Mechanical Engineering: Design, Theory and Methodologies

Mechanical Engineering: Design, Theory and Methodologies

Edited by Michelle Vine

CLANRYE INTERNATIONAL
www.clanryeinternational.com

Clanrye International,
750 Third Avenue, 9th Floor,
New York, NY 10017, USA

ISBN: 978-1-63240-660-6

Cataloging-in-Publication Data

Mechanical engineering : design, theory and methodologies / edited by Michelle Vine.
p. cm.
Includes bibliographical references and index.
ISBN 978-1-63240-660-6
1. Mechanical engineering. 2. Engineering. I. Vine, Michelle.
TJ145 .M43 2018
621--dc23

For information on all Clanrye International publications
visit our website at www.clanryeinternational.com

CLANRYE
INTERNATIONAL

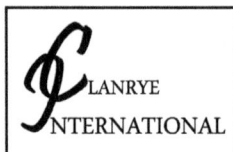

www.ingramcontent.com/pod-product-compliance
Lightning Source LLC
Chambersburg PA
CBHW050457200326
41458CB00014B/5220
* 9 7 8 1 6 3 2 4 0 6 6 0 6 *

P
Particle Phase Modelling, 4
Phase Change Materials (PCMS), 36
Pressurization Configuration Used, 59

Q
Quick Differencing Scheme, 41

R
Recrystallization, 106-107, 109, 111
Reducing Conservatism, 143
Reverse Fault Movement, 28
Rigid Body Configurations, 99
Rigidity Detection, 163-165, 170-171
Robot Calibration, 113, 123
Rp Chains, 125

S
Salter-harris Fracture, 50
Spiral Coil, 86-87, 90-95
Static Equilibrium, 114
Stefan Number, 36, 46-48
Stress-strain Curves, 106
Structural Integrity, 28
Substrate Materials, 1
Successive Dynamic Programming, 14-15
Synthesis, 96, 105, 124-125, 127-130, 163-164, 172-173

T
Time Optimal Path Tracking, 11-12, 19
Topological Graphs, 96-97
Transient Behavior, 86, 90

V
Velocity, 1-2, 4-10, 12-15, 17, 19, 41, 48, 96, 101, 105, 137-138, 153
Virtually Constrained Modes, 156
Von Mises Stress Values, 21

W
Wheeled Mobile Robots (WMRS), 150

Index

A

Abaqus, 29
Absolute Positioning Accuracy, 121
Aluminum Alloys, 106-107
Ansys Workbench Finite Element Analysis (FEA) Software, 21

B

Biomechanical Behaviors, 21, 26
Bonding, 1
Buckling Behavior, 28, 35

C

Capillary Bridge, 67, 69-70, 73-75
Cartilaginous Endplate (CEP), 58
Coil Orientation, 86, 90-91
Cold Spray, 1-2, 9-10
Computational Fluid Dynamics, 1
Computer-aided Numerical Analysis, 21
Converging-diverging Nozzle, 1
Convex Optimization, 11
Coupled Dynamic Model, 131-132, 139
Coupled Simulation, 5
Cut-body Approach, 102
Cut-joint Approach, 96, 101
Cutting Speed Parameters, 106

D

Deform-3d Software, 50, 55
Deposition Efficiency, 1, 9
Diameter-thick Ratio, 28-29, 33, 35
Drilling Process, 50-51, 53-56, 107, 111
Drive Flexibilities Limit, 113
Dynamic Equation, 132
Dynamic Modeling, 114, 131-132
Dynamic Programming (DP), 11

E

Earthquakes, 28
End-effector (EE), 11

F

Fault Displacement, 28-31, 33-35
Fault Movement, 28, 35
Femoral Neck Fracture, 21, 23-24, 26-27
Finite Element Method, 28-29, 50-51
Finite Element Model, 29, 35, 50, 57, 178
Flexible Manipulators, 131
Fluent, 9

G

General Solution Strategy, 14
Geomagic Studio Software, 21
Geometric Calibration, 113, 120, 123
Geometric Model, 113, 116, 120, 122
Geometric Path Planning, 12
Gradients, 1, 6, 15-17, 19

H

Human Femoral Model, 21

I

Industrial Robots, 11-12, 20, 113, 122
Internal Pressure, 28, 35, 62, 70, 75
Intervertebral Disc (IVD), 58

K

Kinematic Loop Constraints, 101
Kinematic Modeling, 96, 162
Kinematic Topology, 96, 103
Kirchoff's Law, 96

L

Lhts Systems, 37
Linearized Dynamic Model, 114

M

Material Model, 21, 50, 54-55
Mechanism, 37, 58, 78, 80, 84, 87, 96-101, 103-105, 124-125, 127, 130-132, 139-140, 163-166, 168-169, 171-172, 181
Mechanism Configuration, 96, 100
Motion Segment Collection, 59
Motor-driven Flexible Manipulator Systems (MDFMSS), 131

N

Novel Technology, 1
Numerical Integration, 11
Numerical Simulations, 9, 53, 56

O

Offset Panel Technique (OPT), 77
Oil-gas Pipelines, 28
Origami-based Product, 77

Contents

Preface........VII

Chapter 1 **An analysis of the particulate flow in cold spray nozzles........1**
M. Meyer, R. Lupoi

Chapter 2 **Successive dynamic programming and subsequent spline optimization for smooth time optimal robot path tracking........11**
M. Oberherber, H. Gattringer, A. Müller

Chapter 3 **Biomechanical comparison using finite element analysis of different screw configurations in the fixation of femoral neck fractures........21**
K. Gok, S. Inal

Chapter 4 **Numerical simulation of buckling behavior of the buried steel pipeline under reverse fault displacement........28**
J. Zhang, Z. Liang, C. J. Han, H. Zhang

Chapter 5 **Thermal analysis of PCM containing heat exchanger enhanced with normal annular fines........36**
M. J. Hosseini, M. Rahimi, R. Bahrampoury

Chapter 6 **Three-dimensional finite element model of the drilling process used for fixation of Salter-Harris type-3 fractures by using a K-wire........50**
A. Gok, K. Gok, M. B. Bilgin

Chapter 7 **The intradiscal failure pressure on porcine lumbar intervertebral discs: an experimental approach........58**
A. R. G. Araújo, N. Peixinho, A. Pinho, J. C. P. Claro

Chapter 8 **Shape and force analysis of capillary bridge between two slender structured surfaces........67**
Z. F. Zhu, J. Y. Jia, H. Z. Fu, Y. L. Chen, Z. Zeng, D. L. Yu

Chapter 9 **Towards developing product applications of thick origami using the offset panel technique........77**
Michael R. Morgan, Robert J. Lang, Spencer P. Magleby, Larry L. Howell

Chapter 10 **Experimental study on transient behavior of embedded spiral-coil heat exchanger........86**
E. I. Jassim

Chapter 11 **Representation of the kinematic topology of mechanisms for kinematic analysis........96**
A. Müller

Chapter 12 **Effect of cutting speed parameters on the surface roughness of Al5083 due to recrystallization........106**
Elias Rezvani, Hamid Ghayour, Masoud Kasiri

Chapter 13 **A two-stage calibration method for industrial robots with joint and drive flexibilities** 113
M. Neubauer, H. Gattringer, A. Müller, A. Steinhauser, W. Höbarth

Chapter 14 **Synthesis of PR-/RP-chain-based compliant mechanisms - design of applications exploiting fibre reinforced material characteristics** 124
U. Hanke, E.-C. Lovasz, M. Zichner, N. Modler, A. Comsa, K.-H. Modler

Chapter 15 **Coupled dynamic model and vibration responses characteristic of a motor-driven flexible manipulator system** 131
Y. F. Liu, W. Li, X. F. Yang, Y. Q. Wang, M. B. Fan, G. Ye

Chapter 16 **B-spline parameterized optimal motion trajectories for robotic systems with guaranteed constraint satisfaction** 141
W. Van Loock, G. Pipeleers, J. Swevers

Chapter 17 **Remarks on the classification of wheeled mobile robots** 150
Christoph Gruber, Michael Hofbaur

Chapter 18 **Solving the double-banana rigidity problem: a loop-based approach** 163
Florian Simroth, Huafeng Ding, Andrés Kecskeméthy

Chapter 19 **Multi-objective optimization of a type of ellipse-parabola shaped superelastic flexure hinge** 174
Zhijiang Du, Miao Yang, Wei Dong

Permissions

List of Contributors

Index

Preface

This book has been a concerted effort by a group of academicians, researchers and scientists, who have contributed their research works for the realization of the book. This book has materialized in the wake of emerging advancements and innovations in this field. Therefore, the need of the hour was to compile all the required researches and disseminate the knowledge to a broad spectrum of people comprising of students, researchers and specialists of the field.

Mechanical engineering deals with the design and manufacture of machines. This book on mechanical engineering deals with the latest concepts and theories of machine design and modeling. Computer-aided engineering and design has greatly aided this field in visualizing devices and tools efficiently. It also integrates concepts from all fields of physics and mathematics for optimal design and manufacture. This book studies, analyses and upholds the pillars of mechanical engineering and its utmost significance in modern times. It attempts to understand the multiple branches that fall under this discipline and how such concepts have practical applications. For all readers who are interested in mechanical engineering, the case studies included in this book will serve as an excellent guide to develop a comprehensive understanding.

At the end of the preface, I would like to thank the authors for their brilliant chapters and the publisher for guiding us all-through the making of the book till its final stage. Also, I would like to thank my family for providing the support and encouragement throughout my academic career and research projects.

Editor

1

An analysis of the particulate flow in cold spray nozzles

M. Meyer and R. Lupoi

The University of Dublin, Trinity College, Department of Mechanical & Manufacturing Engineering, Parsons Building, Dublin 2, Ireland

Correspondence to: M. Meyer (meyerm@tcd.ie)

Abstract. Cold Spray is a novel technology for the application of coatings onto a variety of substrate materials. In this method, melting temperatures are not crossed and the bonding is realized by the acceleration of powder particles through a carrier gas in a converging-diverging nozzle and their high energy impact over a substrate material. The critical aspect of this technology is the acceleration process and the multiphase nature of it. Three different nozzle designs were experimented under constant conditions and their performance simulated using Computational Fluid Dynamics tools. The Deposition Efficiency was measured using titanium as feedstock material and it was shown that it decreases with the cross-sectional throat area of the nozzle. Computational results based on a one-way coupled multiphase approach did not agree with this observation, while more sophisticated modelling techniques with two-way couplings can partially provide high-quality outcomes, in agreement with experimental data.

1 Introduction

New required standards and tolerances come along with an increasing demand of enhanced surface properties, making a new generation of coating technologies necessary and capable of applying high quality layers of advanced materials (Li and Yang, 2013) onto substrates of other metals or alloys.

An alternative to conventional deposition technologies, such as Laser Cladding (Partes and Sepold, 2008), Plasma and Flame Spray (Easter, 2008; Pawlowski, 1995) is Cold Spray (CS). This method is free of melting and therefore avoids the unwanted effects of those techniques which operate under high temperature levels (Kuroda et al., 2011). High pressure gas is accelerated in a converging-diverging supersonic nozzle to velocities in the order of $1000\,\mathrm{m\,s^{-1}}$. The coating material is injected as powder into the nozzle and accelerated by the gas flow. As the powder particles strike against a substrate placed at a distance from the nozzle exit, they deform plastically and bond with the substrate material.

The ratio of particle mass that is deposited successfully over the particle mass fed into the nozzle is called Deposition Efficiency (DE). It is evident that DE strongly depends on the impact velocity of the particles (Pattison, 2006). Despite the simple design and working principle, the flow characteristics are very complex, e.g. due to trans- and supersonic velocities, boundary layer instability, turbulence, and particularly the presence of multiple phases. The rapid change that the flow variables undergo from the inlet to the outlet of the nozzle is the most critical factor.

This complexity makes the nozzle dynamics sensitive to manufacturing inaccuracies (Pattison, 2006). In addition, numerical methods are in general not tailored for all the present local flow situations, e.g. the increased particle volume fraction in the throat region or the high pressure gradients in combination with extreme shear flow. Therefore, investigations often focus on specific aspects of the flow field independently.

For example, Lee et al. (2011) published a numerical investigation about the effects that gas operating conditions, particularly pressure and temperature, have on the flow field. Champagne et al. (2010) discussed the outcome of 1-D-nozzle calculation in comparison to 2-D axi-symmetric simulations and found measured velocities in an intermediate range between those theoretical approaches. A more application-related question was asked by Sova et al. (2013), who analysed the gas flow of a jet that impinges onto a mask. They tracked non-interacting particles that were released in the nozzle exit in order to find how they move through the

mask. In contrast, Yin et al. (2011) observed at the impingement region how the impact velocity increases due to the change of the shock system as the substrate size is changed. A work by Park et al. (2010) concentrates on the gas phase as well, which is a good approximation, since it is dealt with nano-particles. Others, like Tang et al. (2014), were interested in numerical nozzle comparisons, but rather depending on the mixing conditions in the pre-chamber where the carrier and process gas streams join. Similarly, Yin et al. (2014) recently examined the effect of a variation of the injection pressure on the gas flow field on the one hand, but also on the dispersion of the particulate phase on the other. It was found, that the dispersion is strongly increased with enhanced injection pressure. It was also found by Sova et al. (2014) that the injection point as well has a significant effect on the particle dynamics, especially on the thermal conditions. Because the temperature has a major influence on the process efficiency, Li et al. (2011b) were engaged with the numerical modelling of heat transfer within the nozzle and substrate. In a similar manner, Zahiri et al. (2014) developed a 3-D-model that is also mainly concerned with the heat transfer between the nozzle, the fluid and the substrate. Interestingly, to point out the importance of turbulence treatment, a k-ε-turbulence model resulted in over-predicted temperatures of the gas, which could be drastically improved by a model-calibration. Numerical works that focus on the particle behaviour were for example conducted by Li et al. (2011a), who were interested in the differences between under-expanded and over-expanded jets, using a variety of particle sizes. Suo et al. (2014) investigated recently how the particle velocity depends on the type of carrier gas, which was calculated with a simple particle tracking technique. The velocity error was of the order of 10 % in this regard. Yin et al. (2009) again considered in a different context the interaction between particles during the deposition process, thus influencing the deformation of the particles upon impact, not considering any fluid dynamic effects.

Summarising, no studies so far deeply consider particle–particle interaction during the injection and acceleration, although the mixing conditions in the dense throat region and particle dispersion are found to be important in several studies. Several studies found the particle size to have important effects, even within not fully coupled phases, nevertheless, no investigation of the particle loading on the velocity distribution was conducted. There are no conclusive studies which link experimentally measured DE against nozzle design and their relationship at theoretical level.

This forms the starting point for the present study. In order to begin with the integration of all important modelling aspects, this work generates a connection between the particle loading and performance parameters, for example depending on different nozzle designs and operating conditions.

In this regard, experiments when depositing titanium onto aluminium tubes, are compared to numerically computed multiphase flows and discussed taking into account the features of the most widely used numerical approach, the 1-way-coupled *Lagrangian* particle tracking. In addition, a more complex approach is discussed, a 2-way-coupled *Lagrangian* method with stochastic particle collisions, and applied to one of the nozzles, thereby comparing the outcomes at different particle loadings. It is found that conventional numerical methods can be inherently limited for the identification of the performance trends.

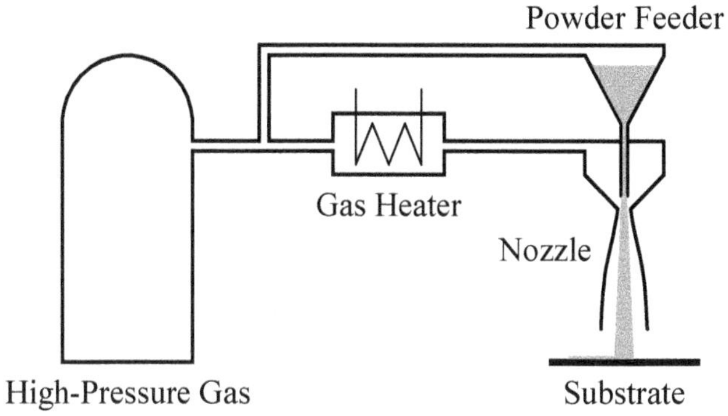

Figure 1. Set-up of the Cold Spray process.

2 Experimental procedure and results

The general set-up for the CS process is shown in Fig. 1. The experiments were conducted utilizing a nitrogen type CS apparatus with an open loop powder feeder. The handling system was capable of delivering a working pressure of up to 3 MPa. A load cell read the powder mass flow rate, while a flow meter measured the gas flow rates in both the powder feeder line and the main line, where a gas heater was installed. This component was used to generate a higher inlet temperature, i.e. nozzle exit speed. Titanium powder (CP-grade 2, $-45\,\mu$m size, spherical) was injected in the subsonic region of the nozzle and deposited onto 50 mm diameter tubes (Al 6082-T6) using three nozzles in order to assess their DE performance.

The geometrical details of the nozzles can be seen in Fig. 2. Correspondent values of the three designs (N1, N2, N3) are summarized in Table 1. A_i and A_e represent the inlet and exit cross-sectional area, respectively. L_c and L_d are the length of the converging and diverging sections of the nozzles and A^* quantifies the cross-sectional throat area. For all test runs the same processing conditions were applied, i.e. the substrate was placed at a stand-off distance of 40 mm from the nozzle exit. The inlet pressure and temperature were set to 3 MPa and 350 °C in order to reach the desired velocity regimes, the powder feed rate was measured to be $55 \pm 9\,\mathrm{g\,min^{-1}}$.

The measured feedstock powder mass flow enables the direct calculation of DE. The respective results are summarised in Table 2. Comparing N1 and N3, the DE of 16.3 % is al-

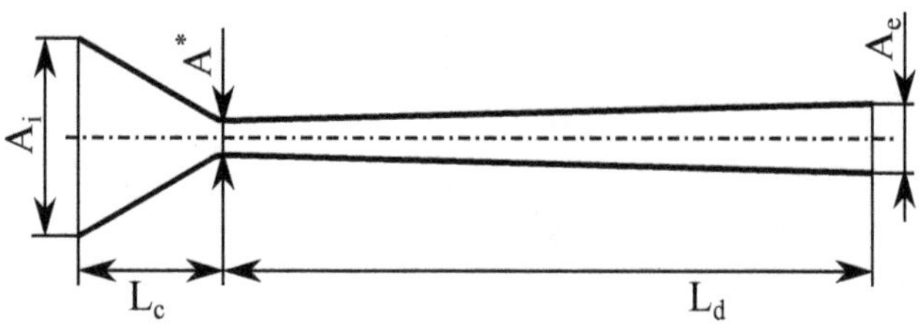

Figure 2. Geometry of the Cold Spray nozzle.

most doubled to a value of 33.3 %, despite the processing conditions remaining constant. Although their overall design is different, nozzle N2 and N3 exhibit similar DE values that correspond to the identical cross sectional throat area. A theoretical analysis was carried out so as to identify the key parameters to unravel the scientific reasons of the experimental outcomes.

3 Uncoupled simulation

In this section, a widely used approach is applied to all three nozzle geometries in order to survey its capabilities regarding an estimate of the experimentally detected behaviour. Therefore, the three cases were simulated with ANSYS Fluent v14.0. An initial analysis of this study was reported by Lupoi (2014).

The operating fluid nitrogen was set to be an ideal gas. The problem was reduced from a three dimensional to an 2-D-axi-symmetric flow. The Navier–Stokes equations for mass, momentum, and energy of the gas phase were solved for a steady state. Moreover, the equations were used in their Reynolds-averaged form and, consequently, extended by a k-ε-turbulence model with standard wall functions. The choice of this type of model was based on its common application in CS and the solution of the variables of interest experienced a negligible change when compared against non-equilibrium wall functions. Gravitational forces were considered negligible. The governing equations are according to FLUENT (2012):

$$\frac{\partial}{\partial x_i}(\rho u_i) = 0 \tag{1}$$

$$\frac{\partial}{\partial x_i}(\rho u_i u_j) = -\frac{\partial P}{\partial x_i} + \frac{\partial \tau_{ij}}{\partial x_i} \tag{2}$$

$$\frac{\partial}{\partial x_i}(\rho e u_i) = -\frac{\partial P u_i}{\partial x_i} + \frac{\partial (u_j \tau_{ij} - q_i)}{\partial x_i} \tag{3}$$

$$\frac{\partial}{\partial x_i}(\rho k u_i) = \frac{\partial}{\partial x_j}\left[\left(\mu + \frac{\mu_t}{\sigma_k}\right)\frac{\partial k}{\partial x_j}\right] + G_k + G_b - \rho\varepsilon - Y_M + S_k \tag{4}$$

Table 1. Geometrical details of the nozzles.

Nozzle	A_i [mm^2]	L_c [mm]	A^* [mm^2]	L_d [mm]	A_e [mm^2]
N1	314	30	3.1	180	28.3
N2	44.2	15.5	5.7	190	47.8
N3	314	20	5.7	190	47.8

Table 2. Comparison of deposition efficiencies.

Nozzle	DE [%]
N1	16.3
N2	32.5
N3	33.3

$$\frac{\partial}{\partial x_i}(\rho \varepsilon u_i) = \frac{\partial}{\partial x_j}\left[\left(\mu + \frac{\mu_t}{\sigma_\varepsilon}\right)\frac{\partial \varepsilon}{\partial x_j}\right] + C_{1\varepsilon}\frac{\varepsilon}{k}(G_k + C_{3\varepsilon}G_b) - C_{2\varepsilon}\rho\frac{\varepsilon^2}{k} + S_\varepsilon \tag{5}$$

In Eqs. (1)–(5), ρ, u, P, e, τ, and q denote the gas density, velocity, static pressure, internal energy, viscous stress tensor, and conductive heat flux. k is the turbulent kinetic energy, ε the eddy dissipation rate, while all other quantities are model-specific constants and source terms. Due to compressibility, a density-based solver was used with a second-order discretisation. The structured mesh was developed to suit the respective flow phenomena with a size of approx. 120 000 elements and tested to provide a mesh-independent solution for the gas phase. It was refined in the near-wall region to capture the boundary layer flow appropriately. The use of standard wall functions requests a wall-adjacent cell hight of no smaller than $y^+ = 15 - 30$ as it should not be placed in the viscous sub-layer. The flow variables, particularly the shear stress and the heat flux, tend to degrade otherwise. However, a sufficient number of grid points are required to resolve the boundary layer. Therefore, the mesh was designed for y^+ values between 20 and 80. The throat radius was resolved with 110 points in the flow direction for a sufficient resolution of the flow gradients. Likewise the resolution at the nozzle exit was kept slightly refined in order to capture the shear layer of the jet. The cell size in the nozzle exit region was tested to be sufficiently fine to capture the shock pattern, i.e. with a change in solution due to mesh refinement less than 1 %. However, adaptive mesh refinement is an option for future work in order to optimise the shock resolution. This is particularly important for smaller particles ($\approx 1\,\mu$m) with shorter response times than in this study. Figure 3 shows the mesh at the nozzle inlet and throat as well as at the exit.

A pressure inlet boundary condition was applied to the nozzle inlet and set to the same values as in the experiments ($p_0 = 3$ MPa, $T_0 = 350\,°$C). The outlet pressure was defined

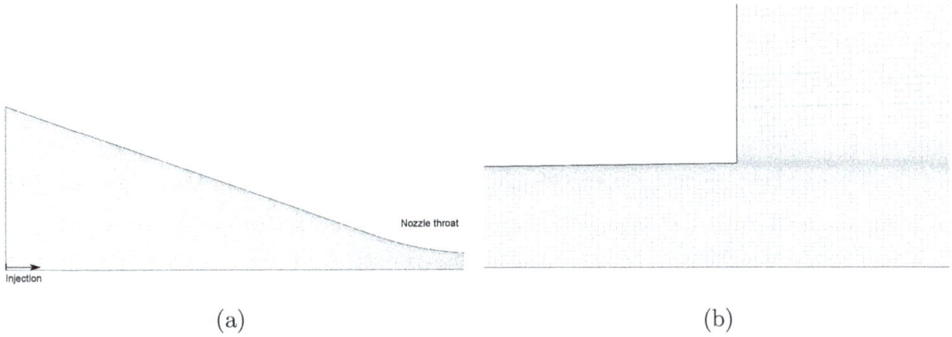

Figure 3. Computational mesh at the nozzle inlet, throat and exit.

Figure 4. Geometry of computational domain and boundaries.

to be atmospheric pressure, sufficiently far downstream from the actual nozzle exit. An adiabatic no-slip boundary condition was applied to the nozzle wall. Figure 4 illustrates the computational domain and boundaries. Table 3 summarises the respective boundary conditions.

Concerning the particle phase modelling, a one-way coupled Lagrangian approach was chosen. In this respect, each particle (45 μm diameter in the model) was released in the inlet zone and further described in a frame of reference that moves with the particle, solving the particle equations based on the local fluid properties. Nevertheless, the change of the gas state variables due to momentum and energy transfer to the particles is not taken into account, as it would require a two-way coupled multiphase model. This one-way coupling is often used in CS simulations, because it provided acceptable results under set conditions (Pattison et al., 2008). Mostly, it is claimed that this simplification is justified due to high Stokes numbers St and low momentum interaction parameters Π_{mom} (Tabbara et al., 2010). In this manner, after obtaining a converged solution for the gas phase, particles are injected into the converging part of the nozzle. Their trajectories are calculated according to the force balance per unit particle mass given by Eq. (6).

$$\frac{d\boldsymbol{u}_p}{dt} = F_D(\boldsymbol{u} - \boldsymbol{u}_p) \qquad (6)$$

Here, the time differential is induced by the motion of the reference frame, but does not imply an unsteady tracking, since the solution is the same for every particle that is exposed to the flow at that specific position. $\boldsymbol{u}$ and $\boldsymbol{u}_p$ are the local velocity vectors of fluid and particle respectively. F_D is a drag force term, that is based on the relative Reynolds number, see Eq. (7), with particle diameter d_p.

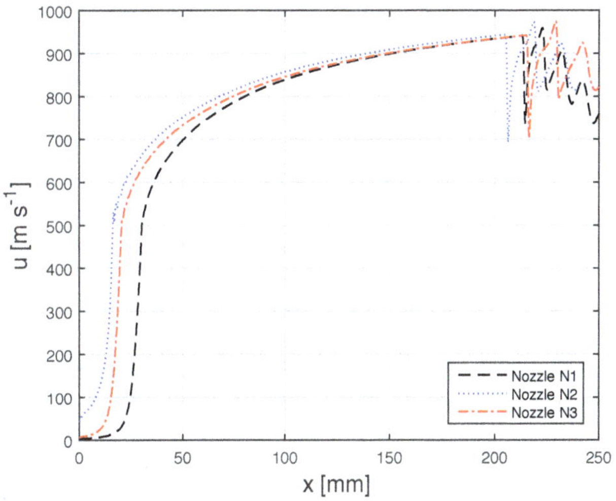

Figure 5. Comparison of nitrogen velocity profiles along the nozzle axis.

$$Re = \frac{\rho d_p \|\boldsymbol{u}_p - \boldsymbol{u}\|}{\mu} \qquad (7)$$

It can be seen, that the modelled acceleration of the particle is mainly influenced by the relative velocity and particle size.

Figure 5 presents the gas velocity profiles along the axial position for the different nozzle designs N1 to N3. The gas phase acceleration is most intense in the transonic region. Each profile shows the typical alternating pattern for over-

Table 3. Boundary conditions for the axisymmetric calculation.

Boundary	Condition type	p_0	$\boldsymbol{v}$	T_0
a–b	Pressure inlet	Specified	$\frac{\partial \boldsymbol{v}}{\partial n}=0$	Specified
b–c	Axial symmetry	$\frac{\partial p}{\partial r}=0$	$\frac{\partial \boldsymbol{v}}{\partial r}=0$	$\frac{\partial T}{\partial r}=0$
c–d, d–e, e–f	Pressure outlet	Specified	$\frac{\partial \boldsymbol{v}}{\partial n}=0$	$\frac{\partial T}{\partial n}=0$
f–g, g–a	Adiabatic no-slip	$\frac{\partial p}{\partial n}=0$	$\boldsymbol{v}=0$	$\frac{\partial T}{\partial n}=0$

expanded flows downstream of the nozzle. In all three nozzle types the gas reaches similar maximum values, although the acceleration in the transonic region differs.

Figure 6 shows a comparison of representative velocity profiles for single particle injections (at the nozzle centre line) in all nozzle design configurations. Since the accelerating drag force is directly related to the relative velocity of the fluid, it increases dramatically in the transonic region and reduces in the diverging section due to the fading gas expansion, as can be seen in Fig. 6. Since particle and gas speeds are still of different levels at the nozzle exits A_e, some slight acceleration is maintained downstream of those points. The shock pattern does not significantly affect the 45 μm particles because of their relatively high inertia. Interestingly, the all particles show very similar profiles and maximal velocities of approximately 595 $m\,s^{-1}$ or 63 % of the carrier gas speed.

Not only the simulated gas phase, but also the particulate material behaves in similar ways regardless of the considered design changes. However, in reality, the deposition performances are entirely different as reported in Table 2. Since the impact velocity is the main driver for DE as experimental conditions were not changed, this fundamental mismatch can only be explained through the fact that the modelling approach neglects important aspects of the process physics: it does not account for any gas-particle and particle–particle interactions. The phase coupling is therefore shown to play a more decisive role in CS nozzle dynamics.

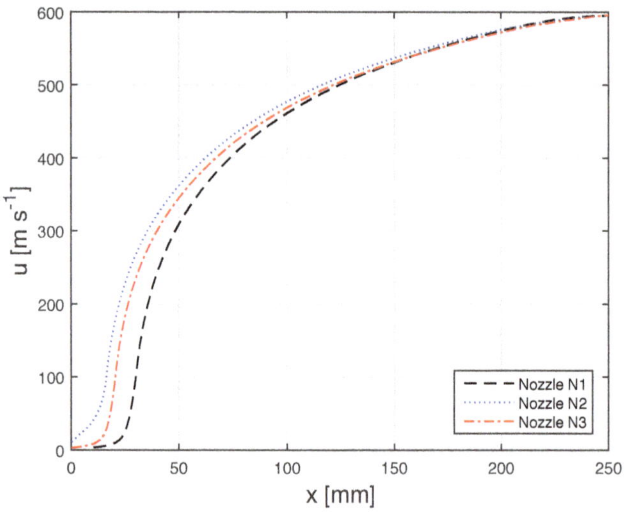

Figure 6. Comparison of titanium particle velocity profiles along the nozzle axis (uncoupled).

4 Coupled simulation

If a significant fluid-particle interaction is present, it must have larger effects in N1 than in N2 and N3. The reason is a higher volume fraction of the particulate phase, originating from the smaller A^* and a lower gas flow rate. A work published by Samareh and Dolatabadi (2008) provides this claim with further theoretical explanation. In this case, the inter-phase relations were modelled in a more sophisticated manner, using an *Eulerian* approach. Accordingly, both the fluid and the particulate phase were modelled as immiscible, interacting continua in the same reference frame. The authors showed a significant decrease in gas velocity at the exit due to the gain in momentum of the particulate phase as the loading was increased. This suggests significant interactions, at least on a theoretical level. A limit of this type of model is the dependency of its validity on relatively high particle density and uniform distribution.

The same authors contributed with another publication (Samareh et al., 2009) that is focused on the simulation of the shock pattern in the jet using a two-way coupled *Lagrangian* approach. It was found that flow patterns could be predicted with high accuracy, including effects of high particle loading in the jet. Using a particle size distribution of mostly small particles (< 10 μm), the calculated exit velocities were within the error range of the measurements. According to the authors, this agreement originated from the complex RSM turbulence model and the two-way phase coupling.

The latter approach is chosen in this study in order to compare different operating conditions. The inaccuracy of previously discussed models go back to the dependency of gas and particle dynamics on local particle loading and volume fraction. Therefore, the respective effects can be investigated if the only parameter that is changed is the particle feed rate, keeping the geometry constant. Since nozzle N1 is the design with the smallest restriction cross-section, it was chosen for this part of the study. It was investigated using the same gas flow conditions but varying particle feed rates from 0 to 16, 32, and through to 64 $g\,min^{-1}$. The numerical mod-

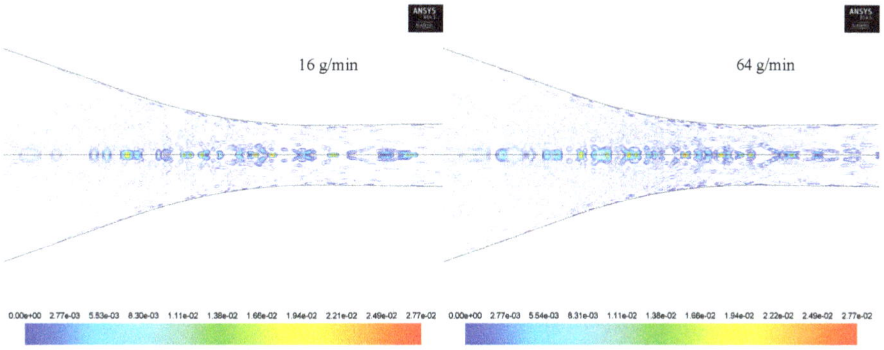

Figure 7. Volume fraction of particles in the throat region.

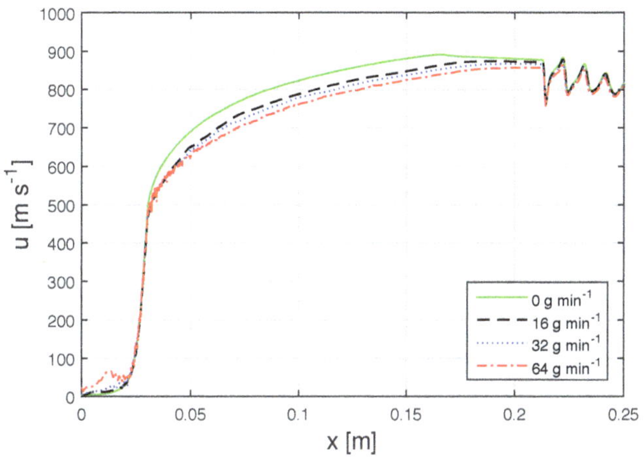

Figure 8. Comparison of nitrogen velocity profiles along axial position (coupled).

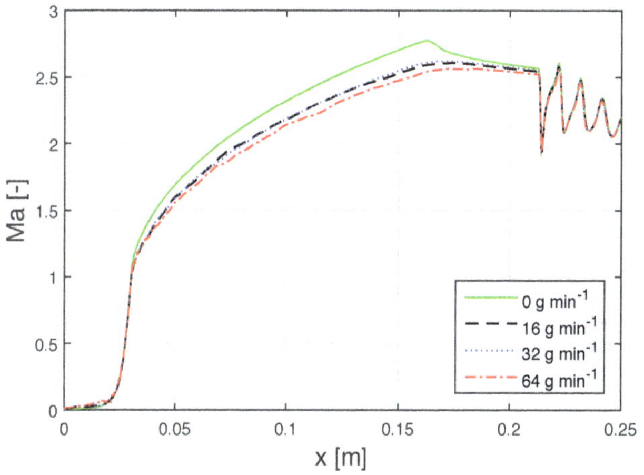

Figure 9. Comparison of gas Mach number along axial position (coupled).

elling was adjusted as follows. The gas phase was modelled and solved as described above, but using a re-normalisation group (RNG) k-ε-turbulence model as it amends the turbulent dissipation at different length scales and was successfully used in gas-particle flows (Fu et al., 2012). In terms of particle injection, apart from the feed rate no changes were made. Nonetheless, the two-way coupling corresponds to an unsteady tracking of particles and produces additional source terms in Eqs. (1)–(5). Likewise, the particle balance Eq. (6) is expanded by an additional force, that depends on the details of the coupling, including the concepts of virtual mass, turbulence coupling, Saffman lift, and stochastic particle–particle collisions, as well as a correction for high pressure gradients which plays a role in the vicinity of shock waves. Consequently, the solution requires an iteration for both the discrete and the continuous phase.

Figure 7 shows the volume fraction distribution in the throat region of nozzle N1 for the two loading cases of 16 and 64 g min^{-1}. Local maximum values are as high as 2.8 % and occur around the centreline, since less space is present for the gas phase. These maxima can be found more frequently in the case of higher loading, but the maximum particle density does not increase measurably. Therefore, higher loadings apparently tend to cause the formation of lump-like spots of high volume fraction if the model accounts for particle particle-interactions. Nevertheless, the higher the particle volume fraction, the more likely collisions become. In this study, it is not observed how this evolution takes place with time while moving downstream through the flow field. This can cause some effect on the velocity observations and motivates a time-dependent solution. A time-averaging of the local particle behaviour could give a more general answer of its impact on the gas velocity.

Figures 8 and 9 show the velocity and Mach number profiles of nitrogen along the nozzle axis for all three particle feed rates and the unladen gas flow. It can be seen, that the velocity profiles are similar in the region of the throat and in the shock-expansion pattern of the jet. However, in the diverging section, the gas velocity decreases considerably with increasing particle feed rate, as more momentum is transferred to the discrete phase. Also, the Mach number profile shows a drastic reduction compared to the pure gas flow, although the differences between the three loading cases are

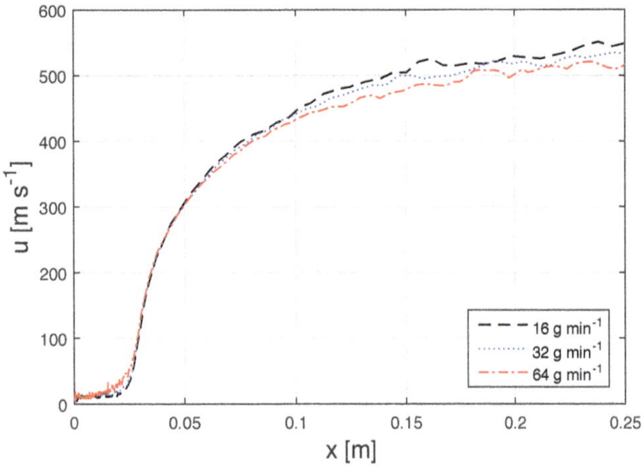

Figure 10. Comparison of titanium particle velocity profiles along axial position (coupled)

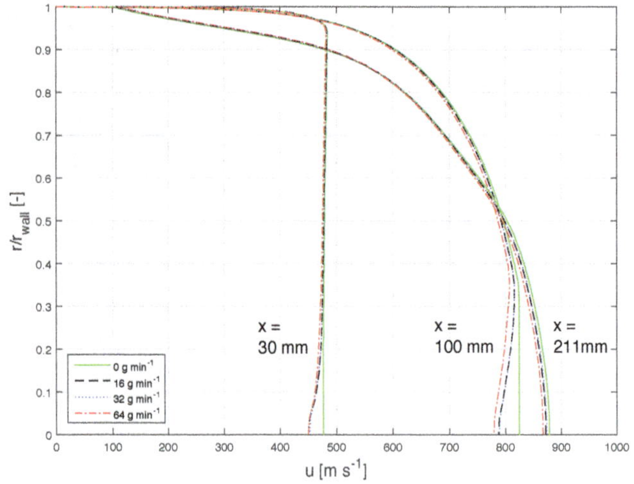

Figure 11. Comparison of gas velocity profiles along radial position (coupled)

comparatively small. A Mach number drop within the last section of the divergent part of the nozzle can be observed for the pure gas flow, which diminishes with increasing feed rate. Interestingly, the gas velocity at the highest particle feed shows a different, more fluctuating trend as compared to the others, especially in regions of dense flow. These fluctuations indicate the importance of the turbulence coupling.

In Fig. 10 the analogous velocity profiles for the particles are shown. Here, the curve depicts the averages over intervals of 100 particles each considering an overlap of 40 %. In this manner, the respective data is reduced to a meaningful representation of the local velocity magnitude. As can be seen, the velocity profiles are similar in the throat region, while the final velocity in fact decreases with increasing feed rate by 8.9 % for the present model. Velocities in the converging nozzle section appear to be the highest for the maximum feed rate which can be explained by a more dominant influence of particle–particle interactions that lead to local velocity maxima. It should be mentioned that the presented analysis neglects information about the velocity direction and hence the particle distribution.

The radial gas velocity profiles for three different axial positions are shown in Fig. 11. The first location with $x = 30\,\mathrm{mm}$ corresponds to the nozzle throat, $x = 100\,\mathrm{mm}$ is a central position of the diverging section and finally $x = 211\,\mathrm{mm}$ is just downstream of the nozzle exit. For this comparison, the radial position is normalised by the local internal nozzle radius. It can be seen how the velocity profiles evolve from an accelerating flow characterised by boundary layers to a fully developed flow. In shock dominated flow at the nozzle exit, the pure carrier gas does not differ much from the loaded cases. However, the gas velocity reduces particularly in the vicinity of the centreline inside the nozzle. The figure shows that increasing the loading does not affect the gas phase as much as injecting $16\,\mathrm{g\,min^{-1}}$ titanium in the first place.

Figures 12 and 13 compare the according radial particle velocity distributions at the nozzle throat and nozzle exit respectively. The different loading cases are compared against each other and represented by both the individual particles in the vicinity of the axial position and a second order polynomial fit. The first set, corresponding to the nozzle throat, shows homogeneously distributed particles for all cases, which agrees with the gas velocity profile. Because of the higher collision rate, the velocity level spreads out with increasing loading and the mean velocity raises. At the nozzle exit, the particles at $16\,\mathrm{g\,min^{-1}}$ exhibit a strong accumulation around the centreline. For the medium loading, this accumulation can still be seen, but the dispersion and the number of low speed particles have increased measurably. At $64\,\mathrm{g\,min^{-1}}$, the particles are spread over all radii with a near-to-constant velocity range, which is slightly lower than in the more dilute cases. The coupled model hence shows an important effect of mass loading on the particle dynamics. These are results at an instant of time, therefore the time-averaging of particle dynamics could possibly show the effects more clearly.

It is difficult to compare these results to the experimental data in default of directly measured velocities. In particular, a model to link the calculated velocities to DE is not available yet, because of a vast amount of practical influences. However, an attempt can be made to compare the calculated changes in velocity to the measured changes in DE as follows. Reducing the cross-sectional throat area of N3 to N1 by half, causes an increase in particulate loading. This is analogous to doubling the particle feed rate in N1 from 32 to $64\,\mathrm{g\,min^{-1}}$. The simulated particle velocity at the substrate stand-off distance ($x = 0.25\,\mathrm{m}$) consequently drops by 4.4 % to $511.9\,\mathrm{ms^{-1}}$. According to Champagne (2007), the DE for aluminium particles in heated air can experience a 65 % decrease, as the particle velocity is reduced by 9.21 %

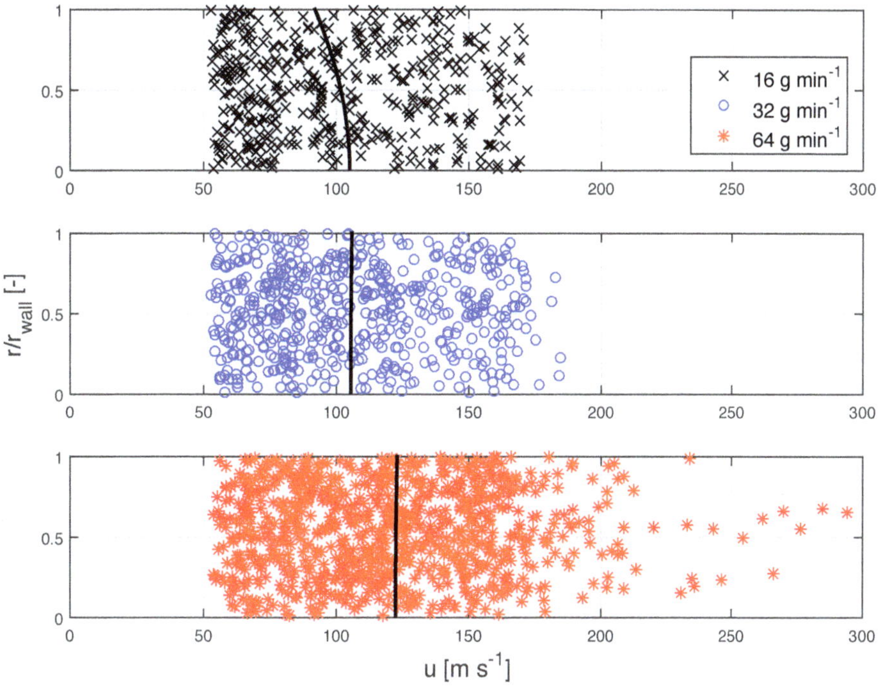

Figure 12. Comparison of titanium particle velocity profiles along radial position at the nozzle throat (coupled)

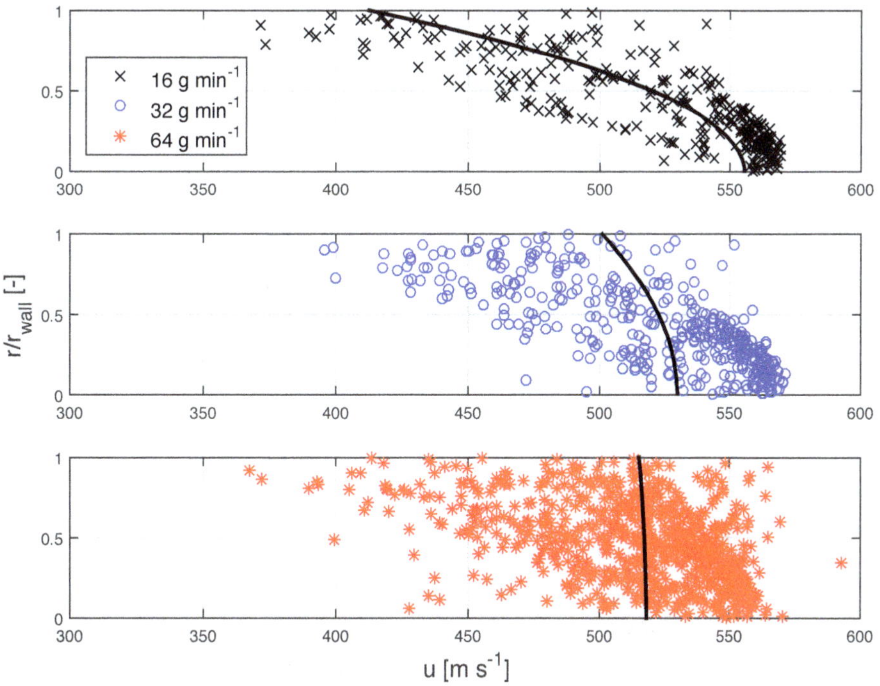

Figure 13. Comparison of titanium particle velocity profiles along radial position at the nozzle exit (coupled)

to 505 ms^{-1}. Transferring this trend to the corresponding simulation results, the 4.4 % decrease in particle velocity can be roughly estimated to cause a drop in DE of 31.1 %. As described in previous sections, the actual reduction of DE is as high as 51.05 %. Therefore, these results show a plausible tendency, however the magnitude is rather underestimated. A reason is the use of data from experiments with a different material and unequal parameters in this argumentation. The enhanced modelling is nonetheless a major improvement in

face of the uncoupled Lagrangian simulation, where the decrease in DE was predicted to be 0 %.

However, in a work by Tabbara et al. (2010) the particle velocities were still considerably overestimated in a validation against experimental data despite a two-way coupling. It was reasoned mainly that this was due to the limits of the k-ε-turbulence model, although the realizable formulation was used that has improved performance for the jet spreading rate. However, particularly the dispersion of the particles was not captured sufficiently. Additionally, Han et al. (2009) showed that the deposition strongly depends on the particle injection process. These aspects represent another direct link between the geometry and the deposition efficiency. Taking the findings of the present study into account, the validity of more elaborately coupled modelling is therefore not only a question of design, gas operating conditions and particle feed rate, but also of local conditions, such as turbulence and particle distribution. This makes further development of more advanced methods and their validation necessary. It could be suggested to make use of a Reynolds-stress model (RSM) that was successfully applied in CS applications in good agreement with experimental observations (Samareh et al., 2009).

5 Conclusions

In this work, the deposition performances of three different De Laval nozzle designs under constant process conditions were investigated and explained by comparing them to numerical results. Titanium was deposited onto aluminium 6082-T6 tubes. It was found that the N1 nozzle, with the smallest throat cross-sectional area, performs the worst in terms of DE. Numerical simulations were performed based on fluid dynamic observations, using steady axisymmetric equations with a k-ε-turbulence model and a one-way coupled discrete phase model. The computed results showed very similar velocity profiles for both phases in all nozzles. The variations in nozzle performance were therefore not numerically reproducible.

The insufficiency of the inter-phase coupling was derived as the main reason, as the comparison with more sophisticated modelling in literature showed. Using a two-way coupled discrete phase model, the effect of increased particle feed rate and hence density on the velocity distributions of both phases was shown to be noticeable for nozzle N1. However, the large number of factors, in relation to the nozzle design, the extreme changes in velocity, and volume fraction makes overall theoretical predictions difficult. Another important factor is the turbulence model, which is derived as another reason for uncertainty. These initial studies will require further development stages in this regard to achieve full validation.

Acknowledgements. The authors wish to express their gratitude to FP7 – Marie Curie (project acronym: SSAM) for the valuable support in developing the work presented in this article.

Edited by: M. Cotterell
Reviewed by: R. Clarke and one anonymous referee

References

Champagne, V.: The cold spray materials deposition process, Woodhead Publishing Limited, Cambridge CB21 6AH, England, 2007.

Champagne, V. K., Helfritch, D. J., Dinavahi, S. P. G., and Leyman, P. F.: Theoretical and Experimental Particle Velocity in Cold Spray, J. Therm. Spray Techn., 20, 425–431, 2010.

Easter, G.: Thermal Spraying – Plasma, Arc and Flame Spray Technology, Wexford College Press, Palm Springs, CA, 2008.

FLUENT: FLUENT Manual, FLUENT Inc., Lebanon, NH, 2012.

Fu, Y., Wang, T., and Gu, C.: Experimental and numerical analyses of gas-solid-multiphase jet in cross-flow, Proceedings of the Institution of Mechanical Engineers, Part G: Journal of Aerospace Engineering, 227, 61–79, 2012.

Han, T., Gillispie, B. A., and Zhao, Z. B.: An Investigation on Powder Injection in the High-Pressure Cold Spray Process, J. Therm. Spray Techn., 18, 320–330, 2009.

Kuroda, S., Watanabe, M., Kim, K., and Katanoda, H.: Current status and future prospects on warm spray technology, J. Therm. Spray Techn., 20, 653–676, 2011.

Lee, M.-W., Park, J.-J., Kim, D.-Y., Yoon, S. S., Kim, H.-Y., James, S. C., Chandra, S., and Coyle, T.: Numerical Studies on the Effects of Stagnation Pressure and Temperature on Supersonic Flow Characteristics in Cold Spray Applications, J. Therm. Spray Techn., 20, 1085–1097, 2011.

Li, C.-J. and Yang, G.-J.: Relationships between feedstock structure, particle parameter, coating deposition, microstructure and properties for thermally sprayed conventional and nanostructured WC Co, Int. J. Refract. Met. H., 39, 2–17, 2013.

Li, S., Muddle, B., Jahedi, M., and Soria, J.: A Numerical Investigation of the Cold Spray Process Using Underexpanded and Overexpanded Jets, J. Therm. Spray Techn., 21, 108–120, 2011a.

Li, W.-Y., Yin, S., Guo, X., Liao, H., Wang, X.-F., and Coddet, C.: An Investigation on Temperature Distribution Within the Substrate and Nozzle Wall in Cold Spraying by Numerical and Experimental Methods, J. Therm. Spray Techn., 21, 41–48, 2011b.

Lupoi, R.: Current design and performance of cold spray nozzles: experimental and numerical observations on deposition efficiency and particle velocity, Surf. Eng., 30, 316–322, 2014.

Park, J.-J., Lee, M.-W., Yoon, S. S., Kim, H.-Y., James, S. C., Heister, S. D., Chandra, S., Yoon, W.-H., Park, D.-S., and Ryu, J.: Supersonic Nozzle Flow Simulations for Particle Coating Applications: Effects of Shockwaves, Nozzle Geometry, Ambient Pressure, and Substrate Location upon Flow Characteristics, J. Therm. Spray Techn., 20, 514–522, 2010.

Partes, K. and Sepold, G.: Modulation of power density distribution in time and space for high speed laser cladding, J. Mater. Process. Tech., 195, 27–33, 2008.

Pattison, J., Celotto, S., Khan, A., and O'Neill, W.: Standoff distance and bow shock phenomena in the Cold Spray process, Surf. Coat. Tech., 202, 1443–1454, 2008.

Pattison, J. A.: Cold Gas Dynamic Manufacturing, Ph.D. thesis, Darwin College, University of Cambridge, Cambridge (UK), 2006.

Pawlowski, L.: The Science and Engineering of Thermal Spray Coatings, Wiley, New York, NY, 1995.

Samareh, B. and Dolatabadi, A.: Dense Particulate Flow in a Cold Gas Dynamic Spray System, J. Fluid. Eng.-ASME, 130, 81702-1–81702-11, 2008.

Samareh, B., Stier, O., Lüthen, V., and Dolatabadi, A.: Assessment of CFD Modeling via Flow Visualization in Cold Spray Process, J. Therm. Spray Techn., 18, 934–943, 2009.

Sova, A., Doubenskaia, M., Grigoriev, S., Okunkova, A., and Smurov, I.: Parameters of the Gas-Powder Supersonic Jet in Cold Spraying Using a Mask, J. Therm. Spray Techn., 22, 551–556, 2013.

Sova, A., Grigoriev, S., Kochetkova, A., and Smurov, I.: Influence of powder injection point position on efficiency of powder preheating in cold spray: Numerical study, Surf. Coat. Tech., 242, 226–231, 2014.

Suo, X., Yin, S., Planche, M.-P., Liu, T., and Liao, H.: Strong effect of carrier gas species on particle velocity during cold spray processes, Surf. Coat. Tech., pp. 1–4, 2014.

Tabbara, H., Gu, S., McCartney, D. G., Price, T. S., and Shipway, P. H.: Study on Process Optimization of Cold Gas Spraying, J. Therm. Spray Techn., 20, 608–620, 2010.

Tang, W., Liu, J., Chen, Q., Zhang, X., and Chen, Z.: The effects of two gas flow streams with initial temperature and pressure differences in cold spraying nozzle, Surf. Coat. Tech., 240, 86–95, 2014.

Yin, S., Wang, X.-F., Li, W.-Y., and Xu, B.-P.: Numerical Investigation on Effects of Interactions Between Particles on Coating Formation in Cold Spraying, J. Therm. Spray Techn., 18, 686–693, 2009.

Yin, S., Wang, X.-F., Li, W.-Y., and Li, Y.: Numerical Study on the Effect of Substrate Size on the Supersonic Jet Flow and Temperature Distribution Within the Substrate in Cold Spraying, J. Therm. Spray Techn., 21, 628–635, 2011.

Yin, S., Liu, Q., Liao, H., and Wang, X.: Effect of injection pressure on particle acceleration, dispersion and deposition in cold spray, Comp. Mater. Sci., 90, 7–15, 2014.

Zahiri, S. H., Phan, T. D., Masood, S. H., and Jahedi, M.: Development of Holistic Three-Dimensional Models for Cold Spray Supersonic Jet, J. Therm. Spray Techn., 2014.

2

Successive dynamic programming and subsequent spline optimization for smooth time optimal robot path tracking

M. Oberherber, H. Gattringer, and A. Müller

Institute of Robotics, Johannes Kepler University Linz, Altenbergerstr. 69, 4040 Linz, Austria

Correspondence to: M. Oberherber (matthias.oberherber@jku.at)

Abstract. The time optimal path tracking for industrial robots regards the problem of generating trajectories that follow predefined end-effector (EE) paths in shortest time possible taking into account kinematic and dynamic constraints. The complicated tasks used in industrial applications lead to very long EE paths. At the same time smooth trajectories are mandatory in order to increase the service life.

The consideration of jerk and torque rate restrictions, necessary to achieve smooth trajectories, causes enormous numerical effort, and increases computation times. This is in particular due to the high number of optimization variables required for long geometric paths. In this paper we propose an approach where the path is split into segments. For each individual segment a smooth time optimal trajectory is determined and represented by a spline. The overall trajectory is then found by assembling these splines to the solution for the whole path. Further we will show that by using splines, the jerks are automatically bounded so that the jerk constraints do not have to be imposed in the optimization, which reduces the computational complexity. We present experimental results for a six-axis industrial robot. The proposed approach provides smooth time optimal trajectories for arbitrary long geometric paths in an efficient way.

1 Introduction

Highly automated production lines with efficient exploitation of the available resources become increasingly important for high wage countries to stay competitive. The utilization of present hardware plays a central role. This can be achieved by an intelligent path planning, using mathematical models of the mechanics to consider technical constraints in an optimization for the determination of optimal motions. It is an established approach to divide the problem into the geometric path planning using a scalar path parameter s and optimization (Bobrow et al., 1985; Pfeiffer and Johanni, 1987). The path planning can for instance be done by combining lines and circles via clothoids. A popular and comfortable way is the definition of the path using polynomials or splines (Geu Flores and Kecskemethy, 2012; Gattringer et al., 2014). In Quang-Cuong (2014) a classification is proposed to divide optimization strategies deriving an optimal trajectory in the 2-D-phase space with coordinates $(s, \dot{s})$ into three families:

1. Dynamic programming (DP): the phase space is divided into a discrete grid. Based on Bellman's optimality principle (Bellman and Dreyfus, 1962) an optimal trajectory is derived on this grid (Shin and McKay, 1985).

2. Numerical integration: the optimal solution for the path parameter time evolution is obtained by integrating maximum and minimum accelerations and finding optimal switching points for this acceleration and deceleration periods in the phase plane (Bobrow et al., 1985; Pfeiffer and Johanni, 1987). Geu Flores and Kecskemethy (2012) used this method to solve the so-called generalized waiter motion problem.

3. Convex optimization: the third method is based on a convex formulation of the optimization problem that can be solved with efficient optimization packages (Verscheure et al., 2009; Ardeshiri et al., 2011).

The tasks to be performed in industrial applications often contain long paths with rough as well as fine contours. Furthermore, the movement should be precise in order to comply with required accuracies on the one hand, and the motor torques should be smooth in order to avoid vibrations and to protect the hardware. Short optimization times are also crucial in order that the saving in execution time is not annihilated by increased calculation times.

To overcome these demands, one may start to optimize the whole path using one approach mentioned above considering kinematic (joint velocity and acceleration) and dynamic (torque) constraints. Smooth trajectories can be achieved by taking jerk or torque rate restrictions into account. In Constantinescu and Croft (2000) and Oberherber et al. (2014) methods to consider them in phase space are presented, while Debrouwere et al. (2013) proposes a sequential convex scheme to solve a nonlinear program. However, for standard six axis industrial robots and long geometric paths the calculation effort is enormous due to the high number of optimization variables and restrictions. This is also caused by the fine discretization, required for a detailed implementation of fine structures of the geometric path. A non-equidistant discretization, where only certain regions of the path are discretized finely, would be a first choice to reduce the size of the problem. But, if the path is very long and contains many finely discretized regions, this approach is not feasible. Another way to reduce the number of optimization variables is a decomposition of the path into segments, performing the optimization for the segments and assembling the solutions to the whole optimal trajectory. In order to achieve a continuous trajectory, terminal conditions have to be defined at the intersection points and regarded in the optimization. We use a DP approach to calculate optimal trajectories for each segment and subsequently for the entire path. For this task we consider restrictions of path velocity, joint velocity and acceleration and also torque limits. Neglecting the jerk restrictions is reflected in a bang-bang behavior of the motor torques.

To obtain smooth robot movements in short optimization times, we approximate the solution provided by the DP approach for each segment of the path with a spline. Subsequently, this approximation is used as initial guess for the optimization of the spline using an active-set solver (Nocedal and Wright, 2006), considering the same restrictions as before. The degree of the splines and the number of control points determine the smoothness of the solution. Since the initial state is near to the global optimum, the calculation times remain low.

As the initial state for the spline optimization is derived by a DP approach, the first part can be assigned to category one. A categorization of the spline optimization is not really possible, since we use an active set solver to optimize the non-convex problem.

The paper is organized as follows: in Sect. 2 the geometric path, describing the task the robot should perform, is defined. Section 3 treats the time optimal path planning problem in the parameter space in a general way, Sect. 4 introduces our solution strategy. Section 5 addresses the used optimization based on DP in detail. An approach to attain smooth trajectories with short calculation times is introduced in Sect. 6. Experimental results, realized on a Stäubli RX130L – a six-axis industrial robot, are presented in Sect. 7.

In this paper we use the abbreviation $\dot{(\,)} = \frac{\mathrm{d}}{\mathrm{d}t}$ for the time derivative of a quantity. The derivative with respect to the path parameter is denoted with $(\,)' = \frac{\mathrm{d}}{\mathrm{d}s}$. Bold lower case letters characterize vectors, while capital letters are used for matrices with the exception of commonly used notations in mechanics. The euclidean norm of the vector $\boldsymbol{x}$ is denoted with $\|\boldsymbol{x}\|$.

2 Geometric path planning

The aim of path planning is the definition of the robot's task. There are several methods to define such a geometric path, like polynomials, combination of lines, circles and clothoids (Müller et al., 2007) or splines (Bobrow, 1988; Gattringer et al., 2014). The latter provide a comfortable way to define the robots motion whether in joint coordinates $\boldsymbol{q}$ or in Cartesian coordinates z_{E}. To consider obstacles in the workspace, it is common to define the geometric path in the Cartesian space as shown in Fig. 1.

A three-dimensional spline curve, describing the end-effector position can be written as

$$\boldsymbol{r}_{\mathrm{E}}(s) = \sum_{l=1}^{n_{\mathrm{D}}} \boldsymbol{d}_l N_l^d(s), \tag{1}$$

with the scalar path parameter $s = [s_{\mathrm{B}}, s_{\mathrm{E}}]$ (s_{B} – begin of the path, s_{E} – end of the path), and the n_{D} control points $\boldsymbol{d}_l$ defining the shape of the curve. These control points can either be defined directly or can be determined using interpolation points $\boldsymbol{p}_l$ as shown in Gattringer et al. (2014). $N_l^d(s)$ denote the B-spline basis functions of degree d. There are different ways to define this basis functions as local or global support functions. For details we refer to De Boor (1978) and Piegl and Tiller (1995). Unit Quaternions $\mathcal{Q} = \left[e_0, \boldsymbol{e}^T\right]^T$ (scalar part e_0 and vector part $\boldsymbol{e}$) are used for the definition of the end-effector orientation $\mathcal{Q}_{\mathrm{E}}$. The evolution of the separate coordinates e_0, e_x, e_y and e_z along the path is again defined via splines with a subsequent normalization. The angular velocity of the end-effector, represented in the end-effector frame, can be calculated to ${}_{\mathrm{E}}\boldsymbol{\omega}_{\mathrm{E}} = 2\left[e_0\dot{\boldsymbol{e}} - \widetilde{\boldsymbol{e}}\dot{\boldsymbol{e}} - \boldsymbol{e}\dot{e}_0\right]$.

3 Time optimal path tracking

3.1 Problem description

In Sect. 2 the geometric path is defined as a function of the scalar path parameter s. The goal of this section is to find an optimal relationship between the path parameter and time:

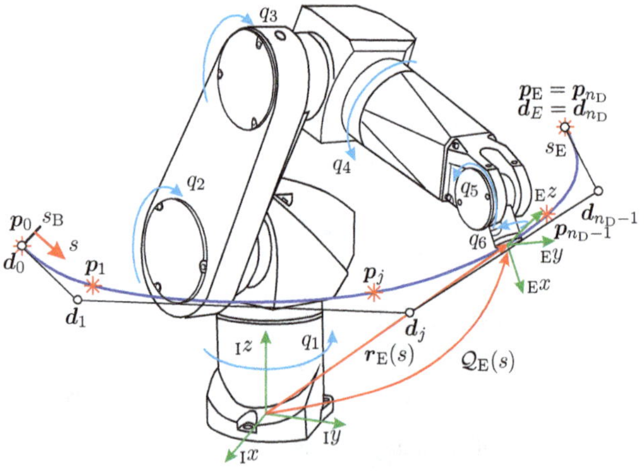

Figure 1. Six-axis industrial robot with geometric path.

$s(t)$. This is accomplished with an optimization considering technical constraints like motor-torque, velocity and acceleration restrictions. The general formulation of the path tracking problem is given by

$$\min_{t_E \in \mathbb{R}^+, \boldsymbol{\tau}(\cdot)} \int_0^{t_E} \left(k_1 + k_2 \boldsymbol{\tau}^T \boldsymbol{\tau}\right) \mathrm{d}t, \tag{2}$$

wherein $\boldsymbol{\tau}$ denotes the vector of motor torques as a function of time. With the coefficients k_1 and k_2 a weighting between time and energy optimality can be achieved. Basically we are able to handle this general case, but in this paper we concentrate on time optimal solutions, therefore the factors $k_1 = 1$ and $k_2 = 0$ are used and the torques vanish from the cost function. The cycle time t_E represents the solution of the optimization and is consequently an unknown quantity. A change of the integration variable from t to s leads to

$$t_E = \int_0^{t_E} 1 \, \mathrm{d}t = \int_{s_B}^{s_E} \frac{1}{\dot{s}(s)} \mathrm{d}s \tag{3}$$

and the optimization problem can be written as

$$\min_{\dot{s}(\cdot)} \int_{s_B}^{s_E} \frac{1}{\dot{s}(s)} \mathrm{d}s. \tag{4}$$

3.2 Technical constraints

3.2.1 Process constraints

For the trajectory optimization several technical constraints should be considered. The first one concerns the end-effector velocity $v_E = \left\| \frac{\mathrm{d}\boldsymbol{r}_E}{\mathrm{d}t} \right\|$, which is a process related restriction. Such constraints can be found in grinding or welding operations. By using the chain rule

$$\frac{\mathrm{d}x}{\mathrm{d}t} = \frac{\mathrm{d}x}{\mathrm{d}s}\frac{\mathrm{d}s}{\mathrm{d}t} = \frac{\mathrm{d}x}{\mathrm{d}s}\dot{s} = x'\dot{s}, \tag{5}$$

the path velocity follows as

$$v_E = \left\| \frac{\mathrm{d}\boldsymbol{r}_E}{\mathrm{d}s}\frac{\mathrm{d}s}{\mathrm{d}t} \right\| = \left\| \boldsymbol{r}'_E(s) \right\| \dot{s} \tag{6}$$

in the parameter range.

3.2.2 Manipulator constraints

Restrictions imposed by the used hardware concern the motor torque, joint velocity, and joint acceleration. In Sect. 2 the path is determined in Cartesian coordinates whereas these restrictions are defined in joint coordinates. Since the end-effector coordinates are calculated with the forward kinematics $\underline{z}_E = [\boldsymbol{r}_E^T \mathcal{Q}_E^T]^T = \boldsymbol{f}(\boldsymbol{q})$ for desired joint positions, the inverse kinematics $\boldsymbol{q}(s) = \boldsymbol{f}^{-1}(\underline{z}_E(s))$ provides the joint angles for desired end-effector coordinates. There are different ways to solve this locally, but not globally unique problem. We use a numerical approach based on the relation

$$\begin{pmatrix} \boldsymbol{v}_E \\ \boldsymbol{\omega}_E \end{pmatrix} = \mathbf{J}(\boldsymbol{q})\dot{\boldsymbol{q}} \, , \tag{7}$$

with the end-effector velocities $\boldsymbol{v}_E, \boldsymbol{\omega}_E$ represented in the inertial frame and the Jacobian

$$\mathbf{J} = \begin{pmatrix} \frac{\partial \boldsymbol{v}_E}{\partial \dot{\boldsymbol{q}}} \\ \frac{\partial \boldsymbol{\omega}_E}{\partial \dot{\boldsymbol{q}}} \end{pmatrix}. \tag{8}$$

With the chain rule Eq. (5), the joint velocity and acceleration follow as

$$\dot{\boldsymbol{q}} = \frac{d\boldsymbol{q}}{ds}\dot{s} = \boldsymbol{q}'\dot{s} \tag{9}$$

$$\ddot{\boldsymbol{q}} = \boldsymbol{q}''\dot{s}^2 + \frac{1}{2}\boldsymbol{q}'(\dot{s}^2)'. \tag{10}$$

The end-effector prime quantities are calculated, using

$$\underline{z}'_E(s) = \left[\left(\boldsymbol{r}'_E(s)\right)^T, \left(\boldsymbol{\omega}_E(s)\right)^T \right]^T,$$

$$\underline{z}''_E(s) = \left[\left(\boldsymbol{r}''_E(s)\right)^T, \left(\boldsymbol{\omega}'_E(s)\right)^T \right]^T$$

and the Jacobian $\mathbf{J}$

$$\boldsymbol{q}' = \mathbf{J}^{-1}\underline{z}'_E \tag{11}$$

$$\boldsymbol{q}'' = \mathbf{J}^{-1}\left[\underline{z}''_E - \mathbf{J}'\boldsymbol{q}'\right]. \tag{12}$$

In order to formulate the torque restrictions, a dynamic model of the robot is necessary. It is derived with the help of the Projection Equation (Bremer, 2008) resulting in the Equations of Motion (EoM)

$$\mathbf{M}(\boldsymbol{q})\ddot{\boldsymbol{q}} + \boldsymbol{g}(\boldsymbol{q}, \dot{\boldsymbol{q}}) = \boldsymbol{\tau}, \tag{13}$$

wherein $\mathbf{M}(\boldsymbol{q})$ is the position dependent, symmetric and positive definite mass matrix. All other generalized forces like

Coriolis-, gravitational-, centrifugal- and friction forces are contained in $\boldsymbol{g}$. The motor torques are represented by $\boldsymbol{\tau}$. As well as the velocity and acceleration restrictions Eqs. (6), (11), (12), also the torque restrictions are required in terms of the path parameter. There exist different ways to express the EoM in terms of the path parameter as those proposed in Johanni (1988) and Geu Flores and Kecskemethy (2012). An analytical formulation is shown in Gattringer et al. (2014), where the parametrized EoM are written as

$$\boldsymbol{\tau} = \boldsymbol{a}(s)(\dot{s}^2)' + \boldsymbol{b}(s)\,\dot{s}^2 + \boldsymbol{c}(s) + \boldsymbol{d}_{\text{v}}(s)\dot{s}. \tag{14}$$

The coefficients $\boldsymbol{a}, \boldsymbol{b}, \mathbf{c}$ and $\boldsymbol{d}_{\text{v}}$ of the parametrized EoM follow directly by rewriting and parametrizing the Projection Equation.

By introducing the variable $z = \dot{s}^2$ the optimization problem in parameter space follows as

$$\min_{z(\cdot)} \int_{s_B}^{s_E} \frac{1}{\sqrt{z(s)}} \mathrm{d}s \tag{15}$$

$$s.t.\ \left\|\boldsymbol{r}'_{\text{E}}(s)\right\| \sqrt{z(s)} \leq v_{\text{E,max}} \tag{16}$$

$$\dot{\boldsymbol{q}}_{\min} \leq \boldsymbol{q}'(s)\sqrt{z(s)} \leq \dot{\boldsymbol{q}}_{\max} \tag{17}$$

$$\ddot{\boldsymbol{q}}_{\min} \leq \boldsymbol{q}''(s)z(s) + \frac{1}{2}\boldsymbol{q}'(s)z'(s) \leq \ddot{\boldsymbol{q}}_{\max} \tag{18}$$

$$\boldsymbol{\tau}_{\min} \leq \boldsymbol{a}(s)z'(s) + \boldsymbol{b}(s)z(s) + \mathbf{c}(s) + \boldsymbol{d}_{\text{v}}(s)\sqrt{z(s)} \leq \boldsymbol{\tau}_{\max}. \tag{19}$$

The values for the motor torque restrictions $\boldsymbol{\tau}_{\min}$, $\boldsymbol{\tau}_{\max}$ and joint velocity restrictions $\dot{\boldsymbol{q}}_{\min}$, $\dot{\boldsymbol{q}}_{\max}$ can be taken from the data sheets of the motors respectively gears. Generally the lower limits are equal to the negative values of the upper limits: $\boldsymbol{\tau}_{\min} = -\boldsymbol{\tau}_{\max}$ and $\dot{\boldsymbol{q}}_{\min} = -\dot{\boldsymbol{q}}_{\max}$. The same applies to the acceleration restrictions $\ddot{\boldsymbol{q}}_{\min} = -\ddot{\boldsymbol{q}}_{\max}$, which are usually defined in the joint controllers. Depending on the process to be performed, the path velocity limit $v_{\text{E,max}}$ is set to the optimal working speed.

The better the robot model matches the real system, the better the limits can be exploited. Primarily parameters that are necessary to simulate the kinematics, like link lengths or distances between axis can be taken from CAD – data. The parameters for the dynamic model – masses, centers of gravity or inertia tensors can usually only be estimated with CAD models. For a good match of the derived robot model with the real system, a parameter identification as shown in Neubauer et al. (2014) and Swevers et al. (1996) is indispensable.

4 General solution strategy

The requirements of smooth time optimal trajectories for long geometric paths and short computation times were discussed in Sect. 1 as well as our idea to overcome this challenge. A reduction of optimization variables for a particular optimization is achieved by dividing the path into sections and performing the optimization for these segments. To get smooth trajectories in acceptable calculation times, we define the evolution of $z(s)$ in the "phase space" $s \times z$ as a smooth spline curve and optimize its shape with an active-set solver. For this kind of problem a suitable initial guess $z_0(s)$ is crucial but is not always easy to define. In Verscheure et al. (2009) a parabola is used. This approach works reasonably well for zero velocity at the begin and end of the path, but is not guaranteed to work in the general case of desired start and end velocities. The division of the path into segments makes their consideration mandatory to achieve a continuous overall trajectory. For that purpose we propose an elegant way to derive the initial states by using a DP approach considering the terminal conditions and approximate its solution with a spline curve. Since the DP approach provides the global optimum, it yields an excellent approximation. However, the discretization for this algorithm can be chosen coarse in order to save computation time, since it only provides the initial states for a further optimization of the spline curve.

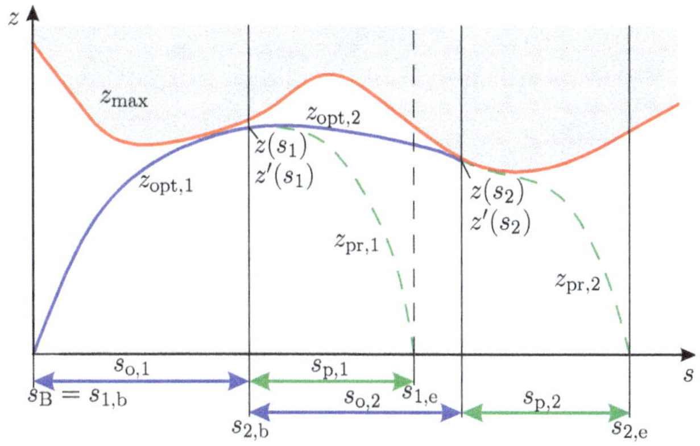

Figure 2. Solution strategy: successive dynamic programming (SDP).

The piecewise optimization procedure is sketched in Fig. 2. It starts with the definition of an optimization s_{o} and a prediction horizon s_{p}. From s_{B} to $s_{\text{o},1} + s_{\text{p},1} = s_{1,\text{e}}$ the optimization is performed and provides an optimal trend, that is split into $z_{\text{opt},1}$ and $z_{\text{pr},1}$. Along the optimization horizon the optimal trajectory is stored and used as a part of the optimal trajectory of the whole path. The prediction horizon is necessary to determine the terminal conditions. At the end of the optimization horizon $z(s_1)$ and $z'(s_1)$ are stored as terminal conditions. Afterwards the horizons are shifted forward by $s_{\text{o},1}$ and the next optimization starts at $s_{2,\text{b}}$ and leads to $s_{2,\text{b}} + s_{\text{o},2} + s_{\text{p},2} = s_{2,\text{e}}$ under consideration of $z(s_1)$, $z'(s_1)$. This procedure is repeated till the end of the path is reached. In further consequence we will call the sequential execution of the algorithm successive dynamic programming (SDP).

The length of the horizons should be chosen carefully. A short optimization horizon in combination with a long prediction horizon leads to unnecessary long calculation times since the solution of the prediction horizon is discarded. Conversely a long optimization horizon and a short prediction

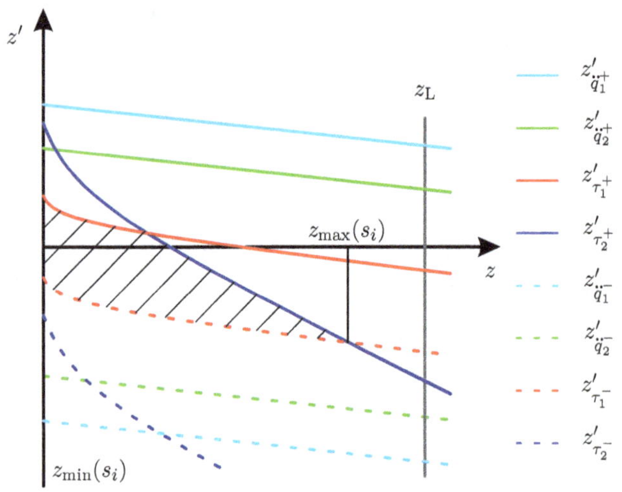

Figure 3. Graphical illustration of technical restrictions in the $z \times z'$ plane.

horizon effects unnecessary deceleration phases. A general choice can not be proposed, rather they have to be chosen problem dependent.

5 Dynamic programming approach to determine an initial solution

5.1 Graphical illustration of restrictions

For a fix point s_i on the path the restrictions in Eqs. (16)–(19) can be graphically illustrated in the $z \times z'$ plane. This is exemplary shown in Fig. 3 for $k = 2$ degrees of freedom and with the assumption that the lower restrictions are equal to the negative upper restriction values.

The velocity restrictions can be squared and combined to

$$z_L = \min \left[\frac{\dot{q}^2_{k,\max}}{q'^2_k}, \frac{v^2_{E,\max}}{\|\boldsymbol{r}'_E\|^2} \right] \tag{20}$$

representing a vertical line in the $z \times z'$ plane. For each joint the acceleration limits follow to

$$z'^{\pm}_{\ddot{q}_k} = \frac{2\left(\pm \ddot{q}_{k,\max} - q''_k z\right)}{q'_k} \tag{21}$$

and thus to linear functions in the $z \times z'$ plane. Due to the consideration of viscous friction (coefficient $\boldsymbol{d}_v$) each joints torque restriction follows to

$$z'^{\pm}_{\tau_k} = \frac{b_k}{a_k} z + \frac{c_k}{a_k} + \frac{d_{v,k}}{a_k}\sqrt{z} \pm \frac{\tau_{k,\max}}{a_k} \tag{22}$$

and thus to nonlinear functions in the $z \times z'$ plane.

The set of curves provide a feasible region (shaded in Fig. 3) for the states. Significant points are given by the minimum and maximum feasible values of z, denoted with $z_{\min}(s_i)$ and $z_{\max}(s_i)$. In Fig. 3 $z_{\min}(s_i)$ is equal to zero. Since a numerical method is used to derive the feasible region it is also possible to consider values $z_{\min}(s_i) > 0$, as they occur for example at the so-called waiter motion problem in Geu Flores and Kecskemethy (2012). $z_{\max}(s_i)$ can either be given by the intersection point of the lowest upper with the highest lower restriction or can concur with the velocity restriction z_L. Within the range $z = z_{\min} \ldots z_{\max}(s_i)$ the feasible region defines the minimum $\underline{z}'(z)$ and maximum $\overline{z}'(z)$ allowed gradients.

5.2 Discretizing the problem

The DP approach is based on Bellman's optimality principle. "[...] An optimal policy has the property that whatever the initial state and the initial decision are, the remaining decisions must constitute an optimal policy with regard to the state resulting from the first decision (Bellman and Dreyfus, 1962) [...]".

With a process running backwards, these decisions can be picked up and optimized with respect to a desired optimality. For that purpose the first step is to discretize the whole path into n segments $s = s_B \ldots s_i \ldots s_E$ with a discretization step size $\Delta s = \frac{s_E - s_B}{n}$ and $i = 0 \ldots n$. Then the optimization horizon s_o with n_o discrete points and prediction horizon s_p with n_p discrete points are chosen as integral multiples of Δs. In the following only one segment $s = s_b \ldots s_e$ is considered, where the start point is denoted with s_b and the endpoint with s_e. Afterwards we evaluate the minimum $z_{\min}(s_i)$ and maximum admissible values $z_{\max}(s_i)$ at all discrete points as shown in Fig. 3 resulting in the limiting curves $z_{\min}(s)$ and $z_{\max}(s)$, that provide the base of operations. Further z is discretized with the step size $\Delta z = \frac{z_{\max}(s_i) - z_{\min}(s_i)}{m}$ into m pieces in the range $z = z_{\min}(s_i) \ldots z_{\min}(s_i) + j\Delta z \ldots z_{\max}(s_i)$ with $j = 0 \ldots m$. The treatment of the path sections leads to a discrete $(n_o + n_p) \times m$ grid instead of a $n \times m$ grid $(n_o + n_p < n)$ in the phase plane. Within this grid also the cost functional from Eq. (15) has to be discretized to

$$W = \sum_{i=n_b}^{n_e} \frac{\Delta s}{\sqrt{z_i}} \tag{23}$$

with $n_b = s_b / \Delta s$ and $n_e = s_e / \Delta s$.

5.3 Successive dynamic programming

The backwards running process starts at the paths end by initializing the cost function W. Starting from a desired end velocity represented by z_e the gradients to all velocity points $z_{e-1,j}$ on the path point s_{e-1} are calculated with

$$z'_{e-1,j} = \frac{z_{e-1,j} - z_e}{\Delta s}. \tag{24}$$

Subsequently they are compared to the minimum $\underline{z}'_{e-1,j}$ and maximum $\overline{z}'_{e-1,j}$ allowed gradients at these points, provided

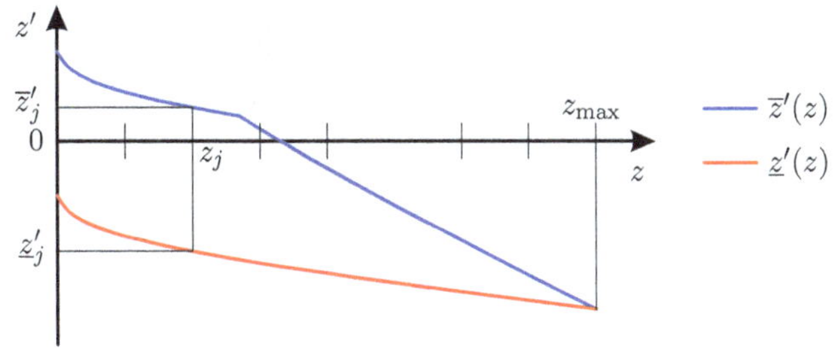

Figure 4. Feasible region in the $z \times z'$ plane with max $\bar{z}'_j$ and min $\underline{z}'_j$ at z_j.

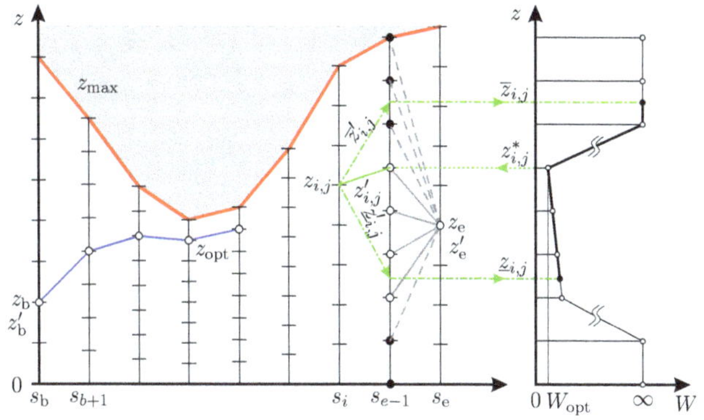

Figure 5. Dynamic programming algorithm.

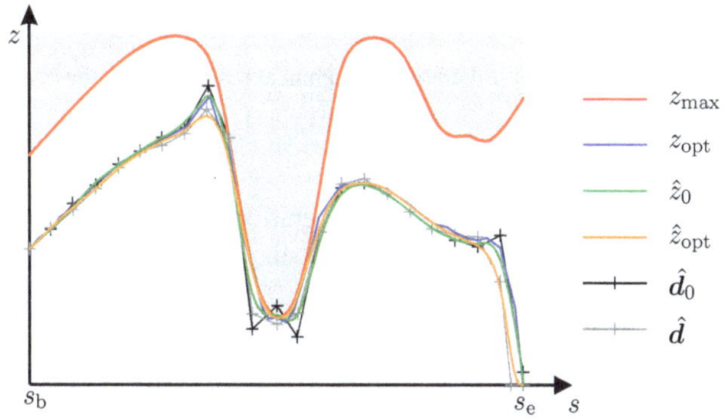

Figure 6. Spline approximation and optimal solution.

by Fig. 4. The cost function is initialized with $W_{e-1,j} = \frac{\Delta s}{\sqrt{z_j}}$ at points with valid gradients (green solid in Fig. 5) and with $W_{e-1,j} \to \infty$ for points with invalid gradients (green dashed). With the cost function on hand the proceeding can be continued for the path points $s_{e-2} \ldots s_b$. At these points the cost function has to be considered in the calculation of the optimal gradients. This is accomplished by determining the minimum and maximum allowed gradients $\underline{z}'_{i,j}$ and $\bar{z}'_{i,j}$ using Fig. 4 of the actual point $z_{i,j}$ and calculating the covered range $[\underline{z}_{i,j}, \bar{z}_{i,j}]$ at the following path point s_{i+1}. Within this range along the z axis the location of the cost functions minimum $z^*_{i,j}$ has to be found. A popular procedure is a minimum search based on the golden ratio. However, we take advantage of its special shape for purely time optimal trajectories $W = \frac{\Delta s}{\sqrt{z}}$. A closer examination shows that the minimum is located at the highest value of z which leads to a feasible value different to ∞. This step is reflected in a strongly reduced computation time. With $z^*_{i,j}$ the optimal gradient follows to $z'_{i,j} = \frac{z^*_{i,j} - z_{i,j}}{\Delta s}$ and the cost function is adapted to $W_{i,j} = W_{i+1,j} + \frac{\Delta s}{\sqrt{z_{i,j}}}$ or $W_{i,j} \to \infty$ if it is out of range. The whole procedure has to be executed for every discrete point at the remaining $n_e - n_b - 2 \times m$ grid. After completion of the optimal gradient determination, the optimal evolution of $z(s)$ can be calculated iteratively with $z_{\mathrm{opt},i+1} = z_{\mathrm{opt},i} + z'_{i,j}\Delta s$ (blue in Fig. 5), starting with a desired start value $z_{\mathrm{opt}}(s_b) = z_b$ and a valid gradient $z'(s_b) = z'_b$ provided by the terminal conditions.

6 Spline based smoothing of the initial solution

6.1 Local spline approximation

The solution provided by the DP approach in Sect. 5 leads to a bang-bang – behavior in the motor torques, resulting in heavy stress for the actuators and the mechanics. In Oberherber et al. (2014) an approach is proposed, that considers torque derivative and joint jerk restrictions in the optimization. This extension of the DP algorithm results in long calculation times and is thus not feasible for long paths. For this reason, we propose a different way in this paper to obtain smooth trajectories. We approximate the optimal evolution of $z_{\mathrm{opt}}(s)$ derived by the dynamic programming algorithm, with a spline curve.

In a first step the trend of $z(s)$ is expressed as a spline

$$\hat{z}_0(s) = \sum_{l=1}^{\hat{n}_D} N_l^d(s)\hat{d}_{0,l} \tag{25}$$

whose $\hat{n}_D$ control points $\hat{\boldsymbol{d}}_0$ follow from a least squares approximation

$$\min_{\hat{\boldsymbol{d}}_0 \in \mathbb{R}^{\hat{n}_D}} \sum_{i=n_b}^{n_e} \left\| \hat{z}_0(s_i) - z_{\mathrm{opt}}(s_i) \right\|^2 \tag{26}$$

minimizing the error between optimal and approximated trend at the discrete points $s = s_b, s_b + \Delta s \ldots s_i, \ldots s_e$ of the DP algorithm. The spline $\hat{z}_0(s)$ is discretized to the originally demanded fine discretization $\Delta\hat{s} = \frac{s_E - s_B}{\hat{n}}$ with $\hat{n} > n$.

6.2 Ensuring consistency

6.2.1 Optimization problem

Since the approximation does not respect any restrictions, local violations are the consequence. Therefore an optimization of the control points, using $\hat{\boldsymbol{d}}_0$ as initial states, is performed

in order to satisfy the restrictions at all discrete points. The optimization problem that has to be solved to fulfill the restrictions is

$$\min_{\hat{d}\in\mathbb{R}^{\hat{n}_D}} \sum_{i=\hat{n}_b}^{\hat{n}_e} \frac{\Delta\hat{s}}{\sqrt{\hat{z}_i}} \tag{27}$$

$$\text{s.t.} \left\|\boldsymbol{r}'_E\right\| \sqrt{\hat{z}} \le v_{E,\max} \tag{28}$$

$$\dot{\boldsymbol{q}}_{\min} \le \boldsymbol{q}'\sqrt{\hat{z}} \le \dot{\boldsymbol{q}}_{\max} \tag{29}$$

$$\ddot{\boldsymbol{q}}_{\min} \le \boldsymbol{q}''\hat{z} + \frac{1}{2}\boldsymbol{q}'\hat{z}' \le \ddot{\boldsymbol{q}}_{\max} \tag{30}$$

$$\boldsymbol{\tau}_{\min} \le \boldsymbol{a}\hat{z}' + \boldsymbol{b}\hat{z} + \mathbf{c} + \boldsymbol{d}_v\sqrt{\hat{z}} \le \boldsymbol{\tau}_{\max} \tag{31}$$

with $\hat{n}_b = s_b/\Delta\hat{s}$ and $\hat{n}_e = s_e/\Delta\hat{s}$. Figure 6 shows the SDP solution z_{opt} and the approximation $\hat{z}_0(s)$ with the control points $\boldsymbol{d}_0$ used as initial states for the optimization. The smooth time optimal trajectory $\hat{z}_{opt}(s)$ follows from the optimal control points $\hat{\boldsymbol{d}}$ provided by the optimization Eqs. (27)–(31).

6.2.2 Gradients

For a fine discretization, the restriction check at every discrete point is computationally expensive. A significant calculation time reduction can be achieved by providing analytical expressions for the gradients of the cost functional and restrictions with respect to the optimization variables. There are actually software tools to compute gradient and Hessians using automatic differentiation. Nevertheless, as the restrictions and the objective are given as analytical functions of the optimization variables, the gradients can easily be calculated analytically. By inserting the spline $\hat{z}_{opt}(s) = \sum_{l=1}^{\hat{n}_D} N_l^d(s)\hat{d}_l$ into the discrete cost functional

$$W = \sum_{i=\hat{n}_b}^{\hat{n}_e} \frac{\Delta\hat{s}}{\sqrt{\hat{z}_{opt,i}}}, \tag{32}$$

its gradient regarding the optimization variables $\hat{d}_l$ follows to

$$\frac{\partial W}{\partial \hat{d}_l} = \sum_{i=\hat{n}_b}^{\hat{n}_e} \frac{-\frac{1}{2}\Delta\hat{s}N_{l,i}^d}{\left(\hat{z}_{opt,i}\right)^{\frac{3}{2}}}. \tag{33}$$

Inserting the spline and its derivative with respect to the path parameter $\hat{z}'_{opt}(s) = \sum_{l=1}^{\hat{n}_D} N_l^{d'}(s)\hat{d}_l$ into the restrictions, their gradients regarding the optimization variables $\hat{d}_l$ follow to

$$\frac{\partial v_E}{\partial \hat{d}_l} = \frac{1}{2}\left\|\boldsymbol{r}'_E\right\| \frac{N_l^d}{\sqrt{\hat{z}_{opt}}} \tag{34}$$

$$\frac{\partial \dot{\boldsymbol{q}}}{\partial \hat{d}_l} = \frac{1}{2}\boldsymbol{q}' \frac{N_l^d}{\sqrt{\hat{z}_{opt}}} \tag{35}$$

$$\frac{\partial \ddot{\boldsymbol{q}}}{\partial \hat{d}_l} = \boldsymbol{q}''N_l^d + \frac{1}{2}\boldsymbol{q}'N_l^{d'} \tag{36}$$

$$\frac{\partial \boldsymbol{\tau}}{\partial \hat{d}_l} = \boldsymbol{a}N_l^{d'} + \boldsymbol{b}N_l^d + \frac{1}{2}\boldsymbol{d}_v \frac{N_l^d}{\sqrt{\hat{z}_{opt}}}. \tag{37}$$

6.2.3 Terminal conditions

To achieve a continuous trajectory a consideration of the terminal conditions for the spline optimization is necessary, as with the DP algorithm. For this purpose the first two control points have to be calculated separately. The first control point is defined by the terminal condition for $z(s_b)$

$$d_1 = z(s_b), \tag{38}$$

while the second control point follows to

$$d_2 = z'(s_b) - d_1 \frac{N_1^{d'}(s_b)}{N_2^{d'}(s_b)} \tag{39}$$

for a transition gradient $z'(s_b)$ and the derivatives of the first two basis functions $N_1^{d'}$ and $N_2^{d'}$ with respect to the path parameter If also transition conditions $z(s_e) = z_e$ and $z'(s_e) = z'_e$ at the end of the path are required, the same procedure also works for the last and second last control point.

The definition of $z(s)$ as spline entails a further advantage namely an easy way to achieve a smooth start and stop. Jerky accelerations at the beginning and decelerations at the end of the path lead to end-effector vibrations which are problematical especial for elastic systems since they need a long time to settle to the desired endpoint. The definition of $z'(s_b) = 0$ leads to a smooth start while $z'(s_e) = 0$ provides a smooth stop.

7 Results

The experiments are realized with a Stäbli RX130L, a six-axis industrial robot. It is controlled by a Bernecker und Rainer system with PD controller and torque feed forward control for each joint. We use a spline curve of degree $d = 4$ in form of our institute logo (a robin), shown in Fig. 7, as geometric path. It is about $l \approx 7.8$ m long and is discretized into $\hat{n} = 2000$ pieces to represent even the fine contours. The end-effector orientation is held constant equal to the initial orientation, so that an observer directly faces the robots endpoint.

For the calculation of the initial solution for the spline optimization with the SDP algorithm the velocity discretization amounts $m = 300$, while the path is discretized into $n = 250$ in the range $s_B = 0 \ldots s_E = 1$. As horizons $s_o = 0.2$ (50 segments) and $s_p = 0.08$ (20 segments) are defined which lead to a subdivision of the total path into five segments.

Figure 8 shows the phase plane $s \times z$ with the liming curve z_{max}, the optimal evolution z_{opt} provided by the SDP approach and the smooth optimal trend $\hat{z}_{opt}$. This spline curve of degree $d = 4$ contains $\hat{n}_D = 40$ control points for each path segment.

A comparison of the motor torques, discontinuous (provided by the SDP approach) and smooth, is given by Fig. 9. The basic behavior looks very similar, with the exception of

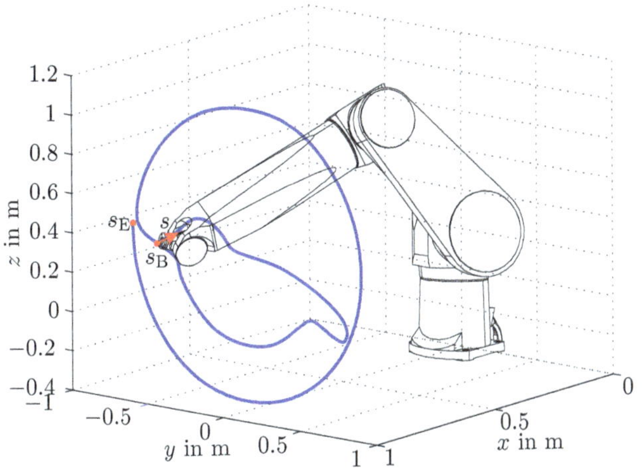

Figure 7. Geometric path – logo of the Institute of Robotics at the JKU Linz.

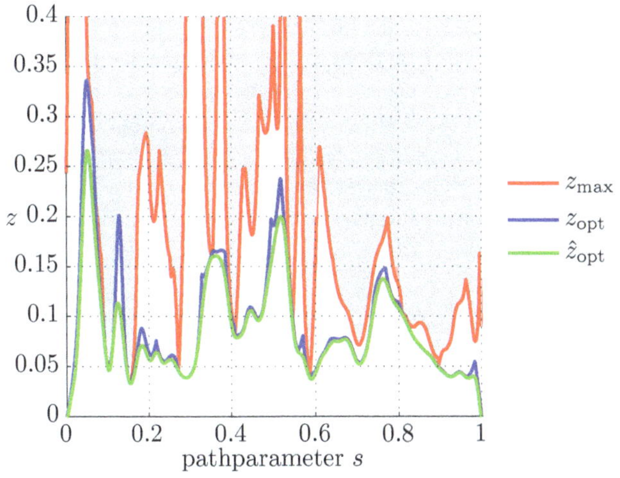

Figure 8. Limiting curve and optimal evolution of the SDP approach and for the smooth solution.

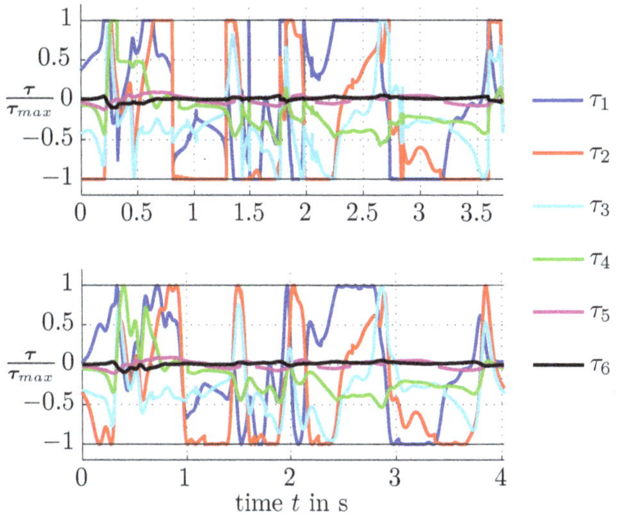

Figure 9. Motor torques, above: torque trends for the SDP solution, below: torque trends for the smooth solution.

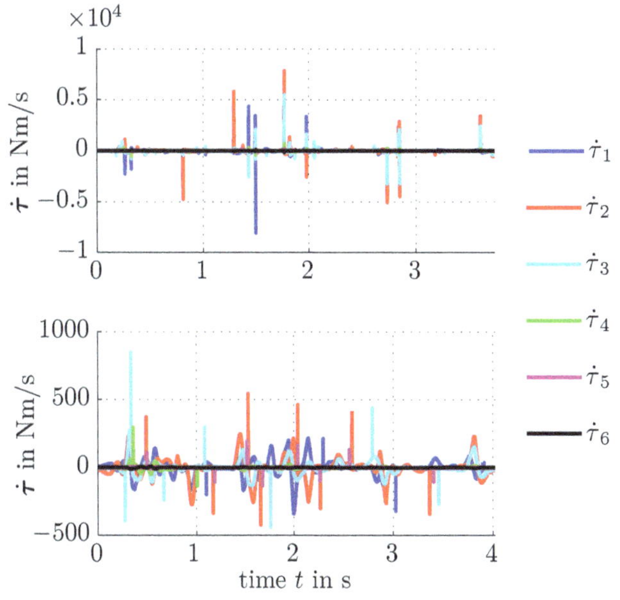

Figure 10. Motor torque rates, above: torque rate trends for the SDP solution, below: torque rate trends for the smooth solution.

a smooth start and stop, where the torques run into the static torques. The difference becomes clearer by looking at the torque rates in Fig. 10 and the joint jerks in Fig. 11. Figure 10 shows, that the torque rates of the discontinuous trajectory are approximately ten times higher then the torque rates of the smooth solution with $\hat{n}_D = 40$ control points for each section.

The smoothness of the trajectory contrasts with an increasing execution time from $t_E \approx 3.72$ s for the bang bang solution to $t_E \approx 4.02$ s for the smooth solution with $\hat{n}_D = 40$ control points for each section. This smooth trajectory was successfully implemented in simulations as well as on the real system, a video clip of the implementation is available on https://youtu.be/c5jllkLE4oU. A reduction of the number of control points to $\hat{n}_D = 25$ leads to a smoother solution, but increases the execution time to $t_E \approx 4.22$ s. Figure 11 shows the joint jerks for the different cases. One can see, that the reduction of the number of control points increases the execution time but reduces the joint jerks.

For validation purposes we implemented an optimization with hard jerk constraints $\dddot{q}_{max} = 3000 \frac{\text{rad}}{\text{s}^3}$, considered in the spline optimization. This optimization converges only for a low number of control points up to $\hat{n}_D = 25$. The calculation time rises to $t_{CPU} \approx 65$ s, while the execution time $t_E \approx 4.13$ s is nearly the same as without jerk restrictions.

Finally we tried to calculate a time optimal trajectory for the whole path in one go, with $\hat{n}_D = 100$ control points. For that purpose we implemented two different approaches to derive an initial guess. A coarse discretized DP approach along

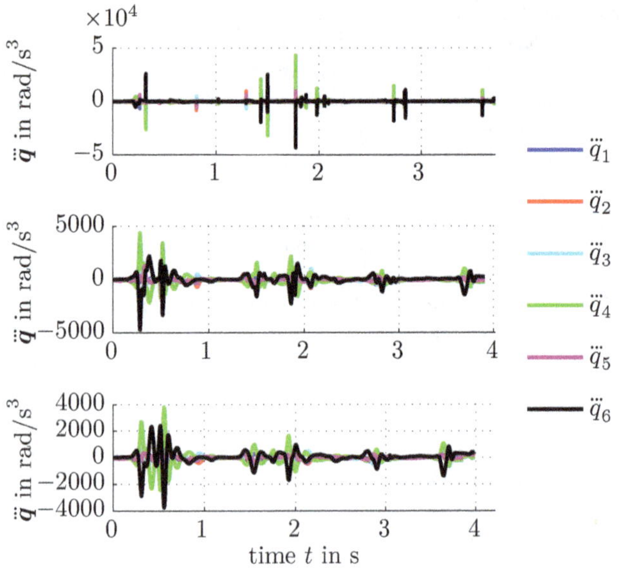

Figure 11. Joint jerks, above: trend of joint jerks for the SDP approach, middle: trend of joint jerks for a smooth solution with $\hat{n}_D = 40$ control points for each section, below: trend of joint jerks for a smooth solution with $\hat{n}_D = 25$ control points for each section.

the whole path, and inspired by Verscheure et al. (2009), a parabola trend of the control points. For that purpose the locations of the control points $\hat{\boldsymbol{d}}$ are defined as

$$\hat{d}_{k+1} = -\min(z_{\max}) \frac{4k\left(-\hat{n}_D + k + 1\right)}{\left(\hat{n}_D - 1\right)^2}, \tag{40}$$

with $k = 1, \ldots, \hat{n}_D$ and $\hat{d}_0 = 0$. The factor $\min(z_{\max})$ ensures, that no velocity restriction is violated by the initial guess. Both approaches require a coarse discretization of the path ($\hat{n} = 1250$) to achieve a convergence of the spline optimization. Despite the coarse discretization, the calculation times increase clearly. The slightly smaller execution times t_E can be attributed to the coarser discretization.

The results of the different methods are listed in Table 1. In this table, SO is the abbreviation for spline optimization and g indicates the usage of analytical gradients. Jerk suggests the consideration of hard joint jerk restrictions in the spline optimization. The last two lines of Table 1 show the results for the optimization in one go with the DP and parabola approach for the initial guess.

The spline optimization was done with the active-set algorithm of the Matlab optimizer fmincon. With the time t_{CPU} we indicate the computation time on a standard PC with a CPU clock of 2.83 GHz. The results in Table 1, clearly indicate the improvements of the presented approach compared to an approach with jerk constraints. Nevertheless, the calculation times are significantly higher compared to the execution times. An improvement could for example be achieved by implementing the optimization not in MATLAB, but in a C-based optimization toolbox.

Table 1. Trajectory execution times t_E and calculation times t_{CPU} for the different methods.

Method	t_{CPU}	t_E	$\hat{n}_D$	$\hat{n}$
	(s)			
SDP	5	3.72	–	2000
SDP and SO	120	4.02	40	2000
SDP and SO and g	25	4.02	40	2000
SDP and SO and g	20	4.22	25	2000
SDP and SO and g and jerk	65	4.13	25	2000
DP and SO and g	58	4.01	100	1250
parabola and SO and g	101	4.01	100	1250

8 Conclusions

This paper presents an approach to derive smooth time optimal trajectories for arbitrary long geometric paths. The main idea is to split the path into sections, to calculate optimal trajectories using terminal conditions, and to assemble the solutions for the individual segments. To achieve smooth trajectories in acceptable calculation times we propose a spline optimization in the phase space. The problem of convenient initial states for the optimization is solved with a DP approach in which terminal conditions can be considered in an easy way. With the spline optimization it is also simple to achieve a smooth start and stop of the robot. The presented approach may be interesting for robot manufacturers which already have algorithms for the path tracking problem and want to extend them to achieve smooth trajectories. Experiments to show the proper functionality of the method are realized on a six-axis industrial robot. An extension of the algorithm to consider jerk and torque rate restrictions in the optimization will be part of future work.

Acknowledgements. This work has been supported by the Austrian COMET-K2 program of the Linz Center of Mechatronics (LCM), and was funded by the Austrian federal government and the federal state of Upper Austria.

References

Ardeshiri, T., Norlöf, M., Löfberg, J., and Hansson, A.: Convex optimization approach for time optimal path tracking of robots with speed dependent constraints, in: Proceedings of the 18th IFAC Congress, Milano, Italy, 28 August–2 September 2011, 14648–14653, 2011.

Bellman, R. E. and Dreyfus, S. E.: Applied Dynamic Programming, Princeton University Press, Princeton, New Jersey, USA, 1962.

Bobrow, J. E.: Optimal Robot Path Planning Using the Minimum-Time Criterion, IEEE Journal of Robotics and Automation, 4, 443–450, 1988.

Bobrow, J. E., Dubowsky, S., and Gibson, J. S.: Time-optimal control of robotic manipulators along specified paths, Int. J. Robot. Res., 4, 3–17, 1985.

Bremer, H.: Elastic Multibody Dynamics: A Direct Ritz Approach, Springer Netherlands, Dordrecht, Netherlands, 2008.

Constantinescu, D. and Croft, E. A.: Smooth and time-optimal trajectory planning for industrial manipulators along specified paths, J. Robot. Syst., 17, 233–249, 2000.

De Boor, C.: A practical guide to splines, Springer Verlag New York, New York, USA, 1978.

Debrouwere, F., Van Loock, W., Pipeleers, G., Tran Dinh, Q., Diehl, M., De Schutter, J., and Severs, J.: Time-Optimal Path Following for Robots with Trajectory Jerk Constraints using Sequential Quadratic Programming, in: Proceedings of IEEE ICRA 2013, 6–10 May 2013, Karlsruhe, Germany, 1916–1921, 2013.

Gattringer, H., Oberherber, M., and Springer, K.: Extending continuous path trajectories to point-to-point trajectories by varying intermediate points, International Journal of Mechanics and Control, 15, 35–43, 2014.

Geu Flores, F. and Kecskemethy, A.: Time-Optimal Path Planning Along Specified Trajectories, in: Multibody System Dynamics, Robotics and Control, edited by: Gattringer, H. and Gerstmayr, J., Springer Vienna, Wien, Austria, 1–16, 2012.

Johanni, R.: Optimale Bahnplanung bei Industrierobotern, VDI Verlag, Düsseldorf, Germany, 1988.

Müller, B., Deutscher, J., and Grodde, S.: Continuous Curvature Trajectory Design and Feedforward Control of Parking a Car, IEEE T. Contr. Syst. T., 15, 541–553, 2007.

Neubauer, M., Gattringer, H., and Bremer, H.: A persistent method for parameter identification of a seven-axes manipulator, Robotica, 1–14, 2014.

Nocedal, J. and Wright, S. J.: Numerical Optimization, 2nd Edn., Springer-Verlag New York, New York, USA, 2006.

Oberherber, M., Gattriger, H., and Springer, K.: Time Optimal Path Planning for Industrial Robots: A Dynamic Programming Approach Considering Torque Derivative and Jerk Constraints, in: Proceedings in Applied Mathematics and Mechanics, 14, 75–76, 2014.

Pfeiffer, F. and Johanni, R.: A Concept for Manipulator Trajectory Planning, IEEE Journal of Robotics and Automation, 3, 115–123, 1987.

Piegl, L. A. and Tiller, W.: The NURBS Book, 2, Springer-Verlag Berlin Heidelberg, Berlin, Germany, 1995.

Quang-Cuong, P.: A General, Fast, and Robust Implementation of the Time-Optimal Path Parameterization Algorithm, IEEE T. Robot., 30, 1533–1540, 2014.

Shin, K. and McKay, N.: Minimum-Time Control of Robotic Manipulators with Geometric Path Constraints, IEEE T. Automat. Control, 30, 531–541, 1985.

Swevers, J., Ganseman, C., De Schutter, J., and Van Brussel, H.: Experimental robot identification using optimised periodic trajectories, Mech. Syst. Signal Pr., 10, 561–577, 1996.

Verscheure, D., Diehl, M., De Schutter, J., and Swevers, J.: Recursive Log-barrier Method for On-line Timeoptimal Robot Path Tracking, in: Proceedings of 2009 American Control Conference, Hyatt Regency Riverfront, 10–12 June 2009, St. Louis, MO, USA, 4134–4140, 2009.

3

Biomechanical comparison using finite element analysis of different screw configurations in the fixation of femoral neck fractures

K. Gok[1] **and S. Inal**[2]

[1]Dumlupinar University, Kütahya Vocational School of Technical Sciences, Germiyan Campus, 43110 Kütahya, Turkey

[2]Dumlupinar University, School of Medicine, Department of Orthopaedic Surgery, Campus of Evliya Celebi, 43100 Kütahya, Turkey

Correspondence to: K. Gok (kadirgok67@hotmail.com)

Abstract. In this research, biomechanical behaviors of five different configurations of screws used for stabilization of femoral neck fracture under axial loading have been examined, and which configuration is best has been investigated. A point cloud was obtained after scanning the human femoral model with a three dimensional (3-D) scanner, and this point cloud was converted to a 3-D femoral model by Geomagic Studio software. Femoral neck fracture was modeled by SolidWorks software for five different configurations: dual parallel, triple parallel, triangle, inverted triangle and square, and computer-aided numerical analysis of different configurations were carried out by ANSYS Workbench finite element analysis (FEA) software. For each configuration, mesh process, loading status (axial), boundary conditions and material model were applied in finite element analysis software. Von Mises stress values in the upper and lower proximity of the femur and screws were calculated. According to FEA results, it was particularly advantageous to use the fixation type of triangle configuration. The lowest values are found as 223.32 MPa at the lower, 63.34 MPa at the upper proximity and 493.24 MPa at the screws in triangle configuration. This showed that this configuration creates minimum stress at the upper and lower proximity of the fracture line. Clinically, we believe that the lowest stress values which are created by triangle configuration encompass the most advantageous method. In clinical practices, it is believed that using more than three screws does not provide any benefit. Furthermore, the highest stresses are as follows: at upper proximity 394.79 MPa in triple parallel configuration, for lower proximity 651.2 MPa in square configuration and for screw 2459 MPa in inverted triangle.

1 Introduction

In our daily life, people can be faced with undesired traumas. As a result of these traumas, femoral neck fractures may occur in the skeletal system. Femoral neck fractures are serious traumas that can lead to pneumonia, pulmonary embolism or death. Therefore, fixing the accuracy and stability of these fractures is necessary. The different configurations of screw fixation are used for stabilization of femoral neck fractures. The question of the best fixation type in surgical treatment of femoral neck fractures is still subject of debate today (Basso et al., 2012; Deneka et al., 1997; Filipov, 2011; Martens et al., 1979; Zdero et al., 2010). In general, although screw fixations of these fractures have been described as appropriate, there are only few studies that contain evidence based on biomechanics regarding which configuration or how many screws result in better stabilization (Audekercke et al., 1979; Cody et al., 2000; Martens et al., 1979; Springer et al., 1991; Swiontkowski et al., 1987). In clinical practice, morbidity related to the fixation of a femoral neck fracture might be due to the configuration of screws that is used; a superior fixation absolutely will create less morbidity (PJ, 1996). Today, biomechanical analyses are performed by using finite element analysis (FEA) in many areas of medicine. In this study,

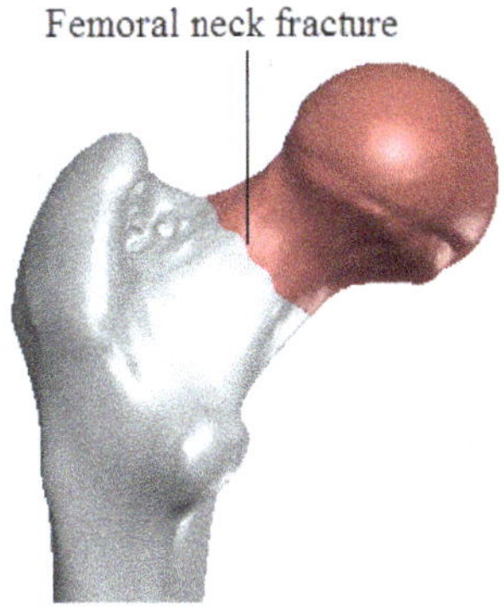

Figure 1. Femoral neck fracture.

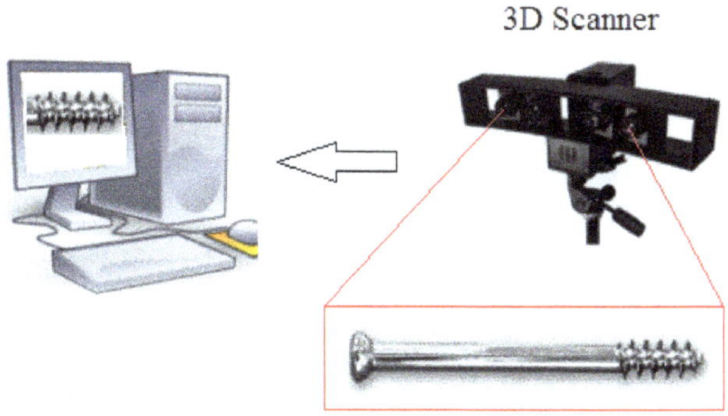

Figure 2. The scanning of screw using a 3-D scanner.

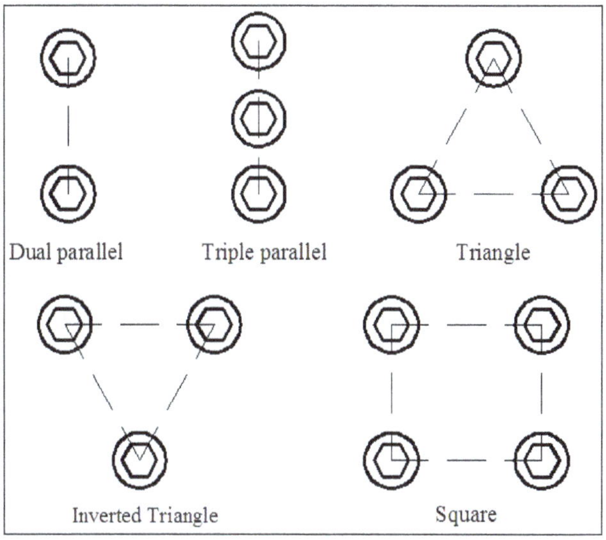

Figure 3. Schematic view of five different configurations for femoral neck fracture.

we aimed to research what the best stable fixation practice might be by applying five different screw configuration types on the femoral neck fracture model with FEA method.

2 Computer-aided design (CAD) and finite element analysis

The human femoral model used in this study was scanned using a 3-D scanner and point cloud was obtained. This process is called reverse engineering (RE). RE process is used to copy the complex shapes and designs by special software and hardware. This method decreases the design process of a product. In particular, it is very important when CAD models of products have been lost. However, in traditional production sequence, reverse engineering typically starts with measuring an existing object so that a solid model can be deduced in order to make use of the advantages of CAD/CAM/CAE technologies (Várady et al., 1997). Later, 3-D model of femur was created using point cloud data by Geomagic Studio 10 program. The scanned data were taken as point cloud data in STL format and converted to Parasolid format using the Rapidform program. The Parasolid format was opened in SolidWorks program and a femoral neck fracture on the 3-D femur model was created as shown in Fig. 1. In order to stabilize the femur neck fracture, $M = 6.5$ screws (cancellous bone screw 6.5 mm diameter and 16 mm thread) were used in this study. Screws were scanned (Fig. 2) using 3-D scanner and modeled in SolidWorks program. Finally, five different configurations – dual parallel, triple parallel, triangle, inverted triangle and square (Fig. 3) – were created using this femur neck fracture model as shown in Fig. 4.

Computer-aided numerical analysis to stabilize the five configurations after fixation was performed using ANSYS Workbench software. The 3-D CAD models of the five configurations (Figs. 3 and 4) were imported into the ANSYS Workbench software to prepare the FEA. Load, boundary conditions and material models were defined in ANSYS Workbench.

Loading and boundary conditions

The mesh process was performed using hex dominant finite element for the FEA modeling after importing five different configurations of 3-D models into the ANSYS Workbench software. The FEA model has 92 552 nodes and 33 117 elements. While the mesh density for the femur was inputted as 4 mm, each screw was inputted as 1 mm, as shown in Fig. 5a. A load of 350 newton (N) in axial direction was applied to the femoral head, and it was fixed from the distal condylar articular face, as shown in Fig. 5b. Contact types the between bone–bone interaction and screw–bone interaction were defined as a frictional contact. Friction coefficients were taken as 0.46 for bone–bone interactions and 0.42 for screw–bone interaction (Goffin et al., 2013). Finally, convergence analysis was conducted as shown in Fig. 6. The force convergence was commonly used in non-linear analyses. If solution is not convergence, there are one or more problems. The purple line on the convergence graph should be acted on the cyan line in order to obtain a good solution. This status depends on

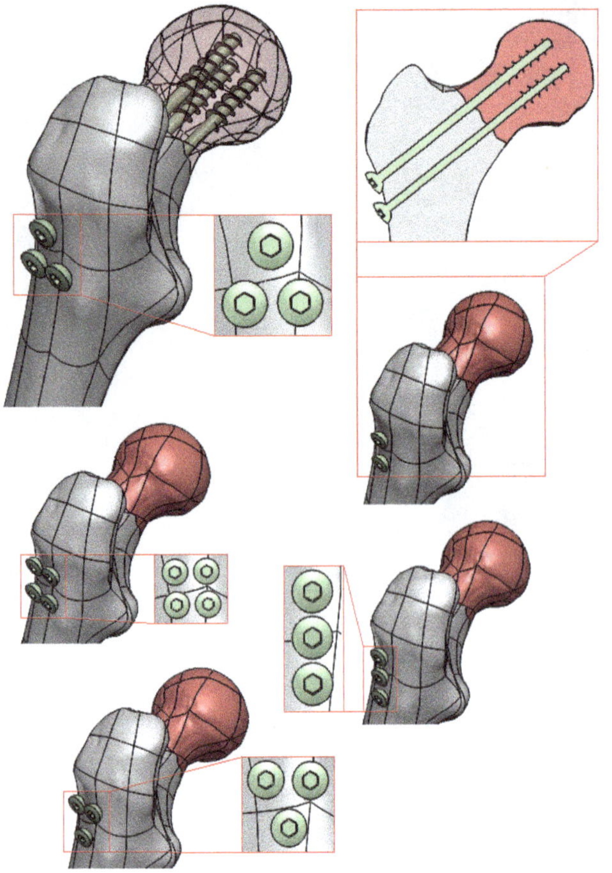

Figure 4. 3-D view of five different configurations for femoral neck fracture.

Table 1. Mechanical properties of bone and screw used in the FEA (Yuan-Kun et al., 2009).

Parameters	Screw (ANSYS Workbench Material Library)	Bone (Yuan-Kun et al., 2009)
Density (kg m^{-3})	4620	2100
Young's modulus (MPa)	96 000	17 000
Yield strength (MPa)	930	135
Ultimate strength (MPa)	1070	148
Poisson's ratio	0.36	0.35

boundary conditions such as friction, contact type and others.

Mechanical properties of bone and screw used in the FEA analyses are given in Table 1. Titanium material was selected for screws, and its mechanical properties were taken from ANSYS Workbench Material Library. Linear isotropic material model was used for mechanical behaviors of bone and screws. The human femur was taken into account as cortical. Müller et al. (1991) mentioned that the anisotropy of bone, i.e., its different mechanical properties along different axes, does not play a major role in internal fixation and will therefore be neglected here.

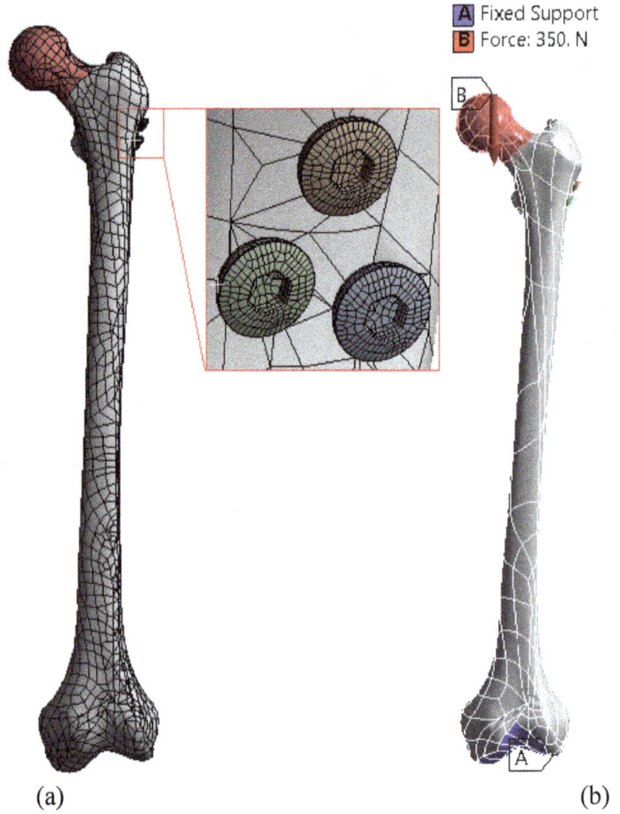

Figure 5. Mesh structure of triangle configuration (**a**), loading and fixing (**b**).

3 Results

After inputting loading and boundary conditions, FEA analysis was solved. According to FEA results, maximum stress values on upper and lower proximity of femoral bone and screws are given in Table 2. These stress values have been evaluated according to the von Mises criteria. According to this criterion, the von Mises stress is an equivalent or effective stress at which yielding is predicted to occur in ductile materials. If the equivalent stress exceeds the yield stress of the material, yielding occurs at that point. In most literature, such a stress is derived using principal axes in terms of the principal stresses (Fig. 7) as in Eq. (1). In latest editions, the von Mises stress with respect to multi-axes stresses (Fig. 8) can also be expressed as in Eq. (2) (Jong and Springer, 2009).

$$\sigma_{\mathrm{Equ}} = \frac{1}{\sqrt{2}}\left[(\sigma_1-\sigma_2)^2+(\sigma_2-\sigma_3)^2+(\sigma_3-\sigma_1)^2\right]^{\frac{1}{2}} \tag{1}$$

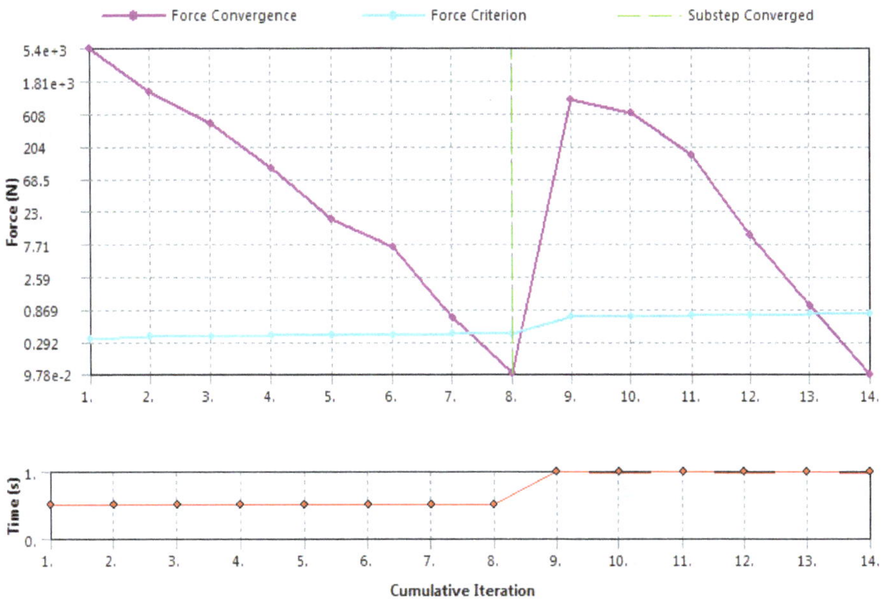

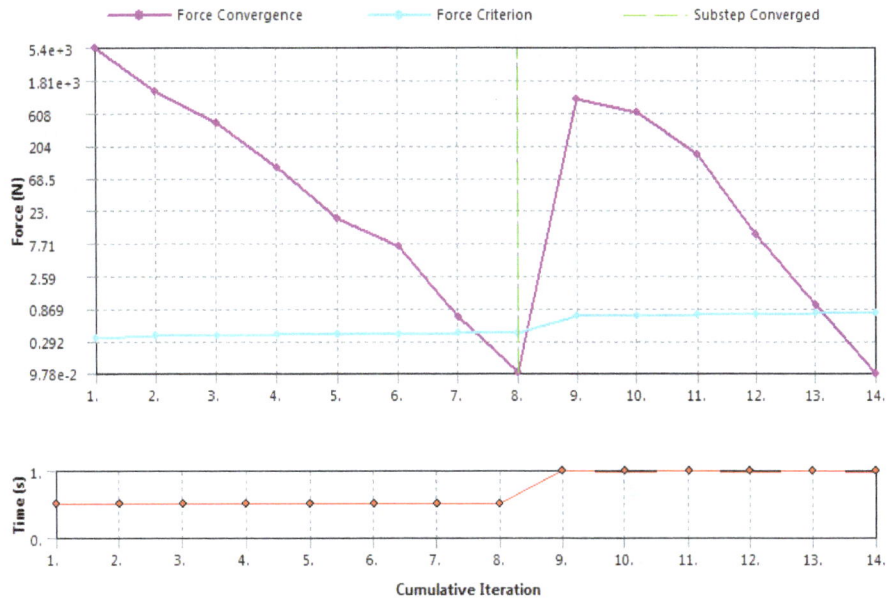

Figure 6. Convergence analysis.

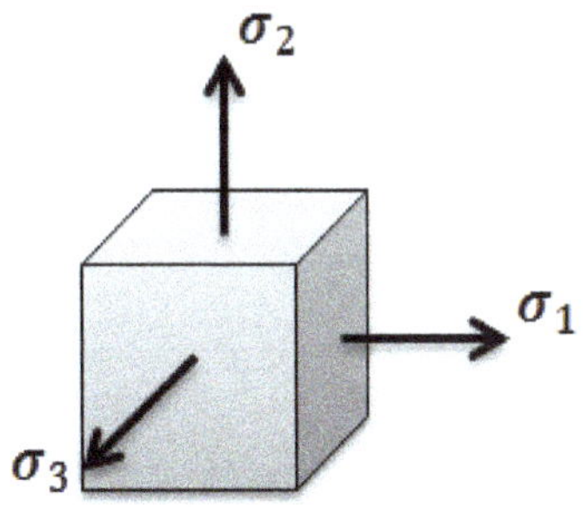

Figure 7. The principal stresses.

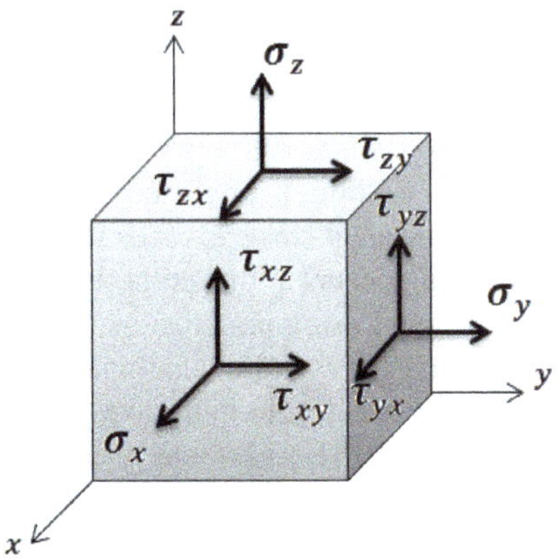

Figure 8. The multi-axial stress state.

$$\sigma_{\text{Equ}} = \frac{1}{\sqrt{2}}\left[\left(\sigma_x - \sigma_y\right)^2 + \left(\sigma_y - \sigma_z\right)^2 + \left(\sigma_z - \sigma_x\right)^2 + 6\left(\tau_{xy}^2 + \tau_{yz}^2 + \tau_{zx}^2\right)\right]^{\frac{1}{2}} \tag{2}$$

Table 2. The stress values in bone and screws.

No.	Fixation type	Stress distributions		
		Screw (MPa)	Upper proximity (MPa)	Lower proximity (MPa)
1	Dual parallel	2157.2	369.49	521.28
2	Triple parallel	1448	394.79	599.88
3	Triangle	493.24	63.34	223.32
4	Inverted triangle	2459	355.32	427.75
5	Square	2121.2	371.75	651.2

As shown in Table 2, it is remarkable under axial loading. The lowest stress values occurring in bone and screws have occurred in triangle configuration and results are as follows: for upper proximity of femur 63.34 MPa, for lower proximity the femur is 223.32 MPa, and for screws 493.24 MPa. The highest values have occurred as follows: for upper proximity of femur is 394.79 MPa with triple parallel configuration, for lower proximity of femur 651.2 MPa with square configuration, and for screws 2459 MPa with inverted triangle configuration. Figure 9 shows the values of stress of axial loading on upper proximity of bone with different configurations, and Fig. 10 shows values for lower proximity bone. Lastly, stress distributions occurring in the screws are shown in Fig. 11.

4 Discussion

Comparison of different configurations of the screws used for stabilization of femoral neck fracture has been studied in biomechanics (Baitner et al., 1999; Holmes et al., 1993; Husby et al., 1987, 1989; Selvan et al., 2004). Reviewing the literature, it is reported in studies that the ideal configuration

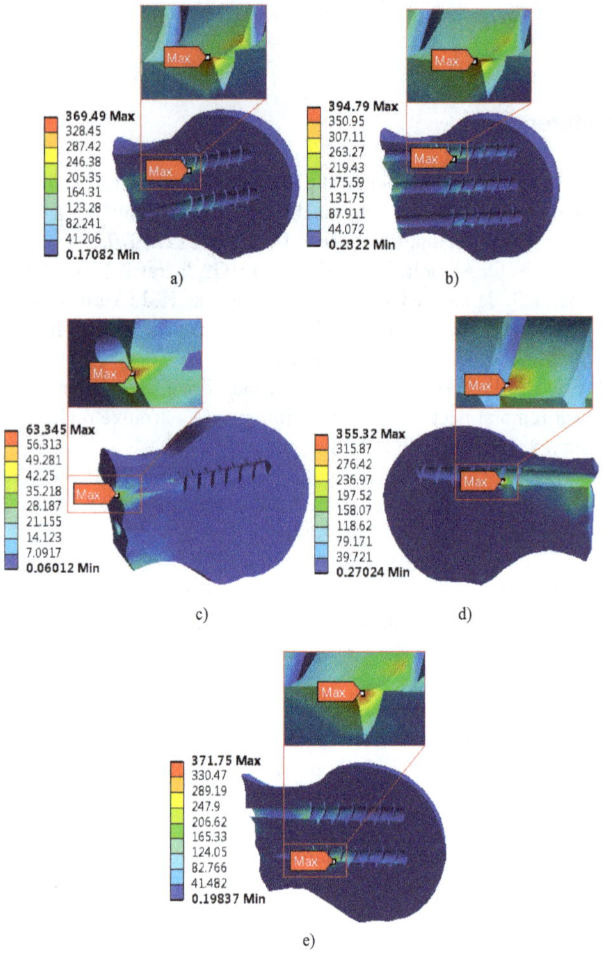

Figure 9. Stress distribution occurring at the upper proximity of femoral bone under axial loading: (**a**) dual parallel, (**b**) triple parallel, (**c**) triangle, (**d**) inverted triangle, and (**e**) square.

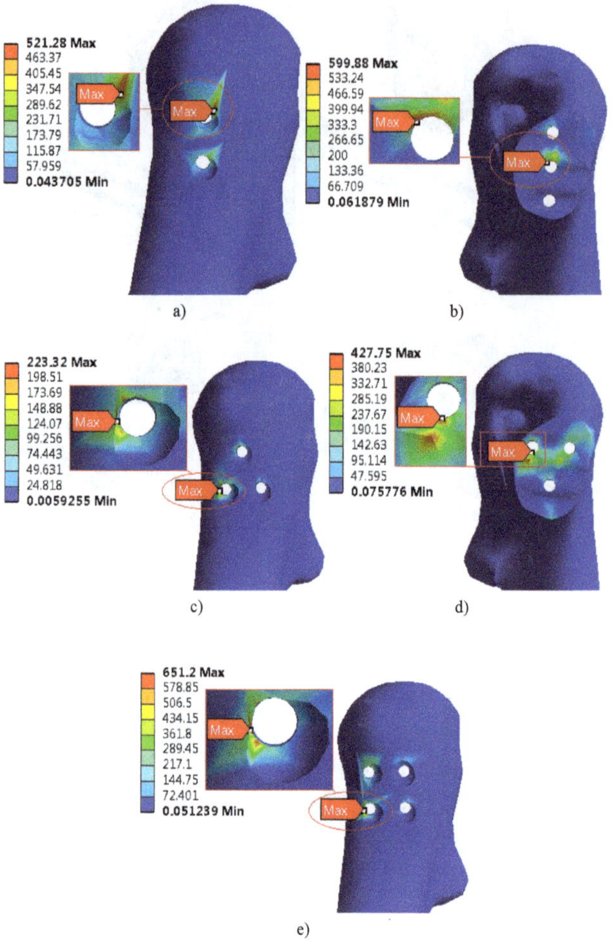

Figure 10. Stress distribution occurring at the lower proximity of femoral bone under axial loading: (**a**) dual parallel, (**b**) triple parallel, (**c**) triangle, (**d**) inverted triangle, and (**e**) square.

of screws is two screws for proximal and one for distal in inverted triangle configuration (Kyle, 2009; Ly and Swiontkowski, 2009; Melvin and Happenstall, 2009; Schmidt et al., 2005). On the contrary, in another study, it is reported that triangle configuration is superior to other configurations (Selvan et al., 2004). It has also been stated that using three or four screws provides similar stability in the femoral head (Kyle, 2009; Lavelle, 2003; Ly and Swiontkowski, 2009; Schmidt et al., 2005; Swiontkowski, 2003). In a review article published by Wu (2010), it is mentioned that inverted triangle configuration is biomechanically superior to traditional triangle configuration. He justifies the superiority of the inverted triangle configuration by explaining the advantage of longer lever arms of two upper screws because during daily activity the loads on the femoral head alternate anteriorly and posteriorly. In our research, we have compared five different configurations: dual parallel, triple parallel, triangle, inverted triangle and square. The stress values show that the triangle configuration during the axial loading has better stabilization, since the distribution of loads occurs firstly on the upper screw and then transfers to lower two screws in a controlled manner in triangle configuration. This situation can be explained biomechanically by pyramid structure. It is thought that as long as the forces are transferred down in a controlled manner in the pyramid, the material is not deformed. One of the most important points is the type of triangle – whether it is an isosceles or equilateral. The stress values with this kind of configurations are equally distributed through the system. Parallel to this situation, triangle configuration created minimum stress values in our study. In clinical practice, low stress means safe fracture line and fixation technique. Stress values have been found high on fracture line and screws in dual parallel, triple parallel, inverted triangle and square configurations. We think that the reason for this situation is asymmetrical and unbalanced distribution of the stress. We believe that triangle configuration is superior in the stabilization of femoral neck fractures, and the biomechanical evidence of this case can be explained with pyramid

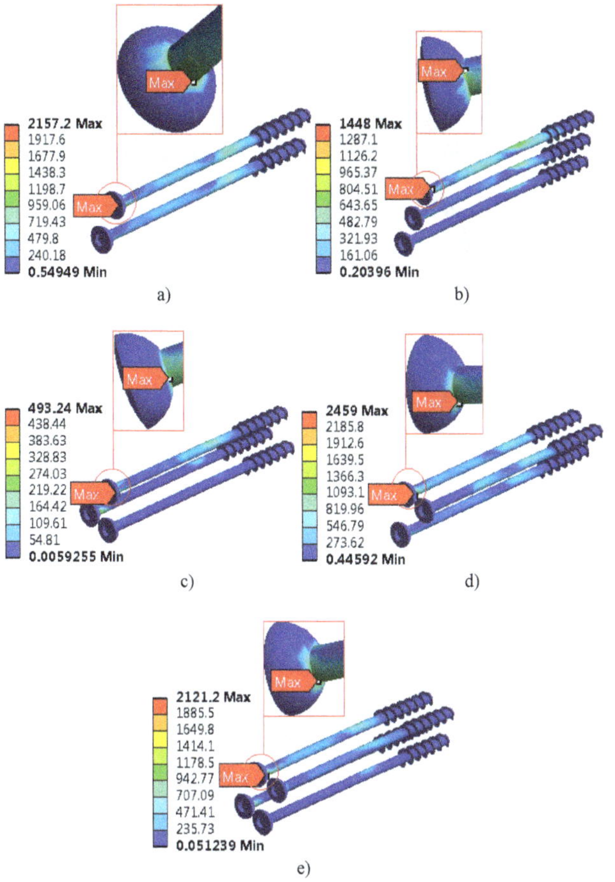

Figure 11. Stress distribution occurring in the screws under axial loading: (**a**) dual parallel, (**b**) triple parallel, (**c**) triangle, (**d**) inverted triangle, and (**e**) square.

theory. In our opinion, using more than three screws would be harmful for stabilization in clinical practices.

5 Conclusions

In this research, biomechanical behaviors of five different screw configurations (dual parallel, triple parallel, triangle, inverted triangle and square) that are applied for the stabilization of femoral neck fracture are investigated. The stress values on the upper and lower proximity of the femur and screws under axial loading have been analyzed, and which configuration is more advantageous has been researched. According to FEA results, it has been found that fixation in triangle configuration is more advantageous. Additionally, in clinical practice, it is thought that the use of more than three screws for stabilization will not be beneficial. We think that further biomechanical studies are needed to improve the safety of stabilization of femoral neck fractures treated with different screw configurations.

References

Audekercke, R. V., Martens, M., Mulier, J. C., and Stuyck, J.: Experimental Study on Internal Fixation of Femoral Neck Fractures, Clin. Orthop. Relat. Res., 141, 203–212, 1979.

Baitner, A. C., Maurer, S. G., Hickey, D. G., Jazrawi, L. M., Kummer, F. J., Jamal, J., Goldman, S., and Koval, K. J.: Vertical Shear Fractures of the Femoral Neck A Biomechanical Study, Clin. Orthop. Relat. Res., 367, 300–305, 1999.

Basso, T., Klaksvik, J., Syversen, U., and Foss, O. A.: Biomechanical femoral neck fracture experiments – A narrative review, Injury, 43, 1633–1639, 2012.

Cody, D. D., Hou, F. J., Divine, G. W., and Fyhrie, D. P.: Femoral structure and stiffness in patients with femoral neck fracture, J. Orthop. Res., 18, 443–448, 2000.

Deneka, D. A., Simonian, P. T., Stankewich, C. J., Eckert, D., Chapman, J. R., and Tencer, A. F.: Biomechanical Comparison of Internal Fixation Techniques for the Treatment of Unstable Basicervical Femoral Neck Fractures, J. Orthop. Trauma, 11, 337–343, 1997.

Filipov, O.: Biplane double-supported screw fixation (F-technique): a method of screw fixation at osteoporotic fractures of the femoral neck, Eur. J. Orthop. Surg. Traumatol., 21, 539–543, 2011.

Goffin, J. M., Pankaj, P., and Simpson, A. H.: The importance of lag screw position for the stabilization of trochanteric fractures with a sliding hip screw: A subject-specific finite element study, J. Orthop. Res., 31, 596–600, 2013.

Holmes, C. A., Edwards, W. T., Myers, E. R., Lewallen, D. G., White, A. A. I., and Hayes, W. C.: Biomechanics of Pin and Screw Fixation of Femoral Neck Fractures, J. Orthop. Trauma, 7, 242–247, 1993.

Husby, T., Alho, A., Hoiseth, A., and Fonstelien, E.: Strength of femoral neck fracture fixation: Comparison of six techniques in cadavers, Acta Orthop., 58, 634–637, 1987.

Husby, T., Alho, A., and Rønningen, H.: Stability of femoral neck osteosynthesis: Comparison of fixation methods in cadavers, Acta Orthop., 60, 299–302, 1989.

Jong, I.-C. and Springer, W.: Teaching von Mises Stress: From Principal Axes To Non-Principal Axes, American Society for Engineering Education, 2009.

Kregor, P. J.: The effect of femoral neck fractures on femoral head blood flow, Orthopedics, 19, 1031–1036, 1996.

Kyle, R. F.: Fractures of the femoral neck, in: Instructional Course Lectures, edited by: Azar, F. M. and O'Connor, M. I., 58. c., American Academy of Orthopaedic Surgeons, USA, 2009.

Lavelle, D. G.: Fractures of hip, in: Campbell's Operative Orthopaedics, edited by: Canale, S. T., Mosby-Year Book, St. Louis, 2003.

Ly, T. V. and Swiontkowski, M. F.: Treatment of femoral neck fractures in young adults, in: Instructional Course Lectures, edited by: Azar, F. M. and O'Connor, M. I., 58. c., American Academy of Orthopaedic Surgeons, USA, 2009.

Martens, M., Audekercke, R. V., Mulier, J. C., and Stuyck, J.: Clinical Study on Internal Fixation of Femoral Neck Fractures, Clin. Orthop. Relat. Res., 141, 199–202, 1979.

Material Library of ANSYS Workbench 14.5 Simulation Software, 2015.

Melvin, J. S. and Happenstall, R. B.: Hip fractures, in: Gowned and Gloved Orthopaedics: Introduction to Common Procedures, edited by: Sheth, N. P. and Lonner, J. H., Saunders Co., Philadelphia, 2009.

Müller, M. E., Perren, S. M., Allgöwer, M., and Osteosynthesefragen, A. f.: Manual of Internal Fixation: Techniques Recommended by the AO-ASIF Group, Springer-Verlag, 1991.

Schmidt, A. H., Asnis, S. E., Haidukewych, Gi, Koval, K. J., and Thorngren, K. G.: Femoral neck fractures, in: Instructional Course Lectures, edited by: Pellegrini, V. D., American Academy of Orthopaedic Surgeons, USA, 2005.

Selvan, V. T., Oakley, M. J., Rangan, A., and Al-Lami, M. K.: Optimum configuration of cannulated hip screws for the fixation of intracapsular hip fractures: a biomechanical study, Injury, 35, 136–141, 2004.

Springer, E. R., Lachiewicz, P. F., and Gilbert, J. A.: Internal Fixation of Femoral Neck Fractures: A Comparative Biomechanical Study of Knowles Pins and 6.5-mm Cancellous Screws, Clin. Orthop. Relat. Res., 267, 85–92, 1991.

Swiontkowski, M. F.: Intracapsular hip fractures, in: Skeletal Trauma, edited by: Browner, B. D., Jupiter, J. B., Levine, A. M., and Trafton, P. G., Saunders Co., Philadelphia, 2003.

Swiontkowski, M. F., Harrington, R. M., Keller, T. S., and Van Patten, P. K.: Torsion and bending analysis of internal fixation techniques for femoral neck fractures: The role of implant design and bone density, J. Orthop. Res., 5, 433–444, 1987.

Várady, T., Martin, R. R., and Cox, J.: Reverse engineering of geometric models – an introduction, Comput. Aided Design, 29, 255–268, 1997.

Wu, C. C.: Using Biomechanics to Improve the Surgical Technique for Internal Fixation of Intracapsular Femoral Neck Fractures, Chang Gung Med. J., 33, 241–251, 2010.

Yuan-Kun, T., Yau-Chia, L., Wen-Jen, Y., Li-Wen, C., You-Yao, H., Yung-Chuan, C., and Li-Chiang, L.: Temperature Rise Simulation During a Kirschner Pin Drilling in Bone, Beijing, 1–4, 11–13 June 2009.

Zdero, R., Keast-Butler, O., and Schemitsch, E. H.: A Biomechanical Comparison of Two Triple-Screw Methods for Femoral Neck Fracture Fixation in a Synthetic Bone Model, J. Trauma Acute Care. Surg., 69, 1537–1544, 1510.1097/TA.1530b1013e3181efb1531d1531, 2010.

4

Numerical simulation of buckling behavior of the buried steel pipeline under reverse fault displacement

J. Zhang, Z. Liang, C. J. Han, and H. Zhang

School of Mechatronic Engineering, Southwest Petroleum University, Chengdu, 610500, China

Correspondence to: J. Zhang (longmenshao@163.com)

Abstract. Reverse fault movement is one of the threats for the structural integrity of buried oil-gas pipelines caused by earthquakes. Buckling behavior of the buried pipeline was investigated by finite element method. Effects of fault displacement, internal pressure, diameter-thick ratio, buried depth and friction coefficient on buckling behavior of the buried steel pipeline were discussed. The results show that internal pressure is the most important factor that affecting the pipeline buckling pattern. Buckling mode of non-pressure pipeline is collapse under reverse fault. Wrinkles appear on buried pressure pipeline when the internal pressure is more than 0.4 P_{max}. Four buckling locations appear on the buried pressure pipeline under bigger fault displacement. There is only one wrinkle on the three locations of the pipeline in the rising formation, but more wrinkles on the fourth location. Number of the wrinkle ridges and length of the wavy buckling increase with the increasing of friction coefficient. Number of buckling location decreases gradually with the decreasing of diameter-thick ratio. A protective device of buried pipeline was designed for preventing pipeline damage crossing fault area for its simple structure and convenient installation. Those results can be used to safety evaluation, maintenance and protection of buried pipelines crossing fault area.

1 Introduction

Fault movement is one of the threats for the structural integrity of buried pipelines caused by earthquakes (Vazouras et al., 2010). Evaluation of the response of buried oil-gas pipelines crossing the faults is among their top seismic design priorities (Karamitros et al., 2007). This is because the axial and bending strains induced to the pipeline by fault may become fairly large and lead to rupture, either due to tension or due to buckling. Leakage of oil-gas pipelines may be results in explosion, poisoning, fire and other accidents. That will lead to huge losses of life and property, or cause social instability. Plastic deformation of buried steel pipeline will reduce the carrying capacity and service life, and pose a potential risk. Generally, there are three types for the fault, they are normal fault, reverse fault and strike-slip fault. The fault motion in this case depends on both the fault dip angle ϕ and the pipeline crossing angle β, which are present in the horizontal and the vertical plane respectively (Shantanu et al., 2011). Mechanical behaviors of the buried steel pipeline under different types of faults are different for the different stratum movement mechanisms.

There are several ways to address the problem. Modern numerical techniques based on finite element method allow a detailed analysis to be performed (Oleg and Vladimir, 2010). The behavior of buried steel pipelines subjected to excessive ground deformation has received significant attention in the pipeline community in the recent year (Vazouras et al., 2012). Wang et al. (2011) analyzed the strain of buried pipes under strike-slip faults. Duan et al. (2011) presented a design method of subsea pipelines against earthquake fault movement, but not considering the buckling morphology of the subsea pipeline cross section. Vazouras et al. (2010, 2012) studied the mechanical behavior of buried pipelines crossing active strike-slip faults. Liu et al. (2008) presented a shell finite element simulation and reported axial strain predictions along the non-pressure pipeline. In fact, buckling modes of non-pressure pipeline and pressure pipeline are different under fault displacement. Due to hanging wall moving towards the foot wall during reverse fault movement, one can expect a

significant amount of axial shortening in the pipeline. Compressive stress may cause buckling of the buried pipeline either in the beam mode or in the shell mode (Shantanu et al., 2011). Some theory analysis methods are established on beam model and rope model, buckling modes of the pipeline cross section cannot be obtained (Liu, 2002). Pipeline is a thin shell structure, when the large deformation appears on the cross section of pipeline, superposition principle cannot be used for the interaction of axial strain and bending strain. And there may be residual stress and stress concentration for the pipeline. The simplified method established by Vazouras considering the S-shape deformed pipeline is only based on geometric deformation of the pipeline, the pipeline-soil was not considered. The bending deformation of the buried pipeline is a non-smooth curve after collapse or wrinkle appears (see farther below). Thus, many simplified methods earlier prove to be inadequate for buckling analysis of pipeline crossing reverse fault, and the finite element method is more suitable.

In this paper, the buckling behaviors of the buried non-pressure and pressure pipeline under reverse fault displacement were investigated by finite element method, considering the soil-pipeline interaction. Effects of buried depth, internal pressure, diameter-thick ratio, fault displacement and friction coefficient on deformation, buckling mode and strain of buried steel pipeline were discussed. And a protective device of buried pipeline was designed for preventing pipeline damage. The results can be used to safety evaluation, maintenance and protection of buried pipelines crossing fault area.

2 Finite element model

The structural response of steel pipeline under reverse fault is examined numerically, using the general purpose finite element program ABAQUS. The nonlinear material behavior of the steel pipeline and the surrounding soil, the interaction between the soil and buried pipeline, as well as the distortion of the pipeline cross-section and the deformation of the surrounding soil are modeled in a rigorous manner, so that the pipeline performance criteria are evaluated with a high-level of accuracy. The pipeline is embedded in an elongated soil prism along the x axis shown in Fig. 1a. Figure 1b shows the buried pipeline mesh and Fig. 1c depicts the mesh of the soil in yz plane. Four-node reduced-integration shell elements (S4R) are employed for modeling the cylindrical pipeline segment, and eight-node reduced-integration elements (C3D8R) are used to simulate the surrounding soil. The fault dip angle $\phi = 45°$, and the pipeline crossing angle $\beta = 0°$.

The fault plane divides the soil in two blocks of equal size (Fig. 1a). The fault hanging wall moves for the reverse fault, and the fault footwall is not moving. The analysis is conducted in two steps as follows, gravity loading is applied firstly, then reverse fault displacement is imposed. The nodes on the bottom boundary plane of the fault footwall (soil nodes) remain fixed in the y directions. The end nodes of the fault footwall remain fixed in the horizontal direction (including the end nodes of the steel pipeline). A uniform oblique displacement due to reverse fault is imposed at the bottom nodes of the fault hanging wall. For the case of pressurized pipelines, an intermediate step of internal pressure application is considered (after the application of gravity and before the fault displacement is activated). The interface between the outer surface of the pipeline and the surrounding soil is simulated with a contact algorithm, which allows separation of the pipeline and soil, and accounts for interface friction. The discretization method of surface to surface used in contact pair can get more accurate contact stress, and reduce the penetration behavior between surfaces. Isotropic coulomb friction is applied through an appropriate friction coefficient μ. In the majority of results reported in the study, μ is considered equal to 0.3.

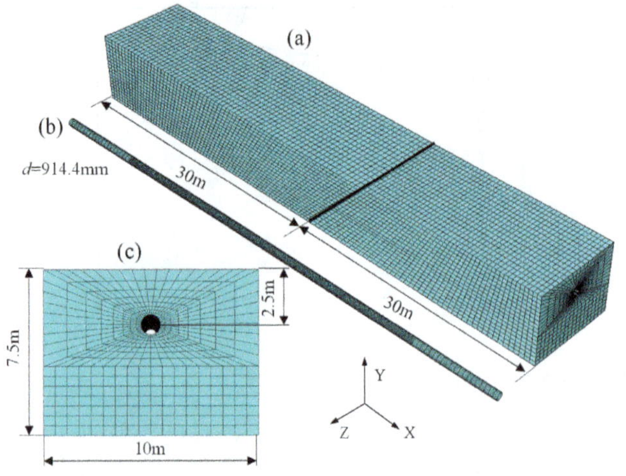

Figure 1. Finite element model. **(a)** The whole model, **(b)** the buried pipeline and **(c)** cross section of the model.

Numerical results are obtained for X65 steel pipelines. The pipeline diameter d is 0.9144 m (36 in), which is a typical size for oil and gas transmission pipeline. The pipeline wall thickness t is considered equal to 8 mm. The pipeline-soil model has dimensions 60 m × 7.5 m × 10 m in directions x, y, z, respectively. Taking loess for example, it has a cohesion $c = 24.6$ kPa, friction angle $\varphi = 11.7°$ (Wang et al., 2010), Young's modulus $E = 33$ MPa, Poisson's ratio $\nu = 0.44$, density $\rho = 1400$ kg m^{-3}. The X65 steel material are typical steel materials for oil and gas pipeline applications, with a nominal stress-strain curve shown in Fig. 2 (Vazouras et al., 2010). The yield stress σ_y of X65 is 448.5 MPa. Young's modulus of steel material equal to 210 Gpa, Poisson's ratio is 0.3, density is 7800 kg m^{-3}. Considering a safety factor equal to 0.72, as suggested in American Society of Mechanical Engineers (2007), and the maximum operating pressure P_{max} of this pipeline given by $P_{max} = 0.72 \times (2\sigma_y t/d)$.

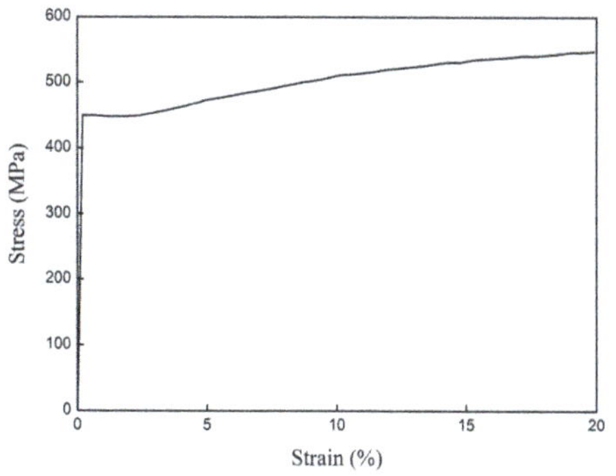

Figure 2. The stress-strain curve of X65.

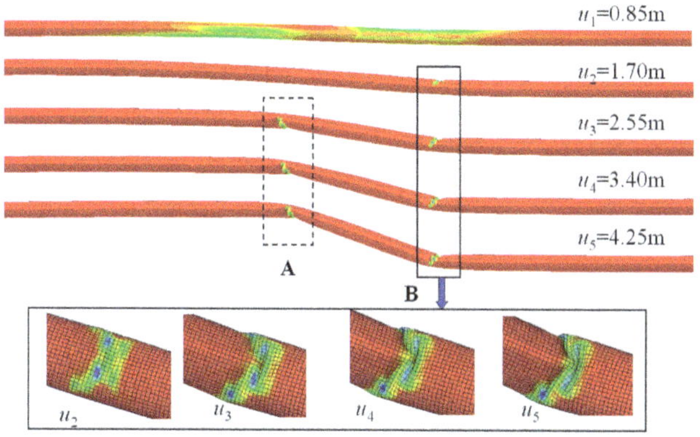

Figure 3. Buckling behaviors of the non-pressure pipeline under different reverse fault displacements.

3 Results and discussions

3.1 Buckling of non-pressure pipeline

When the buried depth $h = 2.5$ m, buckling of the buried non-pressure pipeline under different reverse fault displacements is shown in Fig. 3. Under a small reverse fault displacement, there are two plastic areas on either side of the fault plane. Buckling appears on the section B of the pipeline with the increasing of fault displacement firstly. The buckling mode of non-pressure pipeline is collapse. Then buckling occurs on section A with the fault displacement increases. Buckling modes of the two locations are the same. Deformation shape of the pipeline becomes from S-shape to Z-shape after collapse appears. Local collapse is more serious with the fault displacement.

Figure 4 shows the deformations of buried non-pressure pipeline and surrounding soil when the reverse fault displacement is 3.2 m. A part of the model corresponding to the symmetrical vertical section in the longitudinal direction is shown. Deformations of the two blocks result from the contact interaction with the buried pipeline. Soil arches appear on the top surfaces on both sides of the fault plane. These deformations are caused by the pipeline-soil interaction in the process of the pipeline movement within the surrounding soil. The pipeline is not contact with the hole wall at all. At the two buckling locations, the upper parts of the pipeline contact with the hole wall, while gaps between the pipeline and the surrounding soil under large fault displacement. So, the friction between the soil and buried pipeline is difficult to solve by theoretical analysis method.

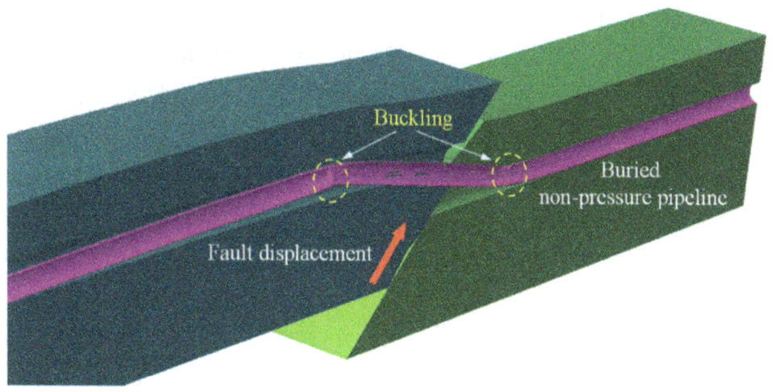

Figure 4. Deformations of the buried non-pressure pipeline and surrounding soil.

3.2 Buckling of pressure pipeline

When the buried depth $h = 2.5$ m, pipeline pressure $P = P_{max}$, buckling of the buried pressure pipeline under different reverse fault displacements is shown in Fig. 5. There are four parts of the buried pipeline with local buckling under reverse fault displacement. When the fault displacement is small, plastic strain appears but with no buckling deformation. With the increasing of the fault displacement, the buckling of section A and D appears firstly, then it appears on section B, and appears on section C lastly. There is only one wrinkle on the section A, B and C, but more wrinkles on section D.

The bending moment increases with the increasing of the fault displacement. For pipeline buckling of section D (Fig. 5a), strain of the lower part is tension strain, while it is mainly compression strain in the upper part. Wrinkle amplitudes increase with the increasing of the fault displacement. For the pipeline buckling of section A, B and C (Fig. 5b), with the increasing of the reverse fault displacement, wrinkle amplitude of section A decreases, wrinkle amplitude of section B increases first and then decreases, while it increases in section C. Because the appear of the wrinkles in section B and C can absorb the energy released by section A. The wrinkle amplitude that far from the fault plane can be eased by the new wrinkle near the fault plane. Wrinkles may reduce the strength of the pipeline and increase the difficulty of pigging. If the plastic strain is bigger than the rupture strain, the leakage will occur.

Variations of axial strain along the two outer generators of the pipeline section D are shown in Fig. 6. At the compression side (Fig. 6a), the outset of local buckling is considered at the stage where outward displacement of the pipeline wall starts at the area of maximum compression. At that stage,

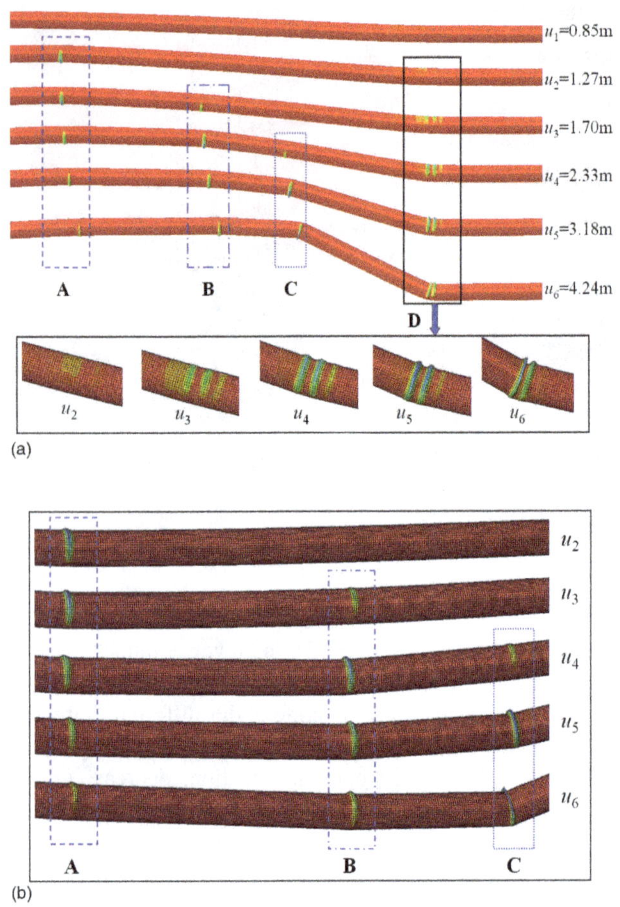

Figure 5. Buckling behaviors of the buried pressure pipeline under different reverse fault displacements. **(a)** Buckling of the pipeline and **(b)** local amplification figure.

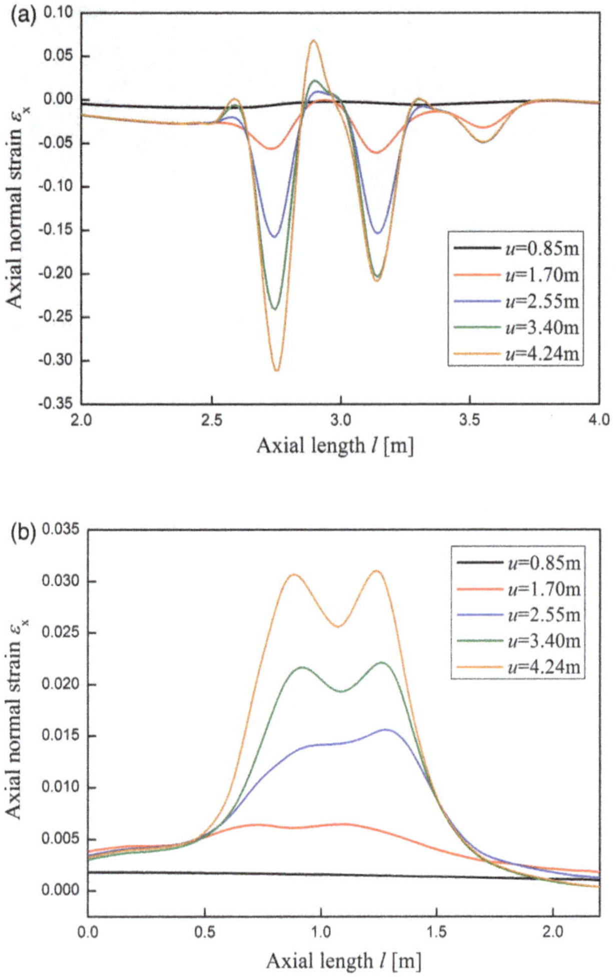

Figure 6. Variation of axial strain of the buckling area on section D under different reverse fault displacements. **(a)** Axial strain at the compression side and **(b)** axial strain at the tension side.

bending strains due to pipeline wall wrinkle development, associated with significant tensile strains at the "ridge" of the buckling. So that the longitudinal compressive strains at this location at the outer surface start decreasing, forming a short wave at this location (Vazouras et al., 2010). At the ridge of the wave, axial strain is the negative maximum value. While at the valley, axial strain is the positive maximum value. The wave amplitude increases with the increasing of the reverse fault displacement. At the tension side (Fig. 6b), the axial strain increases with the increasing of the reverse fault displacement, and the strain curve becomes M-shape under large fault displacement.

Figure 7 shows the bending deformation of the buried pressure pipeline under different fault displacements. Bending deformation of the buried pipeline under bending moment caused by non-uniform deformation of the surrounding soil under reverse fault displacement. With the increasing of fault displacement, bending deformation curve is more serious. Meanwhile, the deformation shape becomes from a smooth curve to a non-smooth curve after the inflection point appears. The inflection points reflect the local buck-

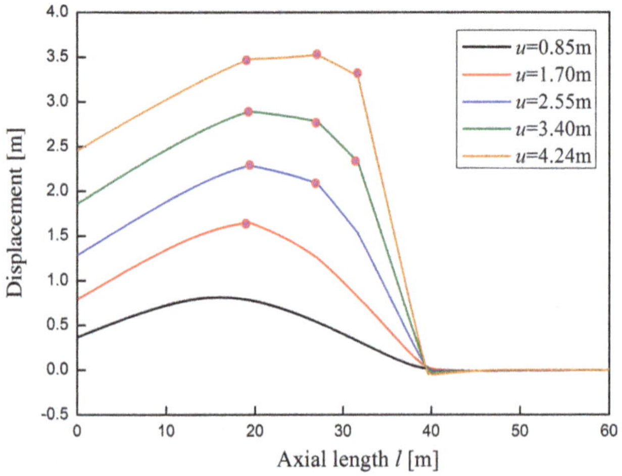

Figure 7. Bending deformations of the buried pipeline under different fault displacements.

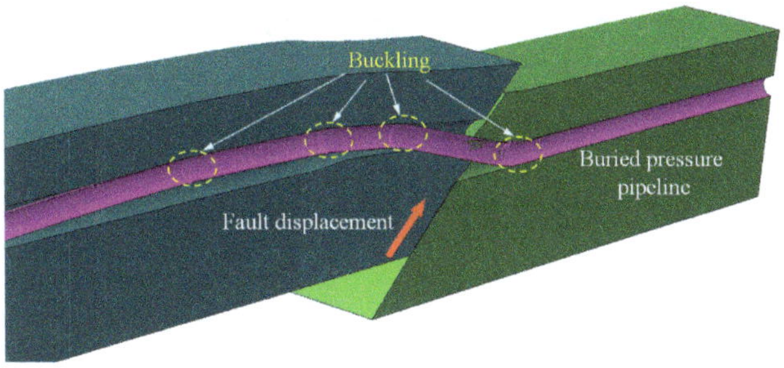

Figure 8. Deformations of the buried pressure pipeline and surrounding soil.

lings. There are three inflection points of the deformation curve in the moving block.

Figure 8 shows the deformations of the pressure pipeline and surrounding soil when the fault displacement is 3.2 m. The pipeline sections near the fault plane contact with the upper and lower part of the hole wall respectively. Deformations of the two block soils are different. Because the top surfaces of the soil can be arched by the bending deformed pipeline, but deformation of the lower part soil is small for the thick stratum.

3.3 Effect of pipeline pressure

When the buried depth $h = 2.5$ m, reverse fault displacement $u = 3.4$ m, friction coefficient $f = 0.3$, buckling behavior of the buried pressure pipeline under different pipeline internal pressures is shown in Fig. 9. When the internal pressure $P \leq 0.4\ P_{max}$, there are only two buckling locations in the pipeline. When $P = 0.5\ P_{max}$, there are three buckling locations. When $P \geq 0.5\ P_{max}$, four buckling locations are on the pipeline. Buckling morphologies of section D pipeline are different under different internal pressures. Collapse buckling appears when the internal pressure is zone or small. While wrinkle appears when the internal pressure is more than 0.4 P_{max}. The internal pressure can enhance the stiffness of the buried pipeline to resistance to bending moment.

Variations of axial strain at the compression side of the pipeline in section D under different internal pressures are shown in Fig. 10. At the compression side, when $P \leq 0.4\ P_{max}$, there is only one crest and valley, the tensile strain is big and the compression strain is small. It illustrates that the buckling pattern is collapse. But with the increasing of the internal pressure, the tensile strain decreases and the compression strain increases, and more than two valleys appear. When $P = P_{max}$, there are three valleys in section D. It illustrates that the buckling pattern become to wavy pattern. Therefore, the internal pressure is the most important factor that affecting the pipeline buckling pattern.

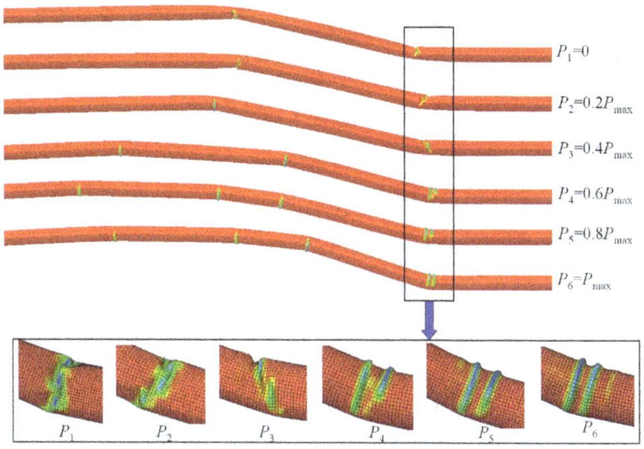

Figure 9. Buckling behaviors of the buried pipeline under different internal pressures.

3.4 Effect of diameter-thick ratio

When the buried depth $h = 2.5$ m, reverse fault displacement $u = 3.4$ m, friction coefficient $f = 0.3$, buckling behavior of the buried pressure pipeline under different diameter-thick ratios is shown in Fig. 11. With the decreasing of the diameter-thick ratio, the buckling locations decrease gradually. The greater diameter-thick ratio can enhance the ability to resistance to bending moment. When $d/t = 45$, there is only one buckling location, and the bending curve of the pipeline is smooth. The amplitude of the wrinkle decreases with the decreasing of the diameter-thick ratio. Therefore, ability to resist damage by reverse fault displacement can be improved by increasing the wall thickness in a dangerous hazard area.

Variations of axial strain at the compression side of the pipeline in section D under different diameter-thick ratios are shown in Fig. 12. At the compression side, with the decreasing of the diameter-thick ratio, wave of the strain curve becomes smooth gradually, but the number of the wave crest increases. The maximum tensile strain and compressive strain decrease with the decreasing of the diameter-thick ratio. When $d/t = 45$, compressive strains of the three valleys are close to each other. It illustrates that the probability of pipeline buckling is smaller with a lower diameter-thick ratio.

3.5 Effect of friction coefficient

In the process of reverse fault displacement, the friction force between surrounding soil and pipeline can be divided into two parts. One is static friction force before the soil yield, the other is the sliding friction force after the soil yield (Liu, 2002). When the axial deformation of the pipeline appears, the surrounding soil will be resistance to the relative movement. When the resistance reaches limit value, the surround-

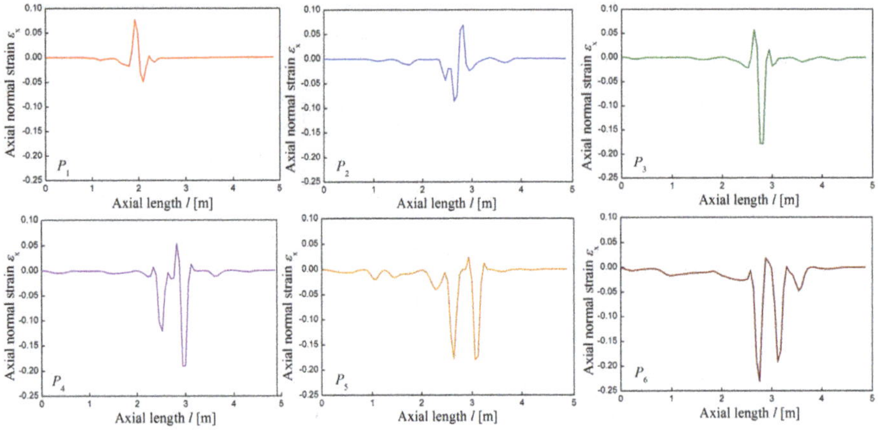

Figure 10. Variations of axial strain at the compression side of the buckling area in section D under different internal pressures.

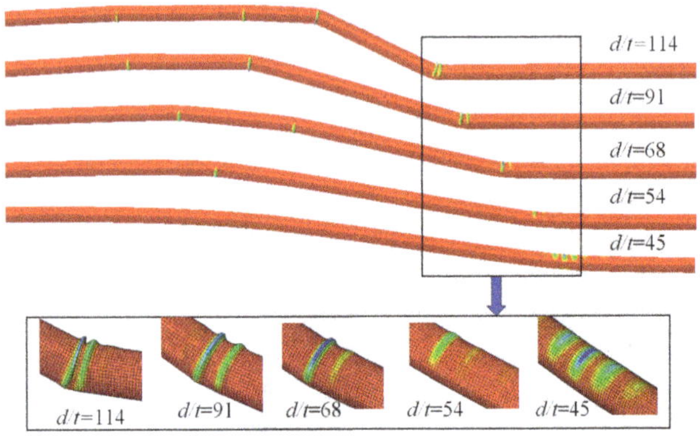

Figure 11. Buckling behaviors of the buried pipeline under different diameter-thick ratios.

ing soil will yield, then relative sliding appears between the pipeline and soil.

When the buried depth h =2.5m, reverse fault displacement $u = 4.2$ m, diameter-thick ratio $d/t = 114$, variations of axial strain at the compression side of the pipeline in section D under different friction coefficients are shown in Fig. 13. At the compression side, the number of the wrinkle ridges increases with the increasing of friction coefficient. The length of the wavy buckling also increases in section D. When friction coefficient $f < 0.4$, the curve shapes are similar. With the increasing of the friction coefficient, the valley number and the maximum compressive strain increase.

3.6 Effect of buried depth

When the reverse fault displacement $u = 2.4$ m, diameter-thick ratio $d/t = 114$, friction coefficient $f = 0.3$, variations of axial strain at the compression side of the pipeline in section D under different buried depths are shown in Fig. 14. At the compression side, the maximum axial tensile strain decreases with the increasing of the buried depth. When the buried depth $h \geq 2.5$ m, the compression strain of the first ridge is the biggest.

4 Protective device design

In order to reduce the failure probability and improve the service life of buried pipeline under reverse fault, protective measures should be designed. In this paper, a protective device of the buried pipeline is designed for preventing pipeline damage crossing fault area. As shown in Fig. 15, it consists of buried pipeline, protective pipeline, sealing ring, baffle ring, flange structure, end cone pipeline, water channel and water. For the reverse fault area, protective pipeline is sleeved on the buried pipeline. Flange structures are used to connect the protective pipeline and end cone pipelines. Cone pipelines were installed in the end. Sealing rings are installed between the buried pipeline and protective pipeline for sealing the fluid. Baffle rings are installed in front of the sealing rings. Water can be pumped into the annular space between the buried pipeline and protective pipeline by the water channel.

Under reverse fault displacement, deformation appears on protective pipeline firstly, and annulus hindered the contact between protective pipeline and buried pipeline. If the protection pipeline is compressed under reverse fault displacement, pressure of the water between the buried pipeline and protective pipeline can act on the buried pipeline. It can avoid the local deformation of buried pipeline caused by large local loads. If buckling appears on protective pipeline, it can be timely repaired and replaced without stopping the transmission of oil and gas. So, this protective device can effectively protect the oil and gas pipeline. And the protective device can be widely used in different locations for its simple structure and convenient installation. However, the reverse fault plane is difficult to be predicted in advance. So, length of the protective device may be longer to reduce the effect of reverse fault.

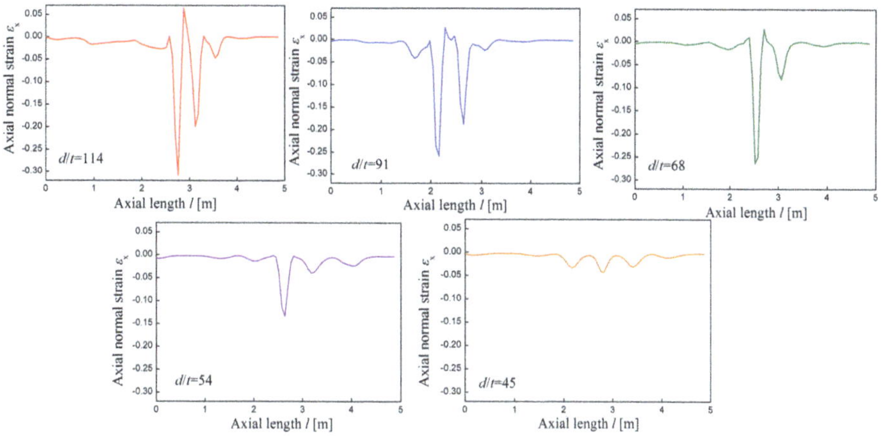

Figure 12. Variations of axial strain at the compression side of the buckling area under different diameter-thick ratios.

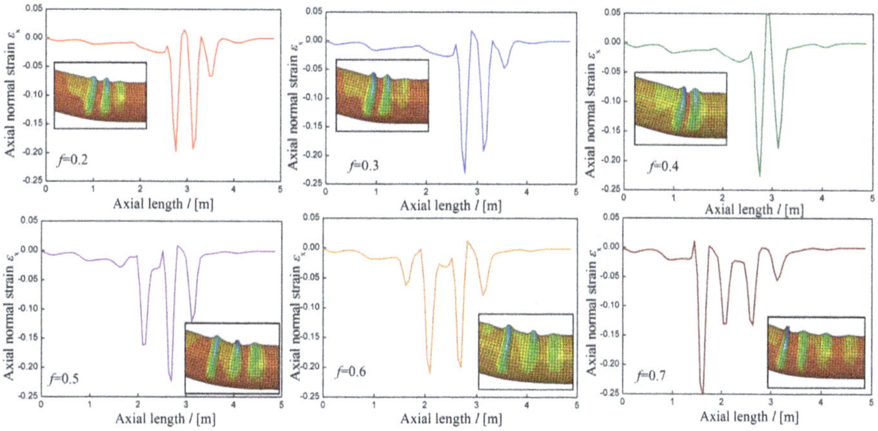

Figure 13. Variations of axial strain at the compression side of the buckling area under different friction coefficients.

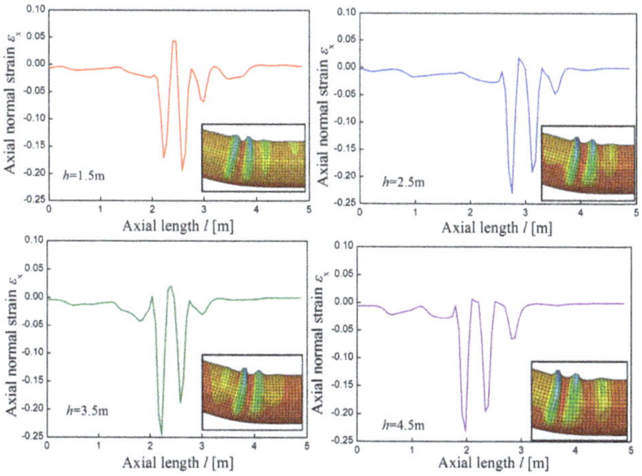

Figure 14. Variations of axial strain at the compression side of the buckling area under different buried depths.

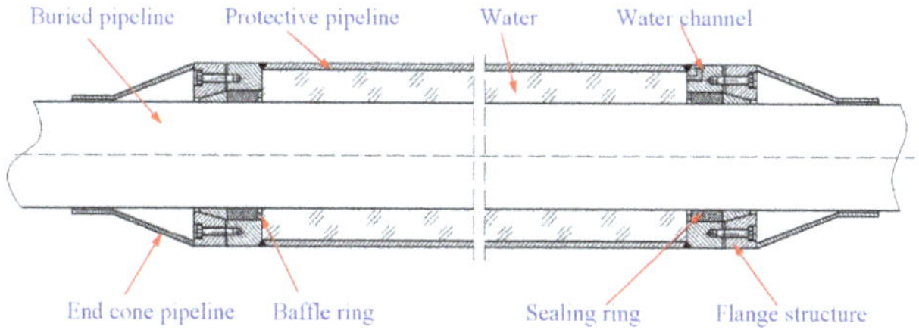

Figure 15. A protective device of the buried pipeline crossing reverse fault area.

Figure 16 shows the buckling behaviors of the pressure pipeline in two conditions when the fault displacement is 1.9 m. In this case, the buried pipeline diameter is 660 mm, the protective pipeline diameter is 762 mm, the annular pressure is 0.3 MPa. Two buckling locations appear on the buried pipeline when there is no protective device. The maximum plastic strain is 0.1418. Under protective device, there is only one plastic zone, and the maximum plastic strain is 0.0053.

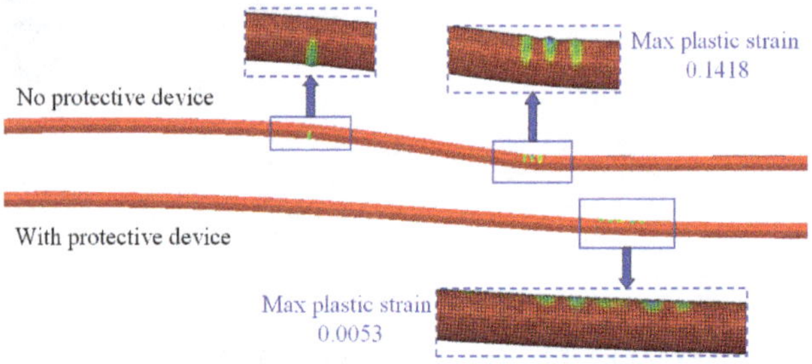

Figure 16. Buckling behaviors of the buried pressure pipeline in two conditions.

Therefore, the protective device can effectively protect the buried pipeline crossing reverse fault area.

5 Conclusions

Numerical simulation of buckling behavior of the buried X65 pipeline under reverse fault displacement in this paper led to the follow conclusions:

1. The buckling mode of non-pressure pipeline is collapse under reverse fault displacement. But wrinkles appear on buried pressure pipeline when the internal pressure is more than 0.4 P_{max}. For pressure pipeline, there are four buckling locations on the buried pipeline under reverse fault displacement. There is only one wrinkle on the three locations of the pipeline in the rising formation, but more wrinkles on the fourth location. Axial strain of the wrinkles section increases with the increasing of the reverse fault displacement. Internal pressure is the most important factor that affecting the pipeline buckling pattern.

2. With the decreasing of the diameter-thick ratio, the buckling locations decrease gradually. Strain curve becomes smooth gradually with the decreasing of the diameter-thick ratio at the compression side. Thick wall pipelines can be laid in the fault areas. Number of the wrinkle ridges and length of the wavy buckling increase with the increasing of the friction coefficient. Buried depth has a great effect on the buckling pattern.

3. A protective device of the buried pipeline is designed for preventing pipeline damage crossing fault area for its simple structure and convenient installation. The protective pipeline and water in the annular space can effectively protect the oil and gas pipeline under stratum deformation.

4. The methodology of deformation evaluation, buckling mode and limit state analysis developed in the paper can be used to safety assessment and prediction of buried pipeline crossing fault area. But a comparison to experimental results or real event data is needed for the verification of the finite element model. The protective device of the buried pipeline is also needed to be tested by experiments, and its performance needs to be evaluated.

Author contributions. J. Zhang completed the numerical simulation of the results and preparation of the manuscript. Z. Liang contributed to the initial design scheme, with assistance from all co-authors. C. J. Han and H. Zhang provided many contributions to the preparation of the manuscript.

Acknowledgements. This research work was supported by the Science and Technology Innovation Talent Engineering Project of Sichuan Province (2015097) and National Natural Science Foundation of China (51474180).

References

American Society of Mechanical Engineers: Gas transmission and distribution piping systems, ANSI/ASME B31.8, New York, 2007.

Duan, M. L., Mao, D. F., Yue, Z. Y., Segen, E., and Li, Z. G.: A seismic design method for subsea pipelines against earthquake fault movement, China Ocean Eng., 25, 179–188, 2011.

Karamitros, D. K., Bouckovalas, G. D., and Kouretzis, G. P.: Stress analysis of buried steel pipelines at strike-slip fault crossings, Soil Dyn. Earthq. Eng., 27, 200–211, 2007.

Liu, A. W.: Response analysis of a buried pipeline crossing the fault based on shell-model, China Seismological Bureau, Institute of Geophysics, Beijing, 2002.

Liu, M., Wang, Y. Y., and Yu, Z.: Response of pipelines under fault crossing, Proceedings of the international offshore and polar engineering conference, Vancouver, BC, Canada, 162–166, 2008.

Oleg, V. T. and Vladimir, P. C.: A semi-analytical approach to a nonlinear stress-strain analysis of buried steel pipelines crossing active faults, Soil Dyn. Earthq. Eng., 30, 1298–1308, 2010.

Shantanu, J., Amit, P., and Arghya, D.: Analysis of buried pipelines subjected to reverse fault motion, Soil Dyn. Earthq. Eng., 31, 930–940, 2011.

Vazouras, P., Karamanos, S. A., and Dakoulas, P.: Finite element analysis of buried steel pipelines under strike-slip fault displacement, Soil Dyn. Earthq. Eng., 30, 1361–1376, 2010.

Vazouras, P., Karamanos, S. A., and Dakoulas, P.: Mechanical behavior of buried steel pipes crossing active strike-slip faults, Soil Dyn. Earthq. Eng., 41, 164–180, 2012.

Wang, B., Li, X., and Zhou, J.: Strain analysis of buried steel pipelines across strike-slip faults, J. Cent. S. Univ. Technol., 18, 1654–1661, 2011.

Wang, S. F., Yin, Y. P., and Men, Y. M.: In-situ test and numerical analysis of skid resistance for micropile to loess landslide, Hydrogeol. Eng. Geol., 37, 22–26, 2010.

5

Thermal analysis of PCM containing heat exchanger enhanced with normal annular fines

M. J. Hosseini[1], M. Rahimi[1], and R. Bahrampoury[2]

[1]Department of Mechanical Engineering, Golestan University, P.O. Box 155, Gorgan, Iran
[2]Department of Mechanical Engineering, K. N. Toosi University of Technology, Tehran, Iran

Correspondence to: M. J. Hosseini (mj.hosseini@gu.ac.ir)

Abstract. In this study, the effect of fins' height and Stefan number on performance of a shell and tube heat exchanger which contains a phase change material is investigated numerically and experimentally. Melting time, solidification time, liquid mass fraction, melting and solidification front and temperature distribution in different directions (longitudinal, radial and angular) are among criteria for the heat exchangers' comparison. In order to generalize the comparison, melting and solidification fronts are studied for different sections of the shell, fin section and mid-section, for different fins' height during charging and discharging processes. The results show that, these two parameters play important roles in the heat exchanger performance. Increasing Stefan number, the melting time reduces; which exhibits a descending trend in rate when the fins are heightened. In addition, investigating both processes, it can be figured out that increasing fins' height influences the solidification time more significantly than melting.

1 Introduction

Due to the increasing gap between the global energy supply and demand, reaching to a thermally efficient and cost optimized thermal energy storage system has received a considerable attention among researchers. There are three methods for storing thermal energy: sensible, latent and thermal–chemical. Among these methods, latent heat thermal storage (LHTS) using phase change materials (PCMs) is known as the most favorable for its high energy storage density with small temperature variation (Mehling and Cabeza, 2007). In other words, PCMs are attractive as they are capable of absorbing and releasing a considerable amount of energy at a nearly constant temperature during melting and solidification processes. Latent heat energy storage systems can be used to store a considerable amount of available thermal energy to be utilized during energy demand period, hereafter providing a promising solution for smoothing the discrepancy between energy supply and demand. Thus, many authors have reported their results of researches on PCM thermal storage during melting and solidification processes in energy storage systems.

Zalba et al. (2003) carried out a review on history of thermal energy storage which deals with phase change materials, heat transfer studies and applications. Sharma et al. (2009) summarized studies on available thermal energy storage systems for different applications.

Temperature and mass flow rate during both melting and solidification are studied for a shell and tube heat exchanger by Hosseini et al. (2012, 2014) in which energy absorption capability of the inserted PCM and thermal characteristic of the PCM storage system are investigated respectively. Experimental results indicated that by increasing inlet HTF temperature to 80 °C, theoretical efficiency in charging and discharging processes rises to 88.4 and 81.4 %, respectively. They also showed that the same inlet temperature increase leads to 37 % reduction in total melting time.

Ezan et al. (2011) experimentally studied melting and solidification of water as a phase change material in a shell and tube system. They focused on investigation of the effect of flow rate, inlet temperature, thermal conductivity of the tube material and shell diameter on the storage capability of the system.

From the results obtained, it is concluded that for both solidification and melting processes, natural convection becomes the dominant heat transfer mechanism after a short interval of heat conduction domination. It is also claimed that during discharging period, the inlet temperature of HTF is more effective on the amount of transferred energy in comparison with flow rate, for selected parameters.

Agyenim et al. (2009, 2010, 2011) and Agyenim and Hewitt (2010) investigated melting and solidification of a PCM in a shell and tube heat exchanger with the HTF passing inside the tube and the PCM filling the shell side, for various operating conditions and geometric parameters.

Fins, or more generally extended surfaces, are used to provide additional heat transfer surface in thermal systems. In LHTS systems, various researchers extensively studied the role of different configurations of fins on the performance improvement characteristics of LHTS systems. Subsequently, different numerical studies looking at the impact of fins on overall PCM melting and solidification can be found in literature (Ogoh and Groulx, 2012; Seeniraj and Narasimhan, 2008; Shatikian et al., 2005); typically those studies still neglect natural convection in the liquid PCM phase. Although the cited numerical studies provides the tool to determine optimum fins geometry and LHTS configuration; the defect in natural convection simulations brings about the need to perform experimental studies.

Mosaffa et al. (2012) studied a two-dimensional numerical model based on an enthalpy formulation for prediction of the solid–liquid interface location in a vertical shell and tube LHTS with radial fins. The results indicate that the PCM solidifies more quickly in a cylindrical shell storage than in a rectangular one. In addition the solid fraction of the PCM increases more quickly when the cell aspect ratio is small.

An experimental study of solidification and melting of a phase change material is investigated in a fin and tube heat exchanger by Rahimi et al. (2014). An experimental apparatus was used to investigate the effect of flow rate, inlet temperature and a geometrical parameter (fin pitch) on charging and discharging processes of the heat exchanger. Experimental results showed that, by increasing inlet temperature from $T_h = 50\,°C$ to $T_h = 60\,°C$, melting time decreases more severally in comparison with the same rise from $T_h = 60\,°C$ to $T_h = 70\,°C$.

Mat et al. (2013) conducted numerical study on melting process in a triplex-tube heat exchanger based on three charging cases/approaches with a phase-change material (PCM), RT82. The study focuses on the PCM melting process in a triplex tube in which 3 different heat transfer methods are considered; (1) internal tube is subjected to heat transfer, (2) outside tube is subjected to heat transfer and (3) both sides are subjected to heat transfer. They also studied the effect of internal, external and internal-external fins on heat transfer enhancement between the PCM and the HTF whereas the consequences of fins' height as an effective parameter were investigated. They found that using a triplex-tube heat exchanger enhanced with internal-external fins, melting time reduces to 43.3 % in comparison with the similar finless heat exchanger.

Liu and Groulx (2014) experimentally investigated phase change heat transfer inside a horizontal cylindrical latent heat energy storage system designed with a central finned copper pipe running the length of the cylindrical container, during charging and discharging operations. It was observed that conduction is the dominant heat transfer mechanism during the initial stages of charging, and natural convection dominates once enough PCM melt is present inside the system. However, conduction dominates during the entire solidification process. Complete melting time is strongly affected by the HTF inlet temperature but it reduces to a much less extent by the HTF flow rate.

The effect of axial fins on enhancement during solidification process has also been examined by Ismail et al. (2001). Their experimental measurements and theoretical predictions obtained from their numerical analysis for different fins' heights, numbers of fins, fins' thicknesses and aspect ratios of annular spaces showed that the use of fins improves the heat transfer rate significantly.

Pakrouh et al. (2015) performed a parametric investigation on pin fin heat sink enhanced with a phase change material. Their results show that an increase in number fins as well as their thickness and height leads to a decrease in base temperature while the operating time reduces.

In the present study, melting and solidification of a specific PCM is explored in a finned shell and tube heat exchanger for two fin heights and three Stephan numbers to study the effect of these two variables on some decision making parameters. These criteria include temperature distribution, melting and solidification front and total melting and solidification time.

2 Geometric details of the heat exchanger

In order to study the heat exchanger, a double pipe heat exchanger is designed in which fines are properly spaced in the shell and around the tube. The scaled section view of the heat exchanger is exhibited in Fig. 1a to show the internal components and the sections under consideration in color. As can be seen, the mounted and welded 8 fins are spaced equally and symmetrically in a way that the tube axis coincides with the fin's surface normal vector and the sections selected for thermal measurement are introduced in white transparent circular plates. The longitudinal distance of the first fin from the cap is half of the fin spacing.

Figure 1b illustrates a photo of the heat exchanger connected to a data logger to record thermal values of the PCM at the considered sections. Complimentary data including the employed materials and shell capacity are given in Table 1.

In order to measure temperature at different positions of the heat exchanger, 3 sections are chosen symmetrically. In each of the sections a number of thermocouples are imple-

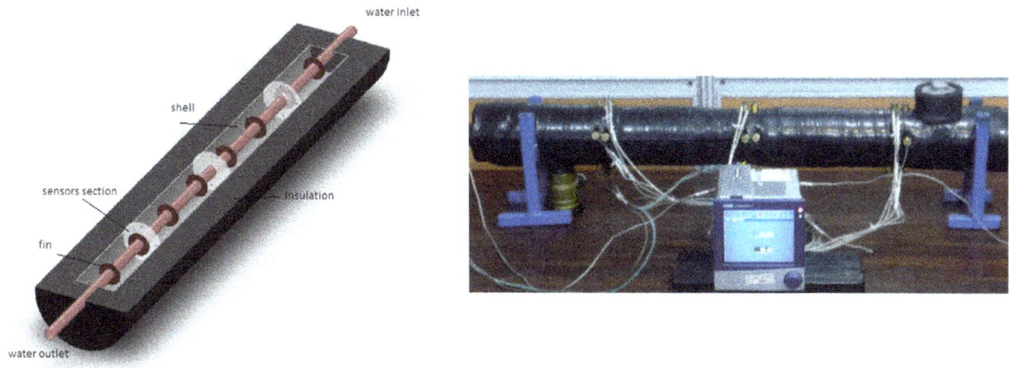

(a) (b)

Figure 1. **(a)** Section view of the heat exchanger and **(b)** real photo of heat exchanger.

Table 1. Complimentary specifications of the heat exchanger.

Specification	Type/value
Fins' material	Copper
Shell's material	Iron
Insulation	Glass wool
Number of fins	8
Shell capacity (m^3)	0.005294
Contact surface (excluding fins) (m^2)	0.06912
Small fins height	13 mm
Small fins' contact surface (m^2)	0.02278
Large fins height	26 mm
Large fins' contact surface (m^2)	0.0673

Table 2. Thermophysical properties of RT50.

Properties	Typical values
T_S [K]	318
T_{Liq} [K]	324
ρ [kg m^{-3}]	780
c_P [J kg^{-1} K^{-1}]	2000
k [W m^{-1} K^{-1}]	0.2
L [J kg^{-1}]	168 000
β [1/K]	0.0006

mented. The exact positions of sensors as well as geometric details of the heat exchanger are shown in Fig. 2.

3 Setup and procedure

Figure 3 illustrates a schematic diagram of the experimental apparatus which basically consist of a flow control system, the heat exchanger unit and the measurement system.

As the study includes both charging and discharging processes, two different loops are designed to provide hot and cold utility for melting and solidification respectively. The hot loop in which hot bath joins to produce the required heat is connected to the cold loop (which includes a cold bath) utilizing appropriate valve to facilitate the selection of each loop to run the corresponding process. As flow rate is an indispensible component of heat transfer rate calculation, a rotameters is employed in the experimental setup. As observed in Fig. 2, three sections in the shell are designated for thermal measurement; the two end sections include 6 thermocouples while the middle one involves 4. The exact position of the section as well as angular position and radial distances of every thermocouple is presented in Fig. 2. Considering these 16 thermocouples as well as the HTF inlet and outlet thermal measuring thermocouples, 18 thermocouples are incorporated in the experimental setup.

4 Methodology of tests

In this study RT50 (Rubitherm GmbH) is selected as PCM to be poured within the shell. The most important characteristics of the material are summarized in Table 2. Additional information about the PCM is given by Hosseini et al. (2014).

After the heat exchanger is filled up with liquid PCM and no leakage was observed, a few runs are made in order to calibrate the system. Then the charging process starts, while the solid PCM is at thermal equilibrium with the conditioned lab temperature (23–25 °C). During the charging process, inlet HTF temperature, T_H, is maintained at a set temperature using a PID controlled hot water bath. After an imposed 150 min of charge time, discharging part of the experiment begins. In this process, cold water which solidifies the PCM is pumped from the bath to the heat exchanger. The experiment is finishes after 180 min of discharge process. During the whole process temperature values are recorded every 1 min.

Using the apparatus and procedures described above, several experiments have been conducted to study the behav-

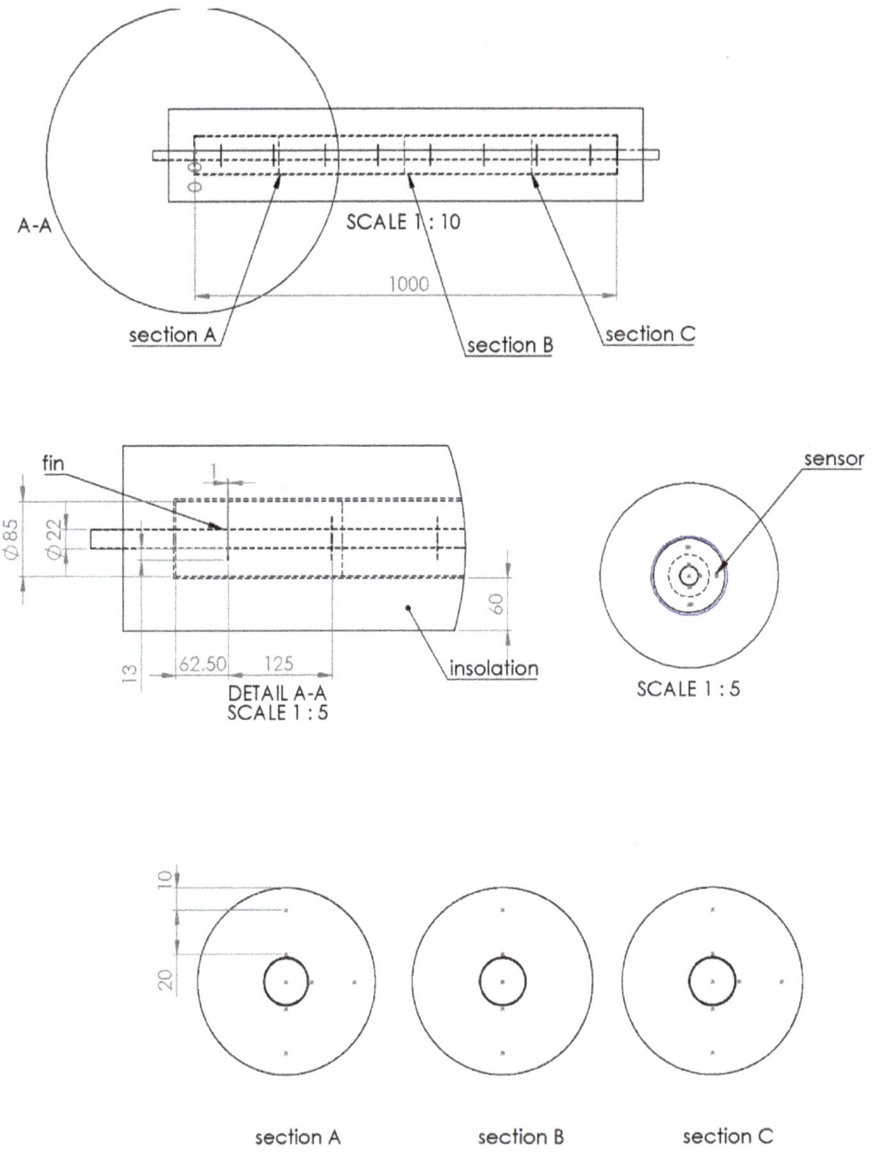

Figure 2. Geometric details of the finned-tub heat exchanger and sensors position.

ior of the PCM during charging and discharging processes. The experiments are performed for different inlet temperatures (70, 75 and 80 °C) of the HTF and fins' height (13 and 26 mm). Corresponding Stephan number (*Ste*) (the ratio of sensible heat to latent heat) as defined in Eq. (1) for each of these hot HTFs are calculated in which the only varying parameter is temperature difference

$$Ste = \frac{C_p(T_{\mathrm{H}} - T_{\mathrm{m}})}{L} \tag{1}$$

where C_p, L, and T_{m} are respectively the specific heat, the latent heat and the mean melting temperature of the PCM and T_{H} is the temperature of hot inlet water. For the experiments presented in this paper, the Stefan numbers for the three melting temperatures are 0.26, 0.32 and 0.38, respectively. Discharging experiments are conducted for the same flow rates but only at $T_{\mathrm{C}} = 25$ °C.

5 Numerical models

In order to simulate phase change phenomenon during melting and solidification in a shell and tube heat exchanger, enthalpy-porosity method Brent et al. (1988) and Gong et al. (1999) is used. In the present study, both PCM and water flows are considered to be unsteady, laminar, incompressible and three-dimensional. The viscous dissipation term is considered negligible so that the viscous incompressible flow and the temperature distribution in annulus space are described by the Navier-Stokes and thermal energy equations, respectively. Also dynamic viscosity considered as a func-

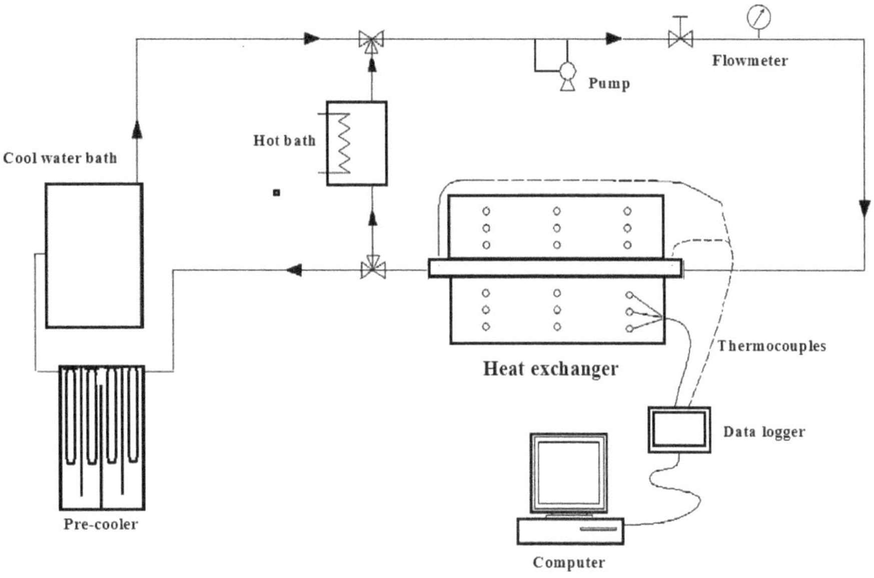

Figure 3. Schematic diagram of experimental setup.

tion of temperature is also used (Hosseini et al., 2014). In addition, the density change within the liquid phase that drives the natural convection is only considered in the body force terms (Boussinesq approximation) in which the variable density is defined as $\rho = \rho_0(\beta(T - T_{\text{Liq}}) + 1)^{-1}$ for $51\,°\text{C} < T < 100\,°\text{C}$ in the liquid state. The initial temperature of the whole system is $T_{\text{Ini}} = 25\,°\text{C}$. Also the lateral surface of the outer tube is assumed to be insulated. Consequently, the continuity, momentum, and thermal energy equations can be expressed as follows:

$$\text{Continuity}: \ \nabla \cdot \boldsymbol{V} = 0 \tag{2}$$

$$\text{Momentum}: \frac{\partial \boldsymbol{V}}{\partial t} + \boldsymbol{V} \cdot \nabla \boldsymbol{V} = \frac{1}{\rho}\Big(-\nabla P + \mu \nabla^2 \boldsymbol{V} + \rho \beta \boldsymbol{g}\,(T - T_{\text{ref}})\Big) + \boldsymbol{S} \tag{3}$$

$$\text{Thermal energy}: \ \frac{\partial h}{\partial t} + \frac{\partial H}{\partial t} + \nabla(\boldsymbol{V} h) = \nabla\left(\frac{k}{\rho c_p} \nabla h\right). \tag{4}$$

The enthalpy of the material is computed as the sum of the sensible enthalpy, h, and the latent heat, ΔH:

$$H = h + \Delta H \tag{5}$$

where

$$h = h_{\text{ref}} + \int_{T_{\text{ref}}}^{T} C_p \mathrm{d}T. \tag{6}$$

The latent heat content can be written in terms of the latent heat of the material:

$$\Delta H = \lambda L \tag{7}$$

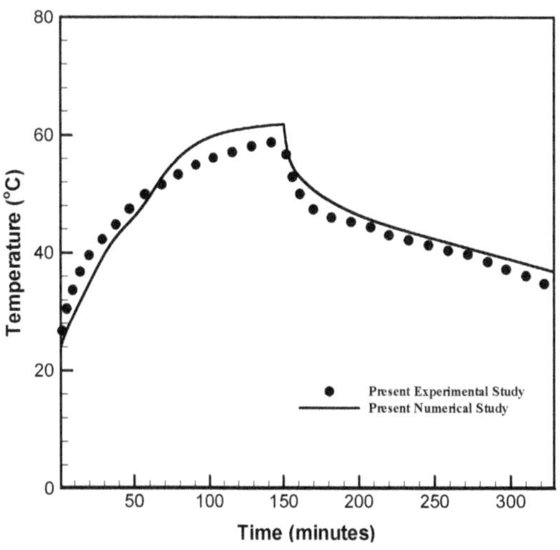

Figure 4. Comparison between numerical and experimental results.

where ΔH may vary from zero (solid) to L (liquid). Therefore, the liquid fraction, λ, can be defined as:

$$\lambda = \begin{cases} \frac{\Delta H}{L} = 0 & \text{if } T < T_{\text{S}} \\ \frac{\Delta H}{L} = 1 & \text{if } T > T_{\text{Liq}} \\ \frac{\Delta H}{L} = \frac{T - T_{\text{S}}}{T_{\text{Liq}} - T_{\text{S}}} & \text{if } T_{\text{S}} < T < T_{\text{Liq}} \end{cases}. \tag{8}$$

In Eq. (3), $\boldsymbol{S}$ is the Darcy's law damping terms (as source term) that is added to the momentum equation due to phase change effects on convective heat transfer which is defined as:

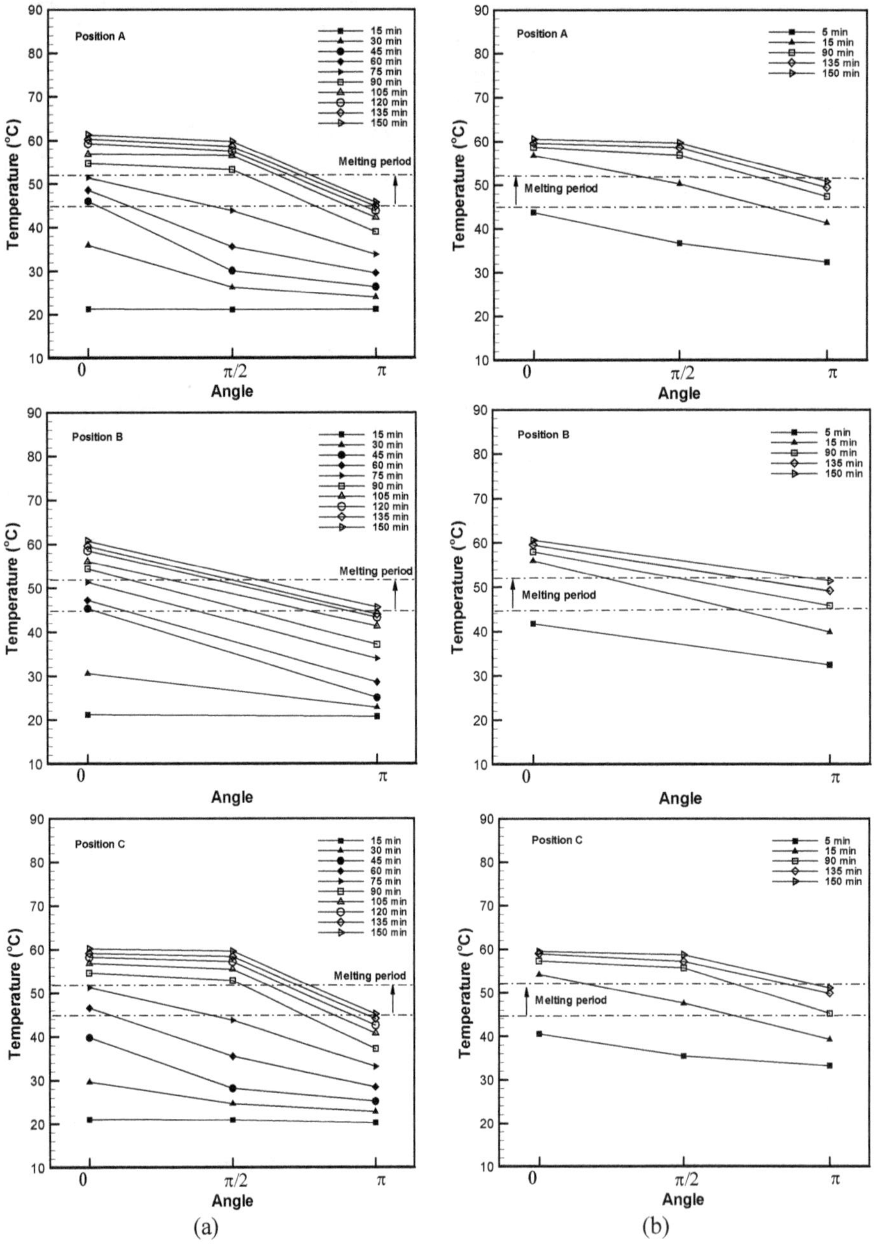

Figure 5. Experimental result of temperature distribution for 13 mm height fins heat exchanger for different angles during charging process for *Ste* = 0.26: **(a)** 32.5 mm radial distance and **(b)** 12.5 mm radial distance.

$$S = \frac{(1-\lambda)^2}{\lambda^3} A_{\text{mush}} V. \tag{9}$$

The coefficient A_{mush} is a mushy zone constant. This constant is a large number, usually 10^4–10^7. In the current study A_{mush} is assumed constant and is set to 10^6.

Computational methodology

The SIMPLE algorithm (Patankar, 1980) whose task is to couple pressure–velocity governing differential equations and QUICK differencing scheme is employed for solving the momentum and energy ones within a 3-D in-house developed code (Hosseini et al., 2012), whereas the PRESTO scheme is adopted for the pressure correction equation. The under-relaxation coefficient for momentum, pressure, density, melting fraction and thermal energy are 0.7, 0.3, 1, 0.9 and 1, respectively. An arrangement of 420 000 grids is found to be sufficient for the present study. The time step in the calculations is as small as 0.05 s and the number of iterations for each time step is 400. The grid size and the time step were chosen after careful examination of the independency of the

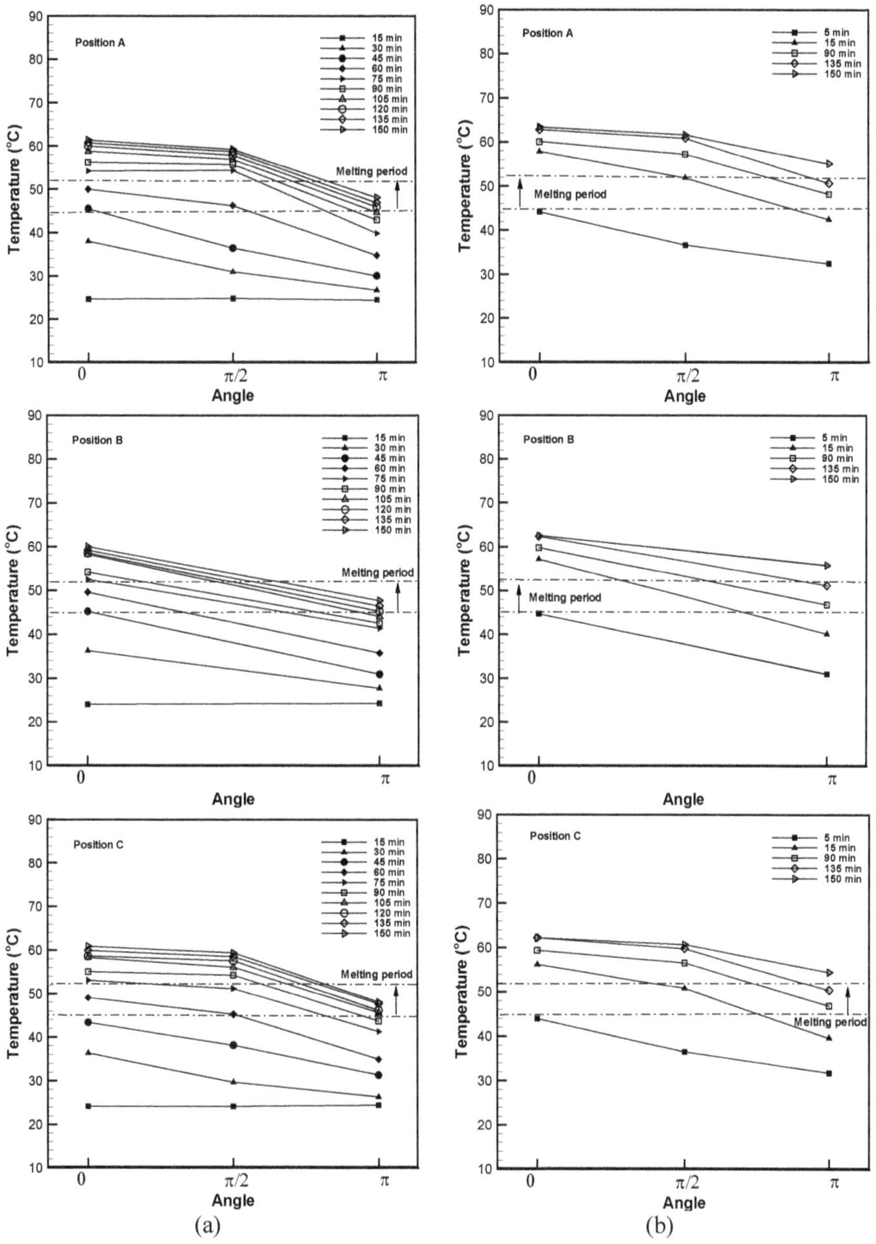

Figure 6. Experimental results of temperature distribution for 26 mm height fins heat exchanger for different angles during charging process for *Ste* = 0.26: **(a)** 32.5 mm radial distance and **(b)** 12.5 mm radial distance.

results to these parameters. The convergence is checked at each time step, with the convergence criterion of 10^{-7} for all variables.

In this section, the presented numerical solution of the simulation of the heat exchanger for fins' height of 13 mm and Stefan number of 0.26 is validated via the current experiment. Figure 4 illustrates the average temperature of the PCM versus time during charge and discharge processes. The figure exhibits a rational compatibility between these two.

6 Results and discussion

6.1 Charging process

As a phase change material is exposed to heat, the melting front appears and develops in a way that the ratio of solid PCM to liquid PCM continuously varies while heat transfer mechanism differs for both fronts; the dominated mechanism of heat transfer in the solid region is conduction through which heat is absorbed from the liquid melt and leads to a rise in the solid PCM temperature. In the other region (melt), convective heat transfer accompanying buoyancy based vor-

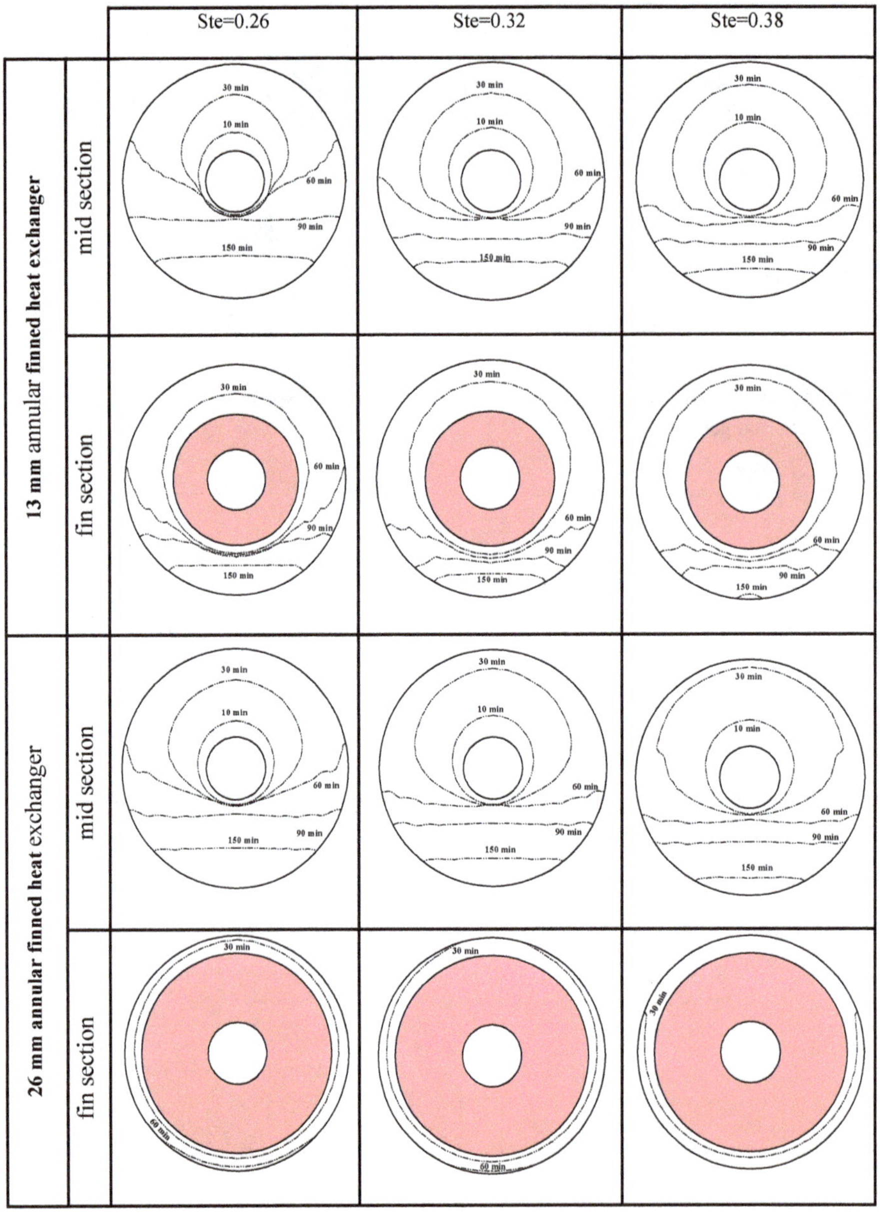

Figure 7. Numerical results showing melting front in shell for different Stefan numbers in 13 and 26 mm fins' height at mid section and fin section.

tices results in superior heat transfer. In this study, in order to investigate temperature variation in every point of these regions and to study the heat transfer mechanisms, local temperatures are recorded via precisely implemented thermocouples. Figure 5 illustrates the temperature distribution in every direction of the enhanced finned heat exchanger including radial, angular and axial for Stephan number of 0.26 and fins' height of 13 mm while the 26 mm counterpart is shown in Fig. 6.

For both of the heat exchangers, the sensor at angle of 0° (above the central tube) senses a higher temperature regardless of time and section. This phenomenon is due to the upward movement of the hot front which originates from buoyancy law. Moreover, for the sections in which 6 thermocouples are employed, the lateral point (90°) is always again hotter than the underside one (180°) due to the same reason for both heat exchangers. Considering both figures, the interesting point here is that, while heating and before the emergence of melt drops, the temperature difference between 0 and 90° is more considerable than this difference between 90 and 180° however, when the melting process initiates and proceeds, the later temperature difference exceeds the first one. Comparing Figs. 5 and 6, it can be concluded that, utilizing higher fins due to farther radial penetration of

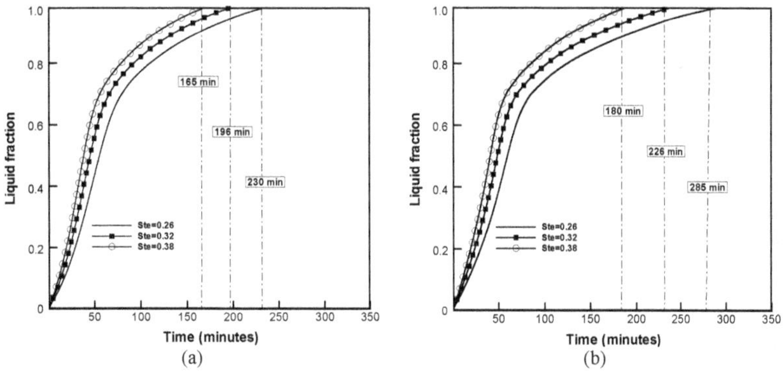

Figure 8. Numerical results showing liquid fraction variation versus time at different Stefan numbers: (**a**) 26 mm annular fin heat exchanger and (**b**) 13 mm annular fin heat exchanger.

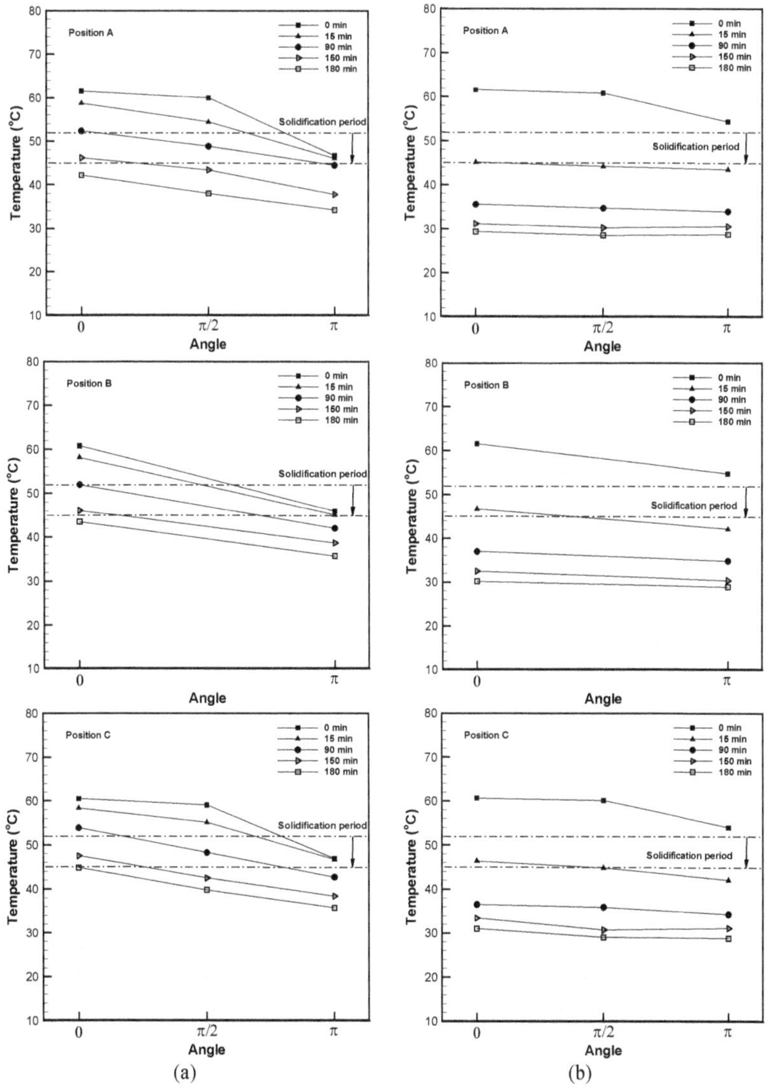

Figure 9. Experimental results of temperature distribution for 13 mm height fins heat exchanger for different angles during discharging process for $Ste = 0.26$: (**a**) 32.5 mm radial distance and (**b**) 12.5 mm radial distance.

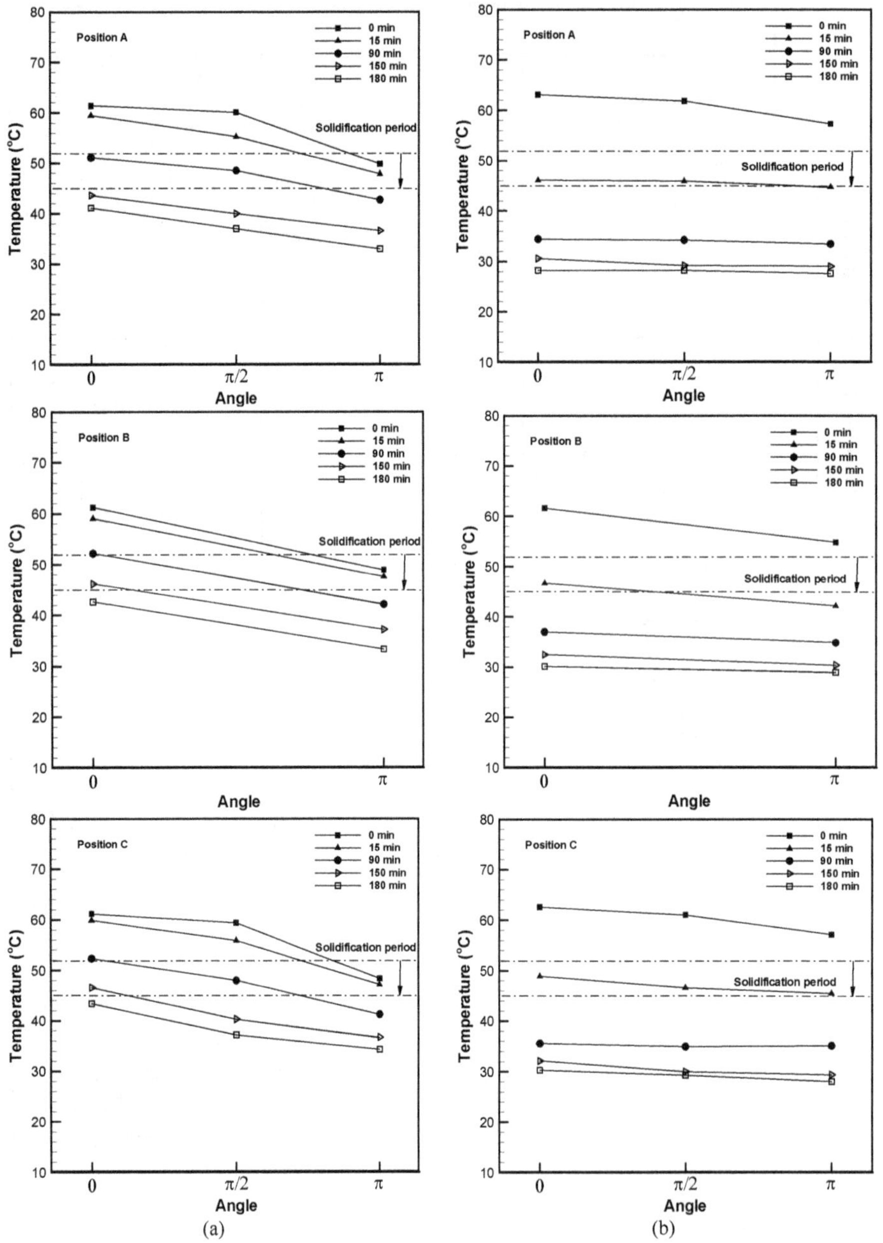

Figure 10. Experimental results of temperature distribution for 26 mm height fins heat exchanger for different angles during discharging process for $Ste = 0.26$: **(a)** 32.5 mm radial distance and **(b)** 12.5 mm radial distance.

heat during initial stages of melting process leads to an accelerated melting process which reduces the melting time. This statement may be verified by comparison of temperature and melting time for both heat exchangers at 32.5 mm radius of the three sections; A, B and C.

In Fig. 7, melting front profiles are illustrated for the tube mid-section and the section of fin for both fins' height and varying Stephan numbers. Considering the profiles of the fin section, it can be seen that melting rate is much more accelerated in these sections in comparison with the mid-sections. The increase is so intense that, for Stephan number of 0.26 and after 60 min of experiment, almost no solid PCM was found around the fin. Implementation of higher fins of 26 mm not only, as stated, increases the rate of melting process at the fin section, but also it accelerates the process in all other sections in comparison with the heat exchanger enhanced with 13 mm fins. In addition, as the melting front is larger for the higher Stephan number (0.38), it is obvious that increasing this dimensionless number lessens the melting time. This effect is so significant that, a less amount of PCM remains in the shell in comparison with other Stephan numbers at a special time.

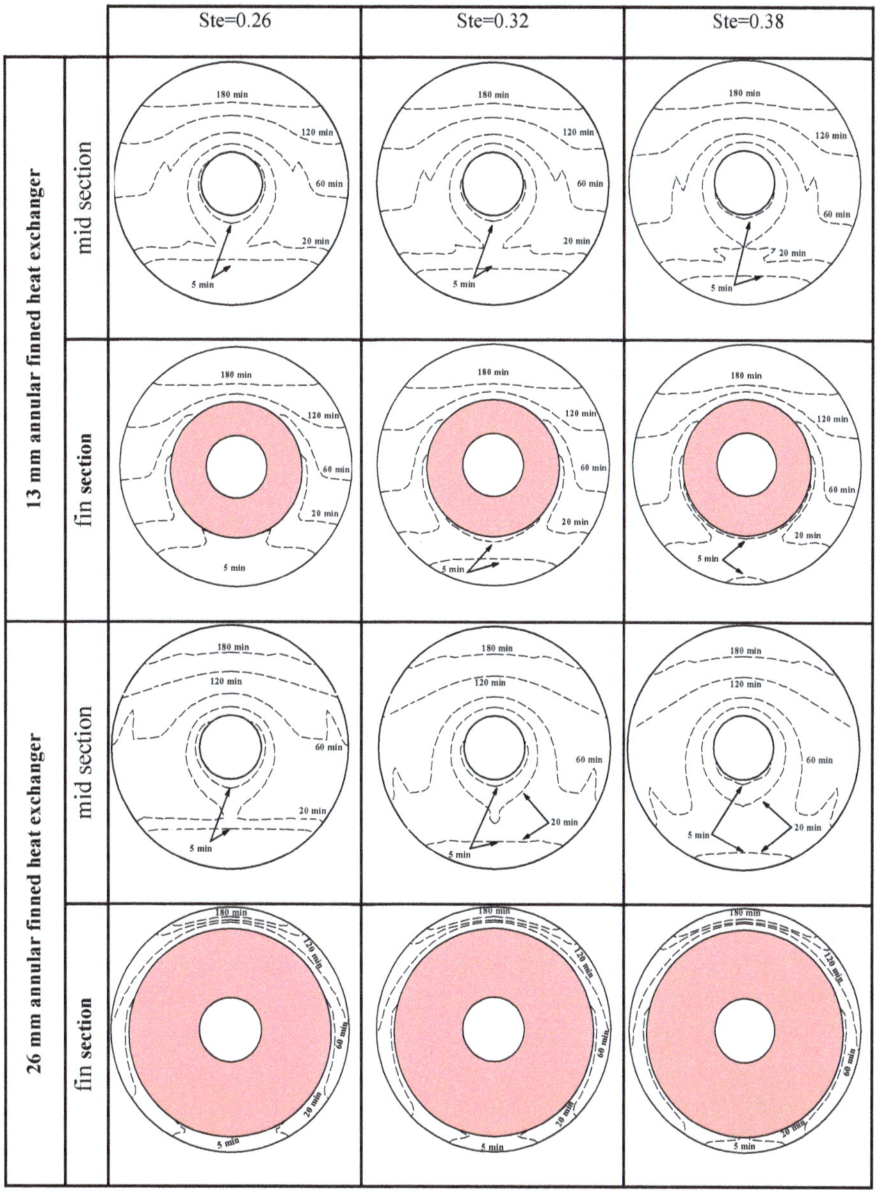

Figure 11. Numerical results of solid front in shell for different Stefan numbers in 13 and 26 mm fins' height at mid section and fin section.

Figure 8 shows the liquid fraction as a function of time. As can be seen in the figure, increasing Stephan number from 0.26 to 0.32 and then to 0.38 of 13 mm-fin heat exchanger, the melting time diminishes 20 and 38 percent respectively. These reductions are 15 and 28 for 26 mm-fin heat exchanger. Comparing Fig. 8a and b it can be implied that by increasing the fins' height from 13 to 26 mm, the total melting time reduces up to 19 %.

6.2 Discharging process

In Figs. 9 and 10 temperature distribution are shown in radial, angular and longitudinal directions for Stefan number of 0.26 during discharge process. Comparing temperature values of these two, it can be figured out that, at the end of melting process for the 26 mm fin, temperature values are more than those of 13 mm fins heat exchanger. When time passes, the fins' presence play a more important role and consequently, the rate of temperature reduction exceeds that of 13 mm fins. In other words, the penetration of fins through the PCM results in less final discharge temperature which is more observable for farther radial distances of the cold carrying tube.

Figure 11 exhibits the solid front formation of the both heat exchangers through the discharge process for the mid-section and the fin section, for different Stephan numbers. The solidification starts with a thin layer of PCM which covers the cold central tube of the heat exchanger. At the fin

sections, due to the fin presence, the solidification process progresses more rapidly. As can be seen in the figure, for 13 mm-fin heat exchangers, after 5 min of the experiment initiation, the solid front fills the bottom of the fin (at 180°). For the 26 mm-fin heat exchanger, the coverage of 180° is completed after 20 min. This figure also shows that, while the Stephan number rises, although the solidification process begins later, the total solidification time stays almost identical for all Stephan numbers. The figure also implies that, at the fin's section, the solidification rate is higher due to the presence of fines. Comparing the solid formation process for both heat exchangers, it is seen that at initial stages of the process, due to the initial condition of higher temperature, the solid formation for the 26 mm-fin heat exchanger occurs more slowly. However, after 60 min of the discharge, the doubling of the fins' height exhibit its influence and the growth of the solid front overtakes the heat exchanger with smaller fins' height. To quantify this phenomenon, 26 mm-fin heat exchanger, for Stephan number of 0.26, reduces the solidification time up to 16 %, comparing the 13 mm-fin heat exchanger (from 460 to 384 min).

7 Conclusions

In this investigation, two parameters including fins' height and Stefan number are studied for a double tube heat exchanger. The results show that:

- PCM temperature at above the tube, regardless of fin size and Stefan number is more than other angular direction around the tube.
- Studying temperature distribution, it is revealed that a point temperature of the PCM in the higher fin heat exchanger is more than the similar point in the shorter fins heat exchanger. This behavior is more obvious for farther radial regions.
- Increasing the Stefan number reduces the melting time for both heat exchangers.
- Increasing fins' height reduces the melting time which is more pronounced for greater Stefan numbers; increasing Stefan number from 0.26 to 0.38 results in melting time reduction to 8.3 and 19.2 respectively for the studied heat exchangers.
- Investigating solidification front, it is clear that, fin's height does not affect the front at mid-section at initial stages of the process progression. While the process advances, solidification front is more influenced by the fins' height in a way that at fins' section, the solidification front progresses in a much faster rate.
- Studying the effect of fins' size on melting and solidification processes, it can be seen that, higher fins influences the solidification time more than melting.

Appendix A

Table A1. Nomenclatures.

c_p	Specific heat capacity (J kg^{-1} K^{-1})
g	Gravity (m s^{-2})
h	Sensible enthalpy (J kg^{-1})
H	Total enthalpy (J)
k	Thermal conductivity (W m^{-1} K^{-1})
L	Latent heat (J kg^{-1})
$\dot{m}$	Mass flow rate (kg s^{-1})
P	Pressure (Pa)
r	Radial distances from the center of HTF tube (m)
$\boldsymbol{S}$	Source term
Ste	Stefan number
T	Temperature (K)
$\boldsymbol{V}$	Velocity vector (m s^{-1})
	Greek symbols
β	Expansion coefficient (K^{-1})
μ	Dynamic viscosity (Pa s)
λ	Liquid fraction
ρ	Density (kg m^{-3})
	Subscripts
ch	Charge
mush	Mushy zone
m	Melting
ref	Reference
H	Hot water
Ini	Initial
Out	Outlet water
PCM	Phase change material
S	Solid
Liq	Liquid

References

Agyenim, F. and Hewitt, N.: The development of a finned phase change material (PCM) storage system to take advantage of off-peak electricity tariff for improvement in cost of heat pump operation, Energy Buildings, 42, 1552–1560, 2010.

Agyenim, F., Eames, P., and Smyth, M.: A comparison of heat transfer enhancement in a medium temperature thermal energy storage heat exchanger using fins, Solar Energy, 83, 1509–1520, 2009.

Agyenim, F., Eames, P., and Smyth, M.: Heat transfer enhancement in medium temperature thermal energy storage system using a multitube heat transfer array, Renewable Energy, 35, 198–207, 2010.

Agyenim, F., Eames, P., and Smyth, M.: Experimental study on the melting and solidification behavior of a medium temperature phase change storage material (Erythritol) system augmented with fins to power a LiBr/H_2O absorption cooling system, Renewable Energy, 36, 108–117, 2011.

Brent, A. D., Voller, V. R., and Reid, K. J.: Enthalpy-porosity technique for modeling convection-diffusion phase change: application to the melting of a pure metal, Numer. Heat Trans. B, 13, 297–318, 1988.

Ezan, M. A., Ozdogan, M., and Erek, A.: Experimental study on charging and discharging periods of water in a latent heat storage unit, Int. J. Therm. Sci., 50, 2205–2219, 2011.

Gong, Z. X., Devahastin, S., and Mujumdar, A. S.: Enhanced heat transfer in free convection-dominated melting in a rectangular cavity with an isothermal vertical wall, Appl. Therm. Eng., 19, 1237–1251, 1999.

Hosseini, M. J., Ranjbar, A. A., Sedighi, K., and Rahimi, M.: A combined experimental and computational study on the melting behavior of a medium temperature phase change storage material inside shell and tube heat exchanger, Int. Commun. Heat Mass Trans., 39, 1416–1424, 2012.

Hosseini, M. J., Rahimi, M., and Bahrampoury, R.: Experimental and computational evolution of a shell and tube heat exchanger as a PCM thermal storage system, Int. Commun. Heat Mass Trans., 50, 128–136, 2014.

Ismail, K. A. R., Alves, C. L. F., and Modesto, M. S.: Numerical and experimental study on the solidification of PCM around a vertical axially finned isothermal cylinder, Appl. Therm. Eng., 21, 53–77, 2001.

Liu, C. and Groulx, D.: Experimental study of the phase change heat transfer inside a horizontal cylindrical latent heat energy storage system, Int. J. Therm. Sci., 82, 100–110, 2014.

Mat, S., Al-Abidi, A. A., Sopian, K., Sulaiman, M. Y., and Mohammad, A. T.: Enhance heat transfer for PCM melting in triplex tube with internal–external fins, Energy Convers. Manage., 74 223–236, 2013.

Mehling, H. and Cabeza, L.: Phase change materials and their basic properties, in: Thermal Energy Storage for Sustainable Energy Consumption, edited by: Paksoy, H. Ö., Springer, Netherlands, 257–277, 2007.

Mosaffa, A. H., Talati, F., Tabrizi, H. B., and Rosen, M. A.: Analytical modeling of PCM solidification in a shell and tube finned thermal storage for air conditioning systems, Energy Buildings, 49, 356–361, 2012.

Ogoh, W. and Groulx, D.: Effects of the number and distribution of fins on the storage characteristics of a cylindrical latent heat energy storage system: a numerical study, Int. J. Heat and Mass Trans., 48, 1825–1835, 2012.

Pakrouh, R. Hosseini, M. J., and Ranjbar, A. A.: A parametric investigation of a PCM-based pin *n* heat sink, Mechanical Science, 6, 65–73, 2015.

Patankar, S. V.: Numerical Heat Transfer and Fluid Flow, Hemisphere, Washington, D.C., 1980.

Rahimi, M., Ranjbar, A. A., Ganji, D. D., Sedighi, K., Hosseini, M. J., and Bahrampoury, R.: Analysis of geometrical and operational parameters of PCM in a fin and tube heat exchanger, Int. Commun. Heat Mass Trans., 53, 109–115, 2014.

Seeniraj, R. V. and Narasimhan, N. L.: Performance enhancement of a solar dynamic LHTS module having both fins and multiple PCMs, Solar Energy, 82, 535–542, 2008.

Sharma, A., Tyagi, V. V., Chen, C. R., and Buddhi, D.: Review on thermal energy storage with phase change materials and applications, Renewable Sustain. Energy Rev., 13, 318–345, 2009.

Shatikian, V., Ziskind, G., and Letan, R.: Numerical investigation of a PCM-based heat sink with internal fins, Int. J. Heat Mass Trans., 48, 3689–3706, 2005.

Zalba, B., Marin, J. M., Cabeza, L. F., and Mehling, H.: Review on Thermal Energy Storage with Phase Change Materials, Heat Transfer Analysis and Applications, Appl. Therm. Eng., 23, 251–283, 2003.

6

Three-dimensional finite element model of the drilling process used for fixation of Salter–Harris type-3 fractures by using a K-wire

A. Gok[1], K. Gok[2], and M. B. Bilgin[1]

[1]Amasya University, Technology Faculty, Department of Mechanical Engineering, 05000 Amasya, Turkey
[2]Dumlupınar University, Kütahya Vocational School of Technical Sciences, Germiyan Campus, 43100 Kütahya, Turkey

Correspondence to: A. Gok (arif.gok@amasya.edu.tr)

Abstract. In this study, the drilling process was performed with Kirschner wire (K-wire) for stabilization after reduction of Salter–Harris (SH) type-3 epiphyseal fractures of distal femur. The study was investigated both experimentally and numerically. The numerical analyses were performed with finite element method (FEM), using DEFORM-3D software. Some conditions such as friction, material model and load and boundary must be identified exactly while using FEM. At the same time, an analytic model and software were developed, which calculate the process parameters such as drilling power and thrust power, heat transfer coefficients and friction coefficient between tool–chip interface in order to identify the temperature distributions occurring in the K-wire and bone model (Keklikolu Plastik San.) material during the drilling process. Experimental results and analysis results have been found as consistent with each other. The main cutting force, thrust force, bone model temperature and K-wire temperature were measured as 80° N, 120° N, 69 °C and 61 °C for 400 rpm in experimental studies. The main cutting force, thrust force, bone model temperature and K-wire temperature were measured as 65° N, 87° N, 91 °C and 82 °C for 800 rpm in experimental studies. The main cutting force, thrust force, bone model temperature and K-wire temperature were measured as 85° N, 127° N, 72 °C and 67 °C for 400 rpm in analysis studies. The main cutting force, thrust force, bone model temperature and K-wire temperature were measured as 69° N, 98° N, 83 °C and 76 °C for 800 rpm in analysis studies. A good consistency was obtained between experimental results and finite element analysis (FEA) results. This proved the validity of the software and finite element model. Thus, this model can be used reliably in such drilling processes.

1 Introduction

A Salter–Harris fracture is a fracture and it involves the epiphyseal plate or growth plate of bone. Salter–Harris fractures are widely seen as an injury type in children, occurring in 15 % of childhood long bone fractures (Salter and Harris, 1963).

Fractures of the distal femur epiphysis have a particularly high risk in terms of growth arrest and other morbidities (Basener et al., 2009; Eid and Hafez, 2002; Mann and Rajmaira, 1990; Peterson et al., 1994). The factors causing this situation are considered to be age, fracture type, degree of displacement, the wavy structure of physics and quality of the fracture reduction (Dahl et al., 2014; Liu et al., 2013; Lombardo and Harvey, 1977). Considering the histology of the situation which is the main cause for angulation and growth complications, physeal bar formation is considered to be the main cause (Dahl et al., 2014; Herring, 2002). In the treatment of Salter–Harris (SH) type-3 epiphysis fractures of distal femur, partially threaded screws are applied in parallel to the articular bone and do not go through the physis line, or Kirschner wires (K-wire) are recommended for fixation (Wheelessonline, 2014; Zionts, 2003). In particular, the technique of fixation with K-wire is widely used.

In our daily lives, any fracture can occur in our musculoskeletal system as a result of any trauma. These fractures

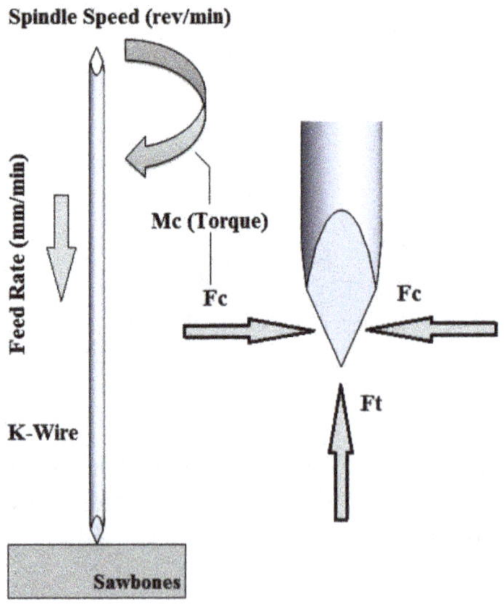

Figure 1. The drilling operation with K-wire and the forces occurring.

are fixed by surgeons using a variety of screws, implants or K-wire. In particular, if the surgeon used a screw to stabilize the fracture, a drill bit suitable for dimensions of the screw would be needed to perform the drilling process. In the literature, bone drilling processes are commonly encountered in order to ensure the broken bone stability. During the drilling process, friction-induced heat emerges due to the temperature difference between drilling tool and bone. This heat results in undesirable thermal damage to the bone and surrounding tissues. The temperature that causes the emergence of heat has a certain critical value. There are many studies on this critical value in the literature. The study by Hillery and Shuaib (1999) showed that the bone undergoes serious damage when the temperature rises above 55 °C in 30 s. Eriksson et al. (1984) studied in vivo and presented that the cortical bone of a rabbit showed thermal necrosis above 47 °C in 60 s. Augustin et al. (2008) reported that the temperature could increase above 47 °C, which causes irreversible osteonecrosis during the bone drilling process.

Optimization of drilling parameters is very important especially bone drilling processes in terms of necrosis in bone and soft tissues. This status requires expensive experimental equipment and additional safety measures to protect from biohazards today. Finite element analysis (FEA) of bone drilling may be an important feature in new surgical techniques. Experimental work eliminates equipment costs as well as potential health risks associated with biological materials. Modeling of the bone drilling process using FEA may be useful for validation of experimental or analytical results. It is considered as a reliable tool to develop new surgical techniques (Alam et al., 2010).

Many researchers also simulated the bone drilling process using the computer-aided FEA tool. Surgical drill bit or K-wires are used in many of these studies and are related to temperature occurring in bone or necrosis caused by them. Some studies are related to optimum drilling parameters. Gok et al. (2015a) developed a new driller system to prevent osteonecrosis, and so they performed optimization of bone drilling processing parameters (Gok et al., 2015b). Yuan-Kun (2008, 2009) developed an elastic–plastic dynamic FEA tool to simulate the effects of processing parameters on the temperature rise during the bone drilling process in which the drill bit and Kirschner pin are used. Yuan-Kun (2011) proposed an empirical equation to calculate the peak bone temperature caused by the applied force and revolutions per minute and compared the results with the FEA simulations. Sezek et al. (2012) measured the inevitable temperature changes that occurred in orthopaedic drilling and studied them to optimize the drilling parameters within a safe drilling temperature lower than 45 °C. Alam et al. (2009) developed a FEA model of the bone drilling process and compared results with experimental results.

In the literature, there are many available studies analyzed by using FEA in bone drilling processes (Gok et al., 2015a, b). But there are no studies to determine the process parameters like heat transfer coefficient, drilling power or machining power and friction coefficient between tool–chip interfaces, which are very important inputs in numeric simulations. Software, analytic model and finite element method (FEM) calculate these process parameters, having been developed for bone drilling simulations by using K-wire in the study.

2 Development of the analytic model and software for the drilling process using K-wire

During the drilling process, moment of rotation occurs due to the rotation and thrust force occurs due to the thrust movement on the drill bit (Fig. 1). The moment of rotation creates the drilling power given in Eq. (1), while thrust force creates the thrust power given in Eq. (2). A large part of the mechanical energy generated during the drilling process is converted into the heat energy given in Eqs. (4) and (5), respectively. Software was developed to calculate the heat transfer coefficient and the thermal conductivity coefficient between bone model, K-wire and ambient using following equations. The software was developed using Visual Basic 6.0 (Fig. 2). In this interface, drilling parameters and heat transfer parameters are input sections. Drilling power, thrust power, total mechanical power, heat transfer between K-wire with bone, heat transfer for K-wire and heat transfer for bone model are

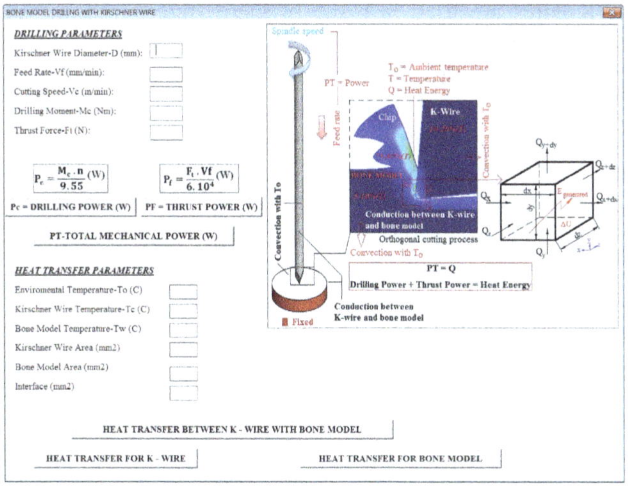

Figure 2. The interface of the developed software.

output sections.

$$P_c = \frac{M_c \cdot n}{9550} \tag{1}$$

$$P_f = \frac{F_t \cdot V_f}{60 \cdot 1000} \tag{2}$$

$$\mathrm{PT} = P_c + P_f \tag{3}$$

In here, P_c is the drilling power, P_f is the thrust power, PT is the total power, n is the spindle speed, M_c is the torque, F_t is the thrust force, and V_f is the feed rate.

$$Q_{\text{convection}} = h_t \cdot A \cdot (T - T_o) \tag{4}$$

$$Q_{\text{conduction}} = -k \cdot A \cdot \left(\frac{\partial T}{\partial n}\right) \tag{5}$$

In here, h_t is the heat transfer coefficient for convection, k is the thermal conductivity coefficient for conduction, A is the contact area, T is the bone model or K-wire temperature, T_o is the ambient temperature, and $\frac{\partial T}{\partial n}$ is the derivative along the outward drawn normal to the surface.

If conservation of energy law (first law of thermodynamics) is applied to this control volume (Fig. 3), Eq. (6) would be obtained; the general equation of heat conduction in Eq. (7) is also obtained by ignoring the higher terms in Taylor series. By writing the m mass, in the internal energy formula in the right side of the Eq. (25) as in Eq. (26), it is then put into place in Eq. (27).

$$E_{\text{input}} - E_{\text{output}} + E_{\text{generated}} - E_{\text{convection}} = \Delta U \tag{6}$$

For the input energy Eq. (6), the input heat amounts in the x, y, and z axes are shown in Eqs. (7), (8), and (9). The areas in these equations are defined in Eqs. (1), (2) and (3) for x, y, and z axes. k, in these equations, is heat conduction coefficient. Because of the contact between the workpiece and the drill bit, a heat transfer with conduction occurs.

$$Q_x = -k \cdot A_x \cdot \frac{\partial T}{\partial x} \tag{7}$$

$$Q_y = -k \cdot A_y \cdot \frac{\partial T}{\partial y} \tag{8}$$

$$Q_z = -k \cdot A_z \cdot \frac{\partial T}{\partial y} \tag{9}$$

$$A_x = \mathrm{d}y \cdot \mathrm{d}z \tag{10}$$

$$A_y = \mathrm{d}x \cdot \mathrm{d}z \tag{11}$$

$$A_z = \mathrm{d}x \cdot \mathrm{d}y \tag{12}$$

For the output energies, if, by using Taylor's series expansion ($Q_{x+\mathrm{d}x}$, $Q_{y+\mathrm{d}y}$ and Q_{z+az}), higher-order terms are neglected, Eqs. (13), (14), and (15) are acquired.

$$Q_{x+\mathrm{d}x} = Q_x + \frac{\partial Q_x}{\partial x} \cdot \mathrm{d}x \tag{13}$$

$$Q_{y+\mathrm{d}y} = Q_y + \frac{\partial Q_y}{\partial y} \cdot \mathrm{d}y \tag{14}$$

$$Q_{z+\mathrm{d}z} = Q_z + \frac{\partial Q_z}{\partial z} \cdot \mathrm{d}z \tag{15}$$

The generated energy in the system is written as volumetric in Eq. (4). Equation (18) can be written for the heat transfer with the environment of workpiece and the drill bit. Equation (19) can be written for the collected energy in the system. Here, mass is m, specific heat capacity is c_p, and the control volume is defined as $\mathrm{d}v$.

$$E_{\text{generated}} = Q \cdot \mathrm{d}v \tag{16}$$

$$\mathrm{d}v = \mathrm{d}x \cdot \mathrm{d}y \cdot \mathrm{d}z \tag{17}$$

$$Q_{\text{convection}} = h_t \cdot \mathrm{d}A_z \cdot (T - T_o) \tag{18}$$

$$\Delta U = m \cdot c_p \cdot \frac{\partial T}{\partial t} \tag{19}$$

Putting these equations above in their places in Eq. (6), the following equations are acquired.

$$Q_x + Q_y + Q_z - \left(Q_{x+\mathrm{d}x} + Q_{y+\mathrm{d}y} + Q_{z+\mathrm{d}z}\right) + Q \cdot \mathrm{d}v - h_t A_z (T - T_o) \cdot \mathrm{d}z = m \cdot c_p \frac{\partial T}{\partial t} \tag{20}$$

$$-\frac{\partial}{\partial x}\left(-k \cdot A_x \cdot \frac{\partial T}{\partial x}\right)\mathrm{d}x - \frac{\partial}{\partial y}\left(-k \cdot A_y \cdot \frac{\partial T}{\partial y}\right)\mathrm{d}y - \frac{\partial}{\partial z}\left(-k \cdot A_z \cdot \frac{\partial T}{\partial z}\right)\mathrm{d}z + Q \cdot \mathrm{d}v - h_t \mathrm{d}A_z (T - T_o)\mathrm{d}z = m \cdot c_p \cdot \frac{\partial T}{\partial t} \tag{21}$$

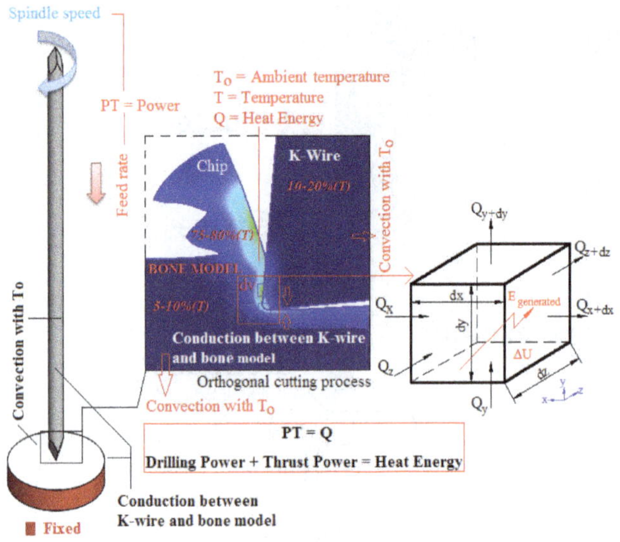

Figure 3. Thermal model of the drilling process using K-wire.

$$\frac{\partial^2 T}{\partial x^2} k \cdot A_x \cdot \mathrm{d}x + \frac{\partial^2 T}{\partial y^2} k \cdot A_y \cdot \mathrm{d}y + \frac{\partial^2 T}{\partial z^2} k \cdot A_z \mathrm{d}z + Q \cdot \mathrm{d}v - h_t \cdot A_z \cdot (T - T_o) \cdot \mathrm{d}z = m \cdot c_p \cdot \frac{\partial T}{\partial t} \tag{22}$$

$$\rho = \frac{m}{\mathrm{d}v}, m = \rho \cdot \mathrm{d}v \tag{23}$$

$$\frac{\partial^2 T}{\partial x^2} \cdot k \cdot \mathrm{d}y \cdot \mathrm{d}z \cdot \mathrm{d}x + \frac{\partial^2 T}{\partial y^2} \cdot k \cdot \mathrm{d}x \cdot \mathrm{d}z \cdot \mathrm{d}y + \frac{\partial^2 T}{\partial z^2} \cdot k \cdot \mathrm{d}x \cdot \mathrm{d}y \cdot \mathrm{d}z + Q \cdot \mathrm{d}x \cdot \mathrm{d}y \cdot \mathrm{d}z - h_t \cdot \mathrm{d}x \cdot \mathrm{d}y \cdot \mathrm{d}z (T - T_o) = \rho \cdot \mathrm{d}v \cdot c_p \frac{\partial T}{\partial t} \tag{24}$$

$$\left(\frac{\partial^2 T}{\partial x^2} + \frac{\partial^2 T}{\partial y^2} + \frac{\partial^2 T}{\partial z^2}\right) + \frac{Q}{k} - \frac{h_t \cdot (T - T_o)}{k} = \frac{\rho \cdot c_p}{k} \cdot \frac{\partial T}{\partial t} \tag{25}$$

In Eq. (27) the general equation of heat conduction is acquired by writing thermal expansion coefficient, α, as in Eq. (26) and Ulutan et al. (2009).

$$\alpha = \frac{k}{\rho \cdot c_p} \tag{26}$$

$$\left(\frac{\partial^2 T}{\partial x^2} + \frac{\partial^2 T}{\partial y^2} + \frac{\partial^2 T}{\partial z^2}\right) + \frac{Q}{k} - \frac{h_t \cdot (T - T_o)}{k} = \frac{1}{\alpha} \frac{\partial T}{\partial t} \tag{27}$$

In here, E_{input} is the input energy to system, E_{output} is the output energy from system, $E_{\text{generated}}$ is the accumulated in the system energy, $E_{\text{convection}}$ is the output energy from system with convection, Q is the heat energy, and α is the thermal diffusivity coefficient.

Table 1. Bone model drilling process parameters for simulation model.

Drilling process parameters	Spindle speed (rpm)	
	400	800
Drilling power (W)	2.51	4.08
Thrust power (W)	0.08	0.05
Total power (W)	2.59	4.14
Friction coefficient	0.53	0.41

3 The bone model drilling processes

A mini 2.5 kW CNC machine was used for the bone model drilling processes. The drilling processes were performed on bone model samples similar to bone by using a K-wire in 3 mm diameter (rake angle −28° and clearance angle +6°) with spindle speeds of 400 and 800 rpm and feed rate of 40 mm min^{-1}. The drilling moment and thrust force which occurred during drilling processes were measured by using a Kistler 9257B dynamometer. Two measurement devices were used for temperature measurements (Fig. 4). While the bone model temperature values were measured using PT100 thermocouple, the K-wire temperature values were measured using non-contact temperature sensor. The PT100 sensor was used to measure the temperature of the bone model sample as illustrated in Fig. 6. As seen, a hole of 6 mm diameter was prepared by drilling in the bone model to place the thermocouple rod with the distance of 0.5 mm to the surgical drill bit to measure the bone model temperature levels during the drilling. The distance to place the thermocouple was used about 1 mm by researchers.

4 Computer-aided finite element analysis

The aim of this study was to calculate the thrust power and drilling power as well as heat transfer coefficients required for numerical simulations of bone model drilling processes using a K-wire. DEFORM-3D finite element-based program was used to perform these bone model drilling processes. The 3-D model of K-wire used in the simulations was created by SolidWorks 2013, and 3-D model of the workpiece was modeled by using the Geometry Primitive feature of DEFORM-3D software.

4.1 Loading and boundary conditions

Firstly, the mesh division for FEA was performed. Tetrahedral element type was selected for the mesh processing. While the mesh structure of bone model consists of 21 498 elements and 4853 nodes, the mesh structure of K-wire consists of 22 958 elements and 5742 nodes. The mesh structure of bone model and drill bit are given in Fig. 5a. In addi-

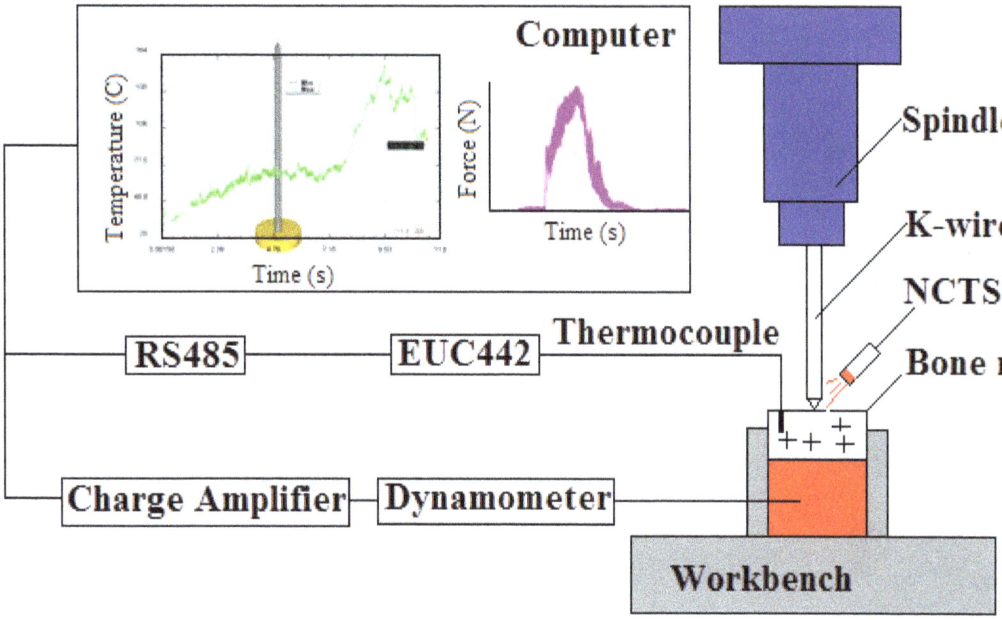

Figure 4. Schematic circuit view of bone model drilling test unit.

tion, the simple cylindrical bone model was fixed from lateral surfaces and bottom. In the drill bit model, different spindle speeds around its axis and constant feed rate in the direction of a drilling axis (Z) are given in Fig. 5b. The contact algorithm between drill bit and bone model was identified as master and slave in the software, where the master element was drill bit and slave element was bone model, respectively. The friction equation of Coulomb was preferred for the friction model of these two elements. Coulomb friction model may be used in low cutting speed. However, high temperatures and high normal stresses occur in high cutting speeds (Arrazola et al., 2008). The friction coefficient between K-wire and chip during the orthogonal cutting process (Fig. 6) is calculated with Eqs. (28) and (29) (Özel, 2006). The forces in Eqs. (28) and (29), respectively, are the main cutting force F_c occurring on the rake face and the thrust force F_t. These forces are measured by using a dynamometer. These forces are, respectively, the shear force (F_s) occurring on the shear plane and the vertical force to it (N_s). Lastly, (F), which occurs on the friction plane, and (N), which is vertical force to it, are calculated by using (F_c) and (F_t), where μ is the friction coefficient, β the friction angle, and γ the rake angle as seen in Fig. 6.

$$\mu = \tan\beta = \frac{F}{N} \tag{28}$$

$$\mu = \frac{F_t + F_c \tan\gamma}{F_c - F_t \tan\gamma} \tag{29}$$

Such a friction model is preferred for low cutting speeds in general. In the simulation study, in which both spindle speeds were used, the drilling process parameters calculated by software developed are given in Table 1. The heat transfer coefficients were calculated for each 1 mm of the drilling process (Fig. 7).

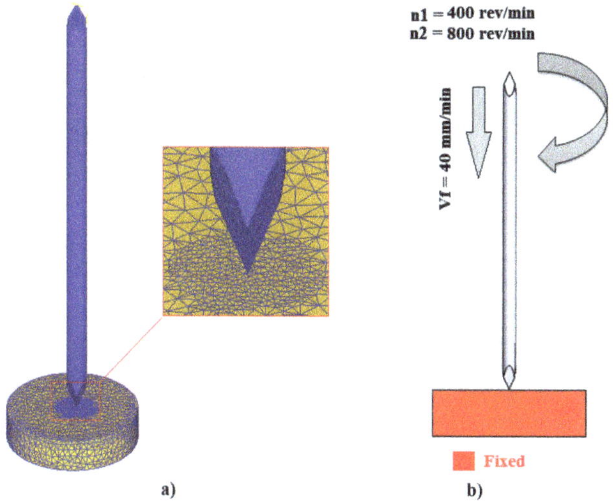

Figure 5. The structure of (**a**) bone model and K-wire as well as (**b**) boundary conditions.

4.2 Material model

A material model including mechanical and thermal properties of the bone model and K-wire was defined in the simulation. The stainless steel (AISI 304) material properties were selected for the K-wire. The mechanical and thermal properties of the AISI 304 and bone model materials are given in Table 2. The mechanical properties of the materials must

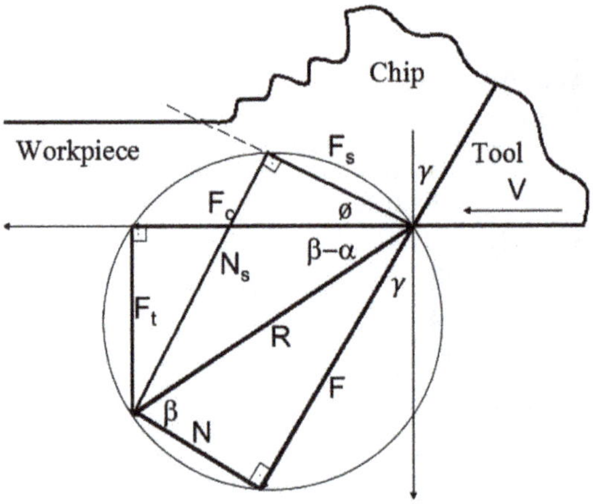

Figure 6. Two-dimensional (2-D) forces system in drilling operations (Merchant's force circle).

have been exactly known since the drilling processes were performed at high temperatures and strain rates. Therefore, the flow stress curve of the bone model material was used. Due to the lack of flow stress curves for all materials, tensile test data were used carefully for the different temperatures and strain rates. The flow stress curves (McElhaney and Byars, 1965) were used in the simulation for the bone model material model defined as a function of strain, strain rate, and temperature as can be seen in Fig. 8. The flow stress $\bar{\sigma}$ in Eq. (30) was selected to exhibit true material behavior as a function of the effective plastic strain ($\bar{\varepsilon}$), effective strain rate ($\dot{\bar{\varepsilon}}$), and temperature (T).

$$\bar{\sigma} = (\bar{\varepsilon}, \dot{\bar{\varepsilon}}, T) \tag{30}$$

5 Results

Bone model drilling experiments were performed using K-wire for two spindle speed values (400–800 rpm) and bone model and K-wire temperatures were obtained separately. Moreover, numerical analyses were performed based on the FEM, using DEFORM-3D software. The main cutting force and thrust force were measured both experimental and via FEM. These two features are very important for drilling operations. Especially, the temperature value occurring in bone during the drilling process is very important in terms of necrosis. If this temperature value exceeds 47 °C, irreversible damage may occur in bone and surrounding tissues. The heat transfer coefficient values were calculated using these temperature values via developed software for drilling simulation. Finally, experimental results and FEA results were compared.

As it can be seen in Fig. 9, it has been observed that the main cutting force and thrust force reduced with increasing spindle speed as a result of experimental and drilling simulations of the bone model drilling processes using a K-wire (Fig. 9a and b). Conversely, the temperature values of K-wire and bone model increased as a result of the increase in spindle speed (Fig. 9c and d). In Fig. 10, the maximum temperature values occurring in bone models at the different spindle speeds were presented.

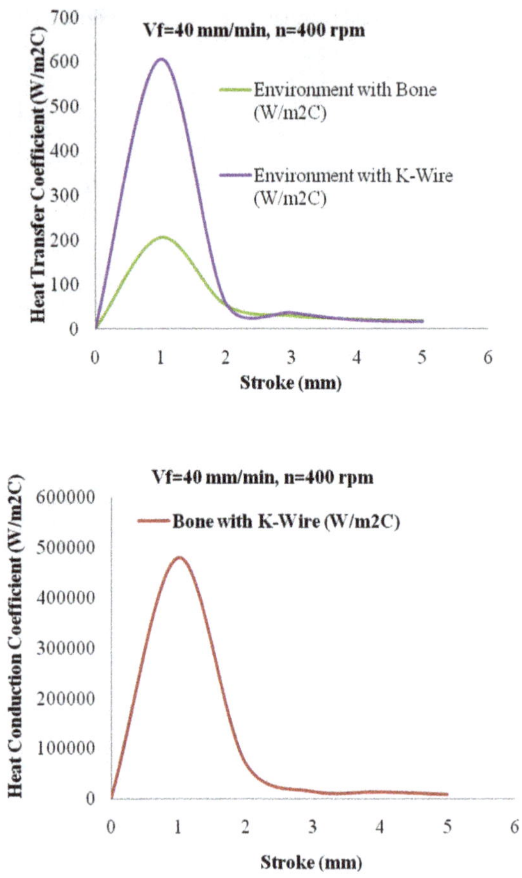

Figure 7. The heat transfer coefficients for each 1 mm of the drilling process.

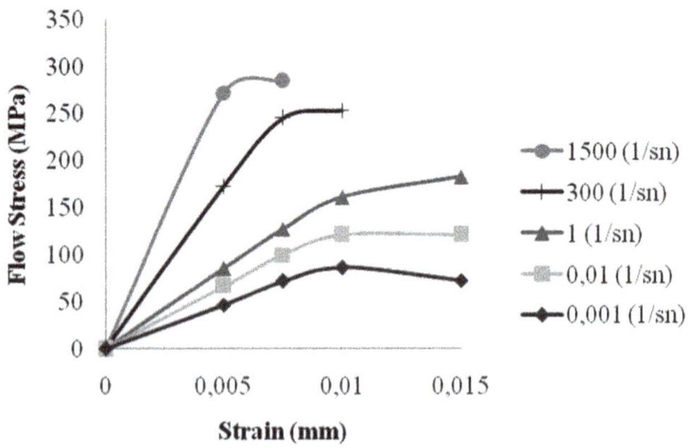

Figure 8. Flow stress of bone (McElhaney and Byars, 1965).

Table 2. Mechanical and thermal properties of the drill bit and bone model materials.

Drill bit material properties (AISI 304) (Deform_Material_Library)(Deform_Material_Library)	
Elasticity modulus (GPa)	20 °C (210)
Poisson's ratio	0.3
Thermal expansion coefficient (10^{-6} °C^{-1})	93.33 °C (1.20×10^{-5})
Thermal conductivity(W/mK)	100 °C (17)
Heat capacity (N mm^{-2} °C)	93.33 °C (2.78)
Emissivity	0.7
The bone model material properties (http://www.matweb.com/)(http://www.matweb.com/)	
Modulus of elasticity (GPa)	17
Poisson's ratio	0.35
Thermal conductivity (W mK^{-1})	0.38
Specific heat (J kg °C)	1260

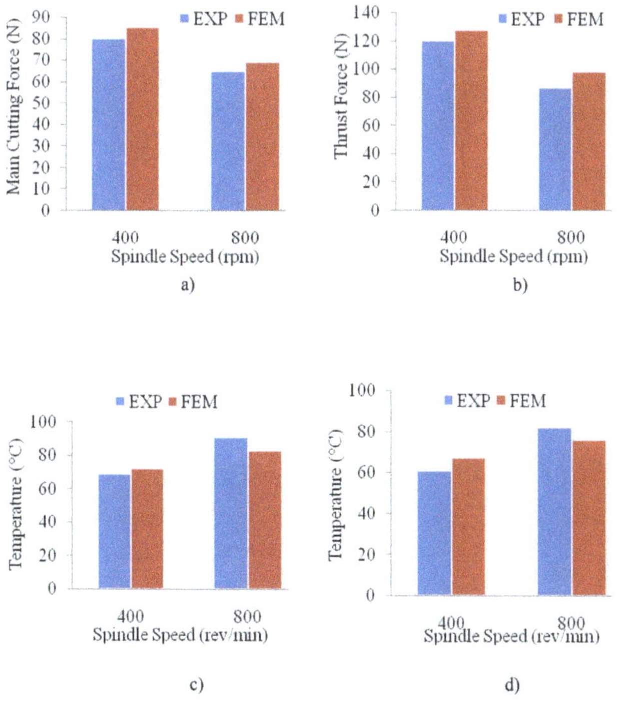

Figure 9. Experimental and FEA results: (**a**) main cutting force, (**b**) thrust force, (**c**) bone model temperature, and (**d**) K-wire temperature in the bone model drilling process.

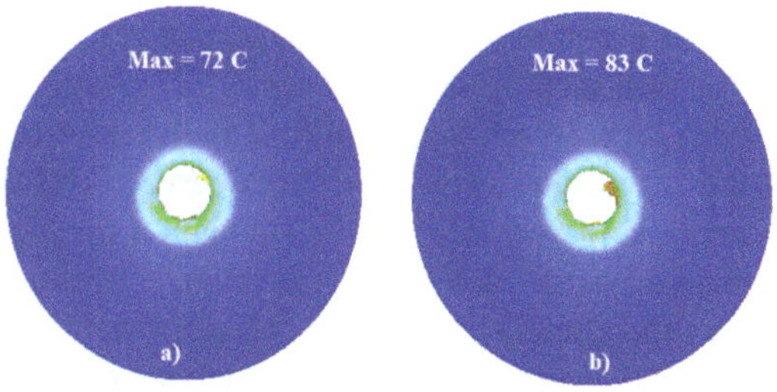

Figure 10. The maximum temperature values occurring in bone model at the different spindle speeds: (**a**) 400 rpm and (**b**) 800 rpm.

6 Discussion

After that bone model drilling processes were performed using K-wire, the main cutting forces, thrust force, bone models and K-wires temperature values were obtained both experimentally and via FEM. The temperature values in bone model samples and K-wire occurred above the critical temperature value (47 °C). This situation is not desired by surgeons too much. To prevent this temperature, there are several methods. One of them is to select the optimum cutting parameters. The other is to cool the drill bit, cutting tool or K-wire from internal or external. This study focused on developing the software, analytic model and FEM model to calculate process parameters using drilling simulations on the bone model samples using K-wire. In studies performed by Gok et al. (2015a, b) bone temperature values were measured less at the lower spindle speed. Additionally, non-contact temperature sensor was used for measure temperature values in K-wire in this study. A more sophisticated measuring device may be used instead of this device in future studies. Additionally, in future, another study may be performed using K-wires of a different type.

7 Conclusions

An analytic model and software was developed to calculate the drilling and thrust power required for chip removal in the numerical simulations of bone model drilling processes, as well as heat transfer mechanisms and coefficients between drill bit, bone model and environment, respectively.

Computer-aided numerical simulation of the bone model drilling process was also performed using DEFORM-3D software. Based on both experimental and FEA results, the following conclusions can be obtained:

- As it can be seen in Fig. 9, it has been observed that the main cutting force and thrust force reduced with increasing spindle speed as a result of experimental and drilling simulations of the bone model drilling processes using a K-wire. Conversely, the temperature values of K-wire and bone model increased as a result of the increase in spindle speed.
- There is a good consistency between experimental results and FEA results. This has proved the validity of the software and finite element model. Thus, this model can be used reliably in such drilling processes.
- The temperature values in bone model samples and K-wire occurred above the critical temperature value (47 °C). Use of lower spindle speeds is recommended.

References

Alam, K., Mitrofanov, A. V., and Silberschmidt, V. V.: Finite element analysis of forces of plane cutting of cortical bone, Comput. Mater. Sci., 46, 738–743, 2009.

Alam, K., Mitrofanov, A. V., and Silberschmidt, V. V.: Thermal analysis of orthogonal cutting of cortical bone using finite element simulations, International Journal of Experimental and Computational Biomechanics, 1, 236–251, 2010.

Arrazola, P. J., Ugarte, D., and Domínguez, X.: A new approach for the friction identification during machining through the use of finite element modeling, Int. J. Mach. Tool. Manu., 48, 173–183, 2008.

Augustin, G., Davila, S., Mihoci, K., Udiljak, T., Vedrina, D., and Antabak, A.: Thermal osteonecrosis and bone drilling parameters revisited, Arch. Orthop. Trauma. Surg., 128, 71–77, 2008.

Basener, C. J., Mehlman, C. T., and DiPasquale, T. G.: Growth disturbance after distal femoral growth plate fractures in children: a meta-analysis, J. Orthop. Trauma., 23, 663–667, 2009.

Dahl, W. J., Silva, S., and Vanderhave, K. L.: Distal Femoral Physeal Fixation: Are Smooth Pins Really Safe?, J. Pediatr. Orthoped., 34, 134–138, doi:10.1097/BPO.0000000000000083, 2014.

Eid, A. M. and Hafez, M. A.: Traumatic injuries of the distal femoral physis. Retrospective study on 151 cases, Injury, 33, 251–255, 2002.

Eriksson, A. R., Albrektsson, T., and Albrektsson, B.: Heat caused by drilling cortical bone. Temperature measured in vivo in patients and animals, Acta. Orthop. Scand., 55, 629–631, 1984.

Gok, K., Buluc, L., Muezzinoglu, U., and Kisioglu, Y.: Development of a new driller system to prevent the osteonecrosis in orthopedic surgery applications, J Braz. Soc. Mech. Sci. Eng., 37, 549–558, 2015a.

Gok, K., Gok, A., and Kisioglu, Y.: Optimization of processing parameters of a developed new driller system for orthopedic surgery applications using Taguchi method, Int. J. Adv. Manuf. Technol., 76, 1437–1448, 2015b.

Hillery, M. T. and Shuaib, I.: Temperature effects in the drilling of human and bovine bone, J. Mater. Process. Tech., 92–93, 302–308, 1999.

Herring, J. A.: Lower extremity injuries, in: Tachdjian's Pediatric Orthopaedics, USA: W.B. Saunders Company, 2327–2334, 2002.

Liu, R. W., Armstrong, D. G., Levine, A. D., Gilmore, A., Thompson, G. H., and Cooperman, D. R.: An Anatomic Study of the Distal Femoral Epiphysis, J. Pediatr. Orthoped., 33, 743–749 doi:10.1097/BPO.0b013e31829d55bf, 2013.

Lombardo, S. J. and Harvey J. P. J. R.: Fractures of the distal femoral epiphyses. Factors influencing prognosis: a review of thirty-four cases, J. Bone Joint Surg., 59, 742–751, 1977.

Mann, D. C. and S, R.: Distribution of physeal and nonphyseal fractures in 2,650 long-bone fractures in children aged 0–16 years, J. Pediatr. Orthop., 10, 713–716, 1990.

McElhaney, J. and Byars, E. F.: Dynamic response of biological materials, American Society of Mechanical Engineers, New York, 150–175, 1965.

Özel, T.: The influence of friction models on finite element simulations of machining, Int. J. Mach. Tool Manu., 46, 518–530, 2006.

Peterson, H. A., Madhok, R., Benson, J. T., Ilstrup, D. M., and Melton, L. J. 3rd: Physeal fractures: Part 1, Epidemiology in Olmsted County, Minnesota, 1979–1988, J. Pediatr. Orthop., 14, 423–430, 1994.

Salter, R. B. and Harris, W. R.: Injuries Involving the Epiphyseal Plate, 587–622, 1963.

Sezek, S., Aksakal, B., and Karaca, F.: Influence of drill parameters on bone temperature and necrosis: A FEM modelling and in vitro experiments, Comput. Mater. Sci., 60, 13–18, 2012.

Ulutan, D., Lazoglu, I., and Dinc, C.: Three-dimensional temperature predictions in machining processes using finite difference method, J. Mater. Process. Tech., 209, 1111–1121, 2009.

Wheeless' Textbook of Orthopaedics, 2014.

Yuan-Kun, T., Hsun-Heng, T., Li-Wen, C., Ching-Chieh, H., Yung-Chuan, C., and Li-Chiang, L.: Finite element simulation of drill bit and bone thermal contact during drilling, Shanghai, China, 16–18 May, 1268–1271, 2008.

Yuan-Kun, T., You-Yao, H., and Yung-Chuan, C.: Finite element modeling of kirschner pin and bone thermalcontact during drilling, Life Sci. J., 6, 23–27, 2009.

Yuan-Kun, T., Wei-Hua, L., Li-Wen, C., Ji-Sih, C., and Yung-Chuan, C.: The effects of drilling parameters on bone temperatures: a finite element simulation, Wuhan, China, 10–12 May, 1–4, 2011.

Zionts, L. E.: Fractures and dislocations about the knee. In: Skeletal Trauma in Children, USA: Saunders, 443–449, 2003.

7

The intradiscal failure pressure on porcine lumbar intervertebral discs: an experimental approach

A. R. G. Araújo, N. Peixinho, A. Pinho, and J. C. P. Claro

Department of Mechanical Engineering, University of Minho, Guimarães, Portugal

Correspondence to: A. R. G. Araújo (angeloaraujo@dem.uminho.pt)

Abstract. The intervertebral disc is submitted to complex loading during its normal daily activities which are responsible for variations of the hydrostatic pressure in its structure. Thus, the determination of the magnitude of failure hydrostatic pressure is essential as a potential for the evaluation of the mechanisms that promote the weakening and the disruption of the annular fibers, commonly linked to herniation process on the spine column. However, few studies include the determination of the failure pressure on discs and the results are widely contradictory. Therefore, the objective of the present work is to determine the values of IDP that promotes the disc disruption. To achieve this goal, the tests were performed using a hydraulic cylinder that inflates the intervertebral disc. The results revealed a mean pressure failure of 0.62 ± 0.08 MPa for lumbar porcine samples ($n = 6$). From this approach it can be concluded that (1) the potential for disc injury may exist at low pressures for lumbar porcine discs when compared several animal and human ones; (2) the rupture of human cervical and porcine lumbar annular fibers could occur for values of intradiscal pressure that are within the physiological range.

1 Introduction

The intervertebral disc (IVD) is a complex and inhomogeneous structure structure composed by an inner gel-like core – the Nucleus Pulposus (NP) surrounded by a layered structure, the Annulus Fibrosus (AF). These structures are limited at the top and bottom by the Cartilaginous Endplate (CEP). The IVD allows successful load-bearing movements due to a synergetic effect of all components. The loads applied on the disc during its normal daily activities are responsible for variations of the internal disc pressure in the NP (Schechtman et al., 2006).

The internal disc pressure or intradiscal pressure (IDP) can be defined as the hydrostatic pressure presented by the NP of an healthy IVD (Claus et al., 2008). The IDP plays a key role on the IVD's ability to withstand the physiological loads (Steffen et al., 1998), being an important parameter to understand the spinal on the disc degeneration. In fact, the IDP data has been essential for prevent the spinal complaints by forming a basis for clinical advice to promote the correct sitting postures. The measurements of IDP also help to clarify the effect of the external loads on the IVD behavior (Claus et al., 2008) and to recognize the mechanism of IDP drop in disc degeneration. In addition, these data form the basis for physiotherapy and rehabilitation programs (Wilke et al., 1999).

From a biomechanical point of view, the IDP is influenced by the axial loads acting on spine (Sato et al., 1999). An increase in the compressive load applied to healthy discs is "converted" into IDP (Schechtman et al., 2006). As the NP can be considered incompressible (Castro et al., 2014), the AF bulges outwardly due to the stretch of annular fibres (Van der Veen et al., 2008) which, together with osmotic phenomenon, promotes a loss in both IVD height and volume.

The importance of IDP is reinforced due to difficult on the assessment of the disc strengthen properties (Schechtman et al., 2006). Although a simple axial compressive overload could not induce damage in an healthy disc, some movements such as compression combined with hyperflexion might generate an IDP beyond what the disc could withstand, promoting several injuries (Nachemson and Elfström, 1970; Schechtman et al., 2006). Previous studies had demonstrated that, before occurring any disc disruption, the com-

pressive overloading provokes the vertebral endplate damage and collapse (Dolan et al., 2013; Schechtman et al., 2006). Thus, the study of the IDP is a subject of deep interest in order to determine its contribution for IVD injury.

The determination of the magnitude of failure IDP is also essential as a potential parameter for the evaluation of the mechanisms that promote the weakening and the disruption of the annular fibers (Iencean, 2000). Once combined with the traditional provocative discography, the IDP monitoring represents an important way to determine the clinical significance internal disc disruption (Menkowitz et al., 2005).

Few studies include the determination of the failure pressure of the IVDs (Iencean, 2000; Menkowitz et al., 2005; Schechtman et al., 2006; Veres et al., 2010). Schechtman et al. (2006) investigated the intrinsic failure strength of the intact bovine caudal disc using an hydraulic inflation actuator. A colored hydrogel was injected into NP under monitored pressure. It was found a mean hydrostatic failure pressure of 18 ± 3 MPa. This method allowed understanding the alterations of the intrinsic disc strength associated with prior loading history or degeneration. However, it does not give information about the microstructural behavior of inner annular fibers after the inflation. Later, Veres et al. (2010) used the same technique performed by Schechtman et al. (2006) to investigate the role of high IDP in the disruption of the annular fibers of the ovine lumbar IVDs. This team included the analysis of the AF damage after pressure insertion by a microstructural investigation. It was found a mean failure pressure of 14.1 ± 3.9. It was also reported that posterior annular region is more susceptible to disruption than the other disc regions, due to its inability to distribute hydrostatic pressures circumferentially.

However, other studies showed that the IVD's injuries could be induced at lower IDPs. Iencean (2000) developed an experimental device for determine the rupture IDP of lumbar intervertebral discs, consisting of a source of pressure, connected to a tube introduced into the IVD through a tunnel drilled in the body of the subjacent vertebra by CEP access. The results revealed that the rupture was reached for IDP ranged from 0.75 to 1.3 MPa for neutral posture and a maximum rupture IDP in anterior flexion of 1.2 MPa. Later, using a 25G needle for the insertion of pressure on cervical discs, Menkowitz et al. (2005), reported a mean intradiscal rupture pressure of 0.28 MPa (range 0.1–1.18 MPa). Both studies have demonstrated that the rupture of human cervical and porcine lumbar annular fibers could occur for IDPs within the physiological range.

Therefore, the objective of the present work is to bring an additional insight about the magnitude of IDP that leads to AF disruption. To achieve this goal, the tests were performed using a hydraulic cylinder that inflates the IVD with glycerin, while a porcine lumbar disc is compressed. The pressure was monitored by a digital manometer and the maximum point of pressure was considered as the rupture point. This inflation method, combined with a pre-defined compression, allows not only inducing a pure hydrostatic loading due to glycerin insertion, but also producing a hydrostatic component and an environment in which the disruption of the disc in-vivo could occur.

2 The materials and methods

2.1 Motion segment collection and preservation

After being collected, the lumbar spines from pigs with 18 months old were immediately sectioned into motion segments (MS) visible in the Fig. 1 (Araújo et al., 2014; Campbell-Kyureghyan et al., 2011). Posteriorly, the MS were sealed in plastic bags and frozen at −20 °C, until the day prior to mechanical testing, minimizing the tissue dehydration. This procedure was adopted since dead and frozen storage presents a negligible effect in mechanical properties of the spine (Adams et al., 1996). Before start any mechanical test, samples were hydrated with 12 h with phosphate buffer saline solution in order to prevent segment desiccation.

2.2 Pressurization configuration used to determine the failure intradiscal pressure

The schematic representation of the apparatus for the inducement of internal disc pressure is presented in Fig. 2.

The pressure is inserted in the inner disc region using a hydraulic cylinder. The pressure generating apparatus consists of a hydraulic cylinder, with a coupled lever that allows controlling the pressure exerted on the system. The injected pressure is assessed by a digital manometer incorporated in the system – the electronic pressure sensor PP7553, from IFM®. This manometer is connected with the LineRecorder® software that allowed registering the pressure acting in the IVD as function of time. The principle of function of each failure pressure test is simple. First, all system was filled with glycerin, which is inserted in the system by the pressure exerted by the hydraulic cylinder. The glycerin was selected as testing fluid since it presents relative higher density and viscosity than water ($1.261\ \mathrm{g\,cm^{-3}}$ and 1499 cP at 20 °C, respectively). To ensure that the entire system was filled with the liquid, and so, the value of failure pressure was not affected by air bubbles in the tubes, a "bleeding" tap was included in the stainless steel basis (Fig. 2). The procedure for air bubbles removal was simple: after the placement of the motion segment on the system, the tap is opened and the glycerin is forced to enter into the system by the suction effect promoted by the lever movement. Then, the fluid was poured by the tap until ensuring that the system presents no air bubbles. Finally, the tap is closed and the system is ready to be submitted to test.

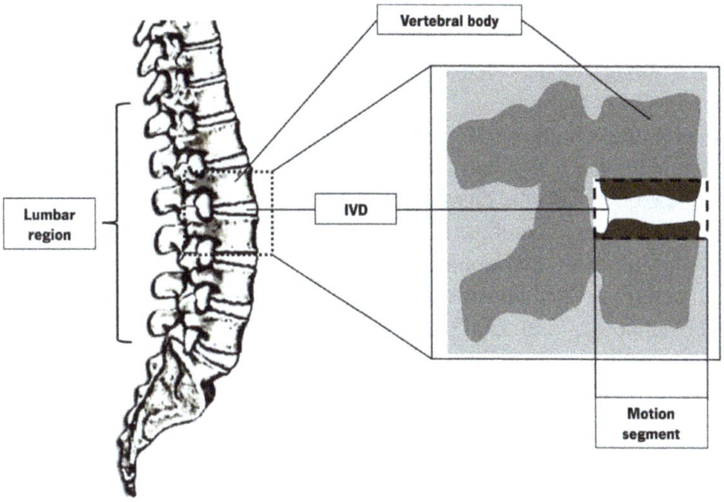

Figure 1. Schematic representation of the motion segment preparation. This structure is highlighted by the dashed rectangle. To obtain a motion segment, two vertebrae are cut parallel and transversely, obtaining an assemble composed by two half vertebra with a disc in between.

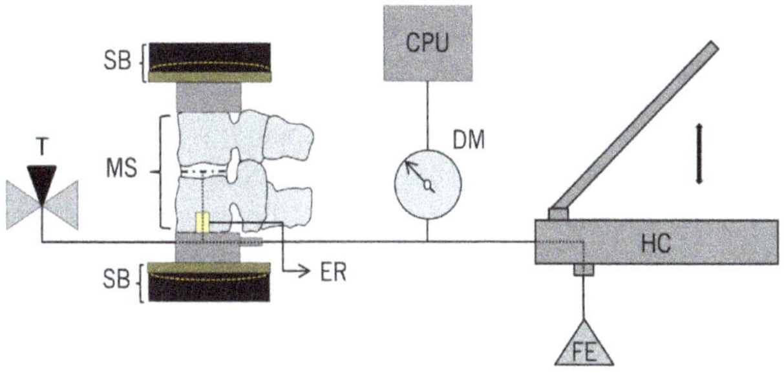

Figure 2. Failure pressure tester. The apparatus is composed by: SB – Spherical Axial Bearing; ER – Epoxy resin; MS – Motion Segment; CPU – Personal Computer; DM – Digital Manometer; T – Tap; HC – Hydraulic cylinder; T – "Bleeding" Tap; FE – Fluid Entrance.

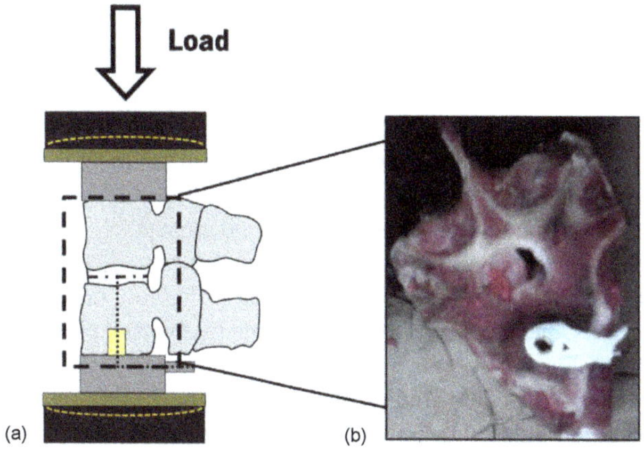

Figure 3. Motion segment attachment. **(a)** The MS was placed between two plates of Instron® 8874, being subjected to compression. The bottom plate is drilled, allowing the fluid passage from pressure apparatus to the hollowed screw. The attachment is done on epoxy resin (yellow region), to provide a better adhesion of the screw; **(b)** top vision of a drilled MS, with the presence of the self-tapping screw on the vertebral body.

2.3 Motion segment attachment

The vertebral bone is a highly porous structure, in which the fluids easily outflow from its inner region to the outer one (Bronner et al., 2010), being extremely hard to tight a screw in it. To effectively overcome this obstacle, a 9 mm diameter and 2 mm height hole as drilled in the top vertebra hole of the MS, in order to fill it with an epoxy resin with fast curing. Consequently, a 4 mm pilot hole was carefully drilled longitudinally through the resin and vertebrae until a sudden change in the structure resistance. This change of resistance indicates the point of contact between CEP and the NP.

The MS was then attached to pressure apparatus, in case to a homemade cylindrical stainless steel bottom plate, by a self-tapping steel screw (Fig. 3). The length of the self-tapping screw was 20 mm, presenting two threads: a section of 10 mm height and 7 mm diameter drywall screw thread to a drill in the vertebra and a region of 10 mm height and 5 mm diameter to attach to the stainless steel plate. This screw also presents a drilled hollow along its entire length, with an internal bore of 1.5 mm diameter that allows the fluid passage. Then, the screw was tightened until reaching the contact point between CEP and the NP. An O-Ring was placed on a cylindrical stainless steel bottom plate (around the self-tapping screw), in order to prevent fluid leakage. The compressive loading was exerted in the top of the other vertebral body. The MS samples were carefully aligned according a pre-defined system of axis, using spherical axial bearing system. This alignment ensures that all discs are placing in the

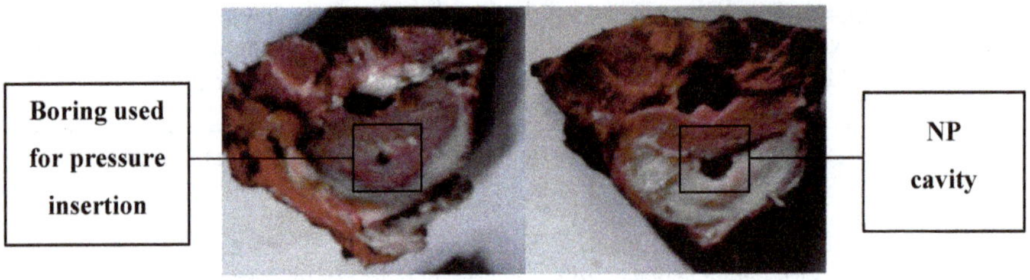

Figure 4. Images of MS sawed transversally after testing. It is visible the presence of a cavity in the NP region, indicating the pressurizing zone.

same position and compressed along the same plane, avoiding errors associated to misplacement of the samples.

The preliminary trials have failed on containing the insertion of glycerin in the nuclear cavity. In fact, during the initial tests, using only fast curing resin as interface between the screw and the MS, the fluid flowed out from the drilled vertebrae motion segment, by the screw insertion. Six motion segments were used during preliminary trials for testing the failure IDP. Then, the approach was improved by using a thin film of silicone in the interface vertebrae-stainless steel base.

2.4 Experimental procedure

A compressive axial displacement of 1 mm was imposed in the top of MS in order to either avoid the longitudinal expansion of the IVD during the pressurization test and to simulate the in-vivo confined conditions (Schechtman et al., 2006). After impose this displacement, the MS is pressurized by the descendent movement of the lever. These samples were pressurized during a period of approximately 10 s, which represents the maximum time that all samples needed to reach failure after the application of an external pressure source. During this period, the Line Recorder® software allowed to monitored and save the values of IDP as function of time. Ten motion segments were used for the monitoring of the failure IDP.

This study had neglected the effect of the pressure drag on the walls during the fluid passage in both tubes and screw. Thus, it is assumed that the pressure read on the pressure sensor corresponds to the real inflation pressure of the IVD.

After all tests, each MS was sectioned transversally in the IVD region, in order to assess if it was pressurized in the nuclear region. The segmented disc was photographed and the final area of each IVD determined using the image processor Image Pro Plus 4.6®.

3 Results

3.1 Visual inspection of the pressurized motion segment

The images were taken to assess the functionality of pressurizing system and, in addition, to validate the results. Six specimens were pressurized in the NP region, which is visible in Fig. 4. Thus, six motion segments were considered for further analysis.

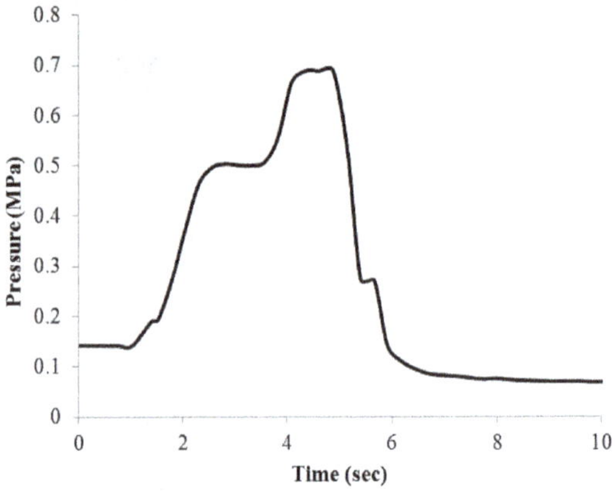

Figure 5. The characteristic response of MS to a failure IDP event (in MPa) as function of time (in s).

In this study only tests revealing a complete containing of the IVD inflation, where no leakage or liquid tear in the vertebral body, were considered. Three specimens were discarded as it was not detected a signal of external NP pressurization. Leakage in the tubes was also found in one sample.

The average disc area and height were $917.2 \pm 107.0\ \text{mm}^2$ and 5.97 ± 0.63 mm, respectively.

3.2 Failure pressure in the motion segment

The criteria of failure in the MS structure, due to the insertion of IDP using an external source, comprises two evident phenomena: (1) the flood of the glycerin from any motion segment region, with exception for the top region of both vertebral bodies, and (2) a significant drop in the pressure detected by the digital manometer. A representative example of the curve of a failure IDP curve as function of time is depicted in Fig. 5.

The values of failure IDP together with the description of the failure in the specimen after visual inspection of the six motion segments that matched the criteria for failure are described in Table 1. In all the cases a substantial drop of pres-

Table 1. Failure pressure (maximum in the curve IDP vs. time) for the 6 motion segments that presents external indicators of rupture, such as sudden pressure drop, leakage in the vertebra region or in spinal cord (fluid outflowing from the spinal cord channel) or tears formed on the annular region or also on the top and bottom parts of annular region (Cartilaginous Endplate region).

Motion segment	Failure pressure (MPa)	Description of the failure on the specimen after visual inspection
1	0.58	– Moderate leakage around the screw region in the vertebra – A little tear formed in the annular region
2	0.69	– Leakage of fluids from the annular region
3	0.69	– Moderate leakage around the screw region in the vertebra – Slight leakage around the spinal cord (fluid outflowing from the spinal cord channel)
4	0.69	– Leakage of fluids from the annular region
5	0.55	– Slight leakage around the screw region in the vertebra – Tears formed on top and bottom parts of annular region (CEP region)
6	0.50	– Slight leakage around the spinal cord (fluid outflowing from the spinal cord channel) – Moderate leakage in the annular region and around the screw region in the vertebra
Mean ± SD	0.62 ± 0.08	

sure was visible after reach the maximum value of IDP that corresponds to the failure pressure.

Several events can be identified during the process of MS failure due to the insertion of an external pressure source. In all discs, an immediate glycerine leakage was visible after reaching the maximum value of pressure. Moreover, there was not possible to identify a pattern between the IDP failure and the place of glycerine outflow, being visible a wide range of rupture regions in the motion segment instead.

On Fig. 6 diverse examples of MS failure are exposed. The images document the different type of rupture occurring in each MS submitted to failure IDP values.

4 Discussion

Despite the panoply of mechanical tests intending to elucidate the contribution of loads to MS collapse, the magnitude of IDP that led to the MS rupture remains still unclear. The measurement of the internal pressure that leads to disc rupture is important not only to understand the mechanisms of IVD failure but also for the design of new implants for NP replacement. In fact, the fiber orientation of the AF is able to withstand the hoop stresses generated hydrostatic pressure in the healthy conditions (Inoue and Espinoza Orías, 2011). When the NP is removed, the outer region of the AF continues to bulge outward during the application of axial loading; conversely, the inner region bulges toward the center of the IVD (Goins et al., 2005; Meakin and Hukins, 2000). Thus, these implants should be able to exert a prescribed pressure on the inner AF walls and this pressure should be able to keep the biomechanical characteristics of the remaining disc, avoiding the disc degeneration.

Based on the method developed by Schechtman et al. (2006) to measure the failure strength on bovine caudal disc, this work determined the failure strength of the porcine MSs under an imposed IDP, using the a cartilaginous endplate access.

The definition of failure of the present work contemplates the flood of the glycerin from any region motion segment together with a significant drop in the pressure detected by the digital manometer. This consideration arises from the fact that, when a failure pressure is imposed in a MS, a chaotic effect is detectable in the whole MS structure. No localized region was detected or identified as a typical region of disc rupture after the insertion of an external pressure, indicating that there was a redistribution of the IDP in the inner region of the multilayered AF, i.e. in the contact zone between NP and AF. Thus, the collapse of the IVD structure is not a local but a generalized event.

Concerning to quantitative results, this study reports a mean pressure failure of 0.62 ± 0.08 MPa for lumbar porcine samples. The comparison between these results with the failure pressure of the annular fibers reported in previous studies for several MS models (Table 2) reveals that the values documented on this report are appreciably lower than the more recent studies, performed by Schechtman et al. (2006) and Veres et al. (2010). In fact, several studies reported the magnitude of the rupture values obtained in this study as within the normal range of physiological IDP for human samples (Claus et al., 2008; Dennison et al., 2008; Sato et al., 1999; Wilke et al., 1999).

Schechtman et al. (2006) investigated the intrinsic failure strength of the intact bovine caudal disc under inflation, injecting a colored gel with a hydraulic actuator: the mean hy-

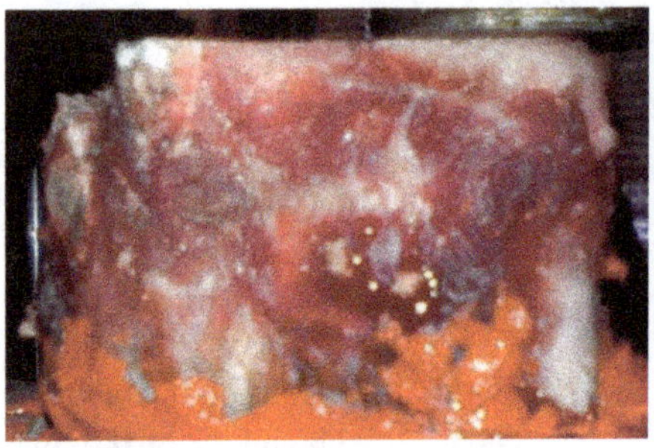

(a) Leakage of fluids from the annular region. Two motion segments failed in this manner.

(a) Severe leakage around the screw region in the vertebrae (bottom vertebrae) and in the annular region. Slight leakage around the spinal cord (fluid outflowing from the spinal cord channel). One motion segment failed in this manner.

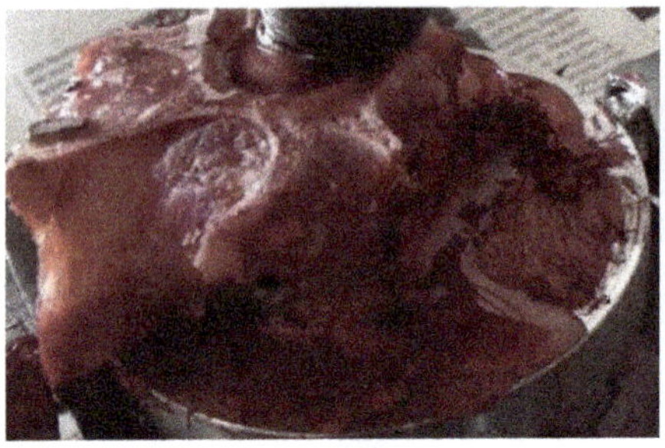

(c) Moderate leakage around the screw region in the vertebrae (bottom vertebrae). Two motion segments failed in this manner.

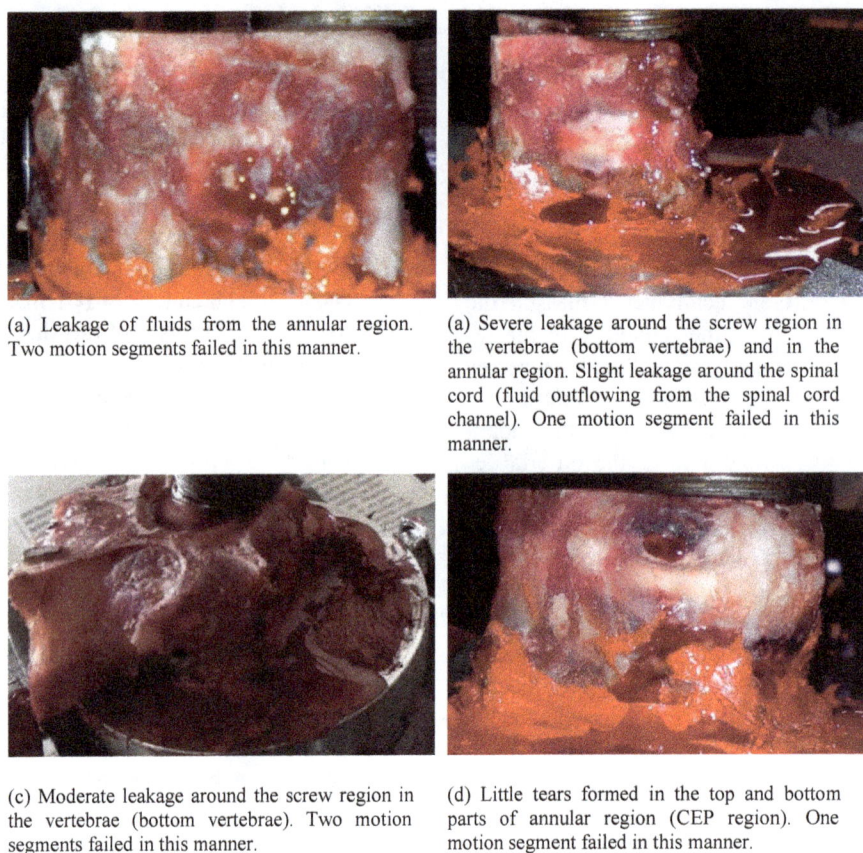

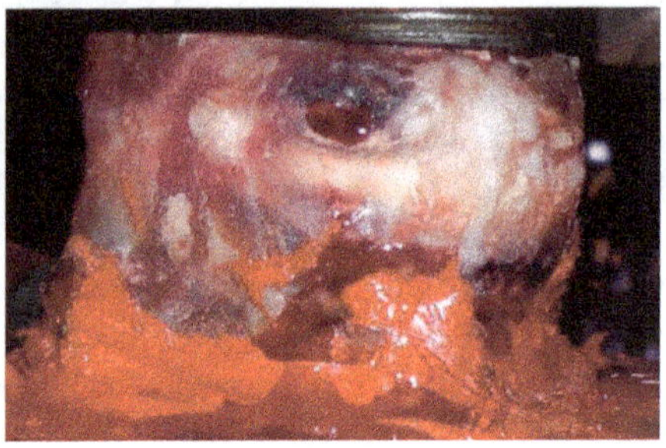

(d) Little tears formed in the top and bottom parts of annular region (CEP region). One motion segment failed in this manner.

Figure 6. Representative images of different configurations of failure in MSs, during the inflation procedure.

Table 2. Recent data about failure pressure data for different IVD models.

Author	Year	Models	Mean failure pressure (MPa)
Iencean	2000	Lumbar human	0.75 to 1.3
Menkowitz	2005	Cervical human	0.28 (min–max: 0.1–1.18)
Schetchman	2006	Caudal bovine	18 ± 3
Veres	2010	Lumbar ovine	14.1 ± 3.9
This study	2015	Lumbar porcine	0.62 ± 0.08

drostatic failure pressure was found to be 18 ± 3 MPa. Later, Veres et al. (2010), using the same technique performed by Schechtman et al. (2006) to investigate the role of high IDP on annular fibers disruption in ovine lumbar IVDs reported a the mean failure pressure of 14.1 ± 3.9. According with the latest developments about the failure pressure of disc, it would be expectable a higher value for the rupture of porcine discs, as the magnitudes presented by these authors are decidedly superior from those documented in this work.

The present results are within the same magnitude of values documented in two works performed with human lumbar (Iencean, 2000) and cervical IVDs (Menkowitz et al., 2005). Iencean (2000), reported a rupture pressure up to 1.3 MPa after inflating the IVD with compress air through a tunnel drilled in the body of the subjacent vertebra via CEP access; Menkowitz et al. (2005) claimed a mean intradiscal rupture pressure of 0.28 MPa (range 0.1–1.18 MPa), using a 25G needle for the insertion of a contrast dye, with IDP monitoring during time. These studies indicate two important facts: (1) the potential for iatrogenic disc injury may exist at low pressures for lumbar porcine IVDs when compared with samples presented in Table 2; (2) the rupture of human cervical and porcine lumbar annular fibers could occur for IDPs within the physiological range, showing that the injury on these structures could be induced at significantly lower pressures.

The disparities in terms of failure pressure, presented in Table 2, could be related to several phenomena. This study had limited the expansion of the MS in terms of height, imposing a permanent 1 mm displacement to each disc sample while the pressurization test occured. The particular case of the use of caudal discs as model for the human lumbar disc is an option that has been questioned at several levels, as they presents different mechanical loading relative to human lumbar spine together with different composition and metabolism and different anatomy (Alini et al., 2008). The

caudal bovine samples presents a nearly cylindrical shape (Schechtman et al., 2006), neglecting the possible existence of critical point of the hydrostatic pressures distribution in the IVD. This misassumption could lead to divergent values concerning to which is expected for human discs. Thus, the present work was conducted with lumbar porcine samples since they were considered was geometrically and morfologically more similar to the human ones, and so, a more suitable choice for mechanical studies on the IVD (Alini et al., 2008). Moreover, the different techniques of measure the failure pressure may compromise the reliability of the results. Several approaches were analyse and their weaknesses should not be neglected when a comparison is made.

This work itself present some limitations as the cadaveric study was limited by sample size or the criteria defined for MS failure. In fact, in this study the real values of rupture could be hidden by the flood of glycerin from the screwed vertebra. However, the samples where the rupture had only ocurred in the screwed vertebra were discarded and, for the quantitative analysis, only the samples with an additional point of rupture were considered. Moreover, it does not contemplate a microstructural analysis of the failure mechanisms. The rupture ocurring in the cited articles neglected the shape differences in samples, in what concerns to the critical points of failure, that are different between specimens. For a complete understanding of the process of IVD failure in compression, the phenomena ocurring on annular region and on the annular wall-endplate must be assessed, as the latter represent a typical critical point for disc rupture in compression (Adams, 2002; Schechtman et al., 2006). In addition, the data provided on this report is a result of an inflation after pre-defined compression of the IVD. Another types of efforts and loadings, such as flexion or rotation must be, together with disc inflation, must be considered for the rupture analysis.

5 Conclusions

In this study, the values of IDP that leads to disc rupture were determined by inflating porcine IVDs with glycerin. The experimental findings provide new insights about the mechanisms of disc failure, bringing an important addition for the validation of the constitutive models as well as to stimulate the development of more reliable solutions to replace the IVD. The main finding of this approach is that the failure could occur for a magnitude of IDP that could be found during daily activities. In light of this date, it can be concluded that the critical IDP pressure is an very specific characteristic, probably depending on compositional factors and a range of features related with specimens particularities. Moreover, this set of new experimental data should be considered on the design of more efficient solutions for the nucleus pulposus replacement, as the maintenance of a prescribed hydrostatic pressure in the inner walls of the AF is essential for keep the physiological conditions of the IVD.

However, care must be taken on the extrapolation of these results to the failure IDP in human samples. Upcoming studies should include numerical or optimization methods, in order to monitoring the factors that could induce divergences in the results, helping to clarify the magnitude of the annular failure strength in the human discs.

Appendix A: Acronyms list

Table A1. List of acronyms.

AF	Annulus Fibrosus
CEP	Cartilaginous Endplate
IVD	Intervertebral Disc
IDP	Intradiscal Pressure
MS	Motion Segment
NP	Nucleus Pulposus

Acknowledgements. This work was funded by the project "NP Mimetic – Biomimetic Nano-Fibre Based Nucleus Pulposus Regeneration for the Treatment of Degenerative Disc Disease", financed by the European Commission under FP7 (grant NMP-2009-SMALL-3-CP-FP 246351).

The authors also express their gratitude to Indústria de Carnes do Minho (ICM) – Primor Group – for the possibility of collecting spine column samples in their facilities.

References

Adams, M. A.: The Biomechanics of Back Pain, Churchill, Livingstone, 2002.

Adams, M. A., McNally, D. S., and Dolan, P.: "Stress" distributions inside intervertebral discs. The effects of age and degeneration, J. Bone Joint. Surg. Br., 78, 965–72, 1996.

Alini, M., Eisenstein, S. M., Ito, K., Little, C., Kettler, A. A., Masuda, K., Melrose, J., Ralphs, J., Stokes, I., and Wilke, H. J.: Are animal models useful for studying human disc disorders/degeneration?, Eur. Spine J., 17, 2–19, 2008.

Araújo, A., Peixinho, N., Pinho, A., and Claro, J. C. P.: A Novel Methodology to Assess the Relaxation Rate of the Intervertebral Disc by Increments on Intradiscal Pressure, Appl. Mech. Mater., 664, 379–383, 2014.

Bronner, F., Farach-Carson, M. C., and Roach, H. I.: Bone and Development, Springer Science & Business Media, Springer-Verlag, London, 2010.

Campbell-Kyureghyan, N. H., Yalla, S. V., Voor, M., and Burnett, D.: Effect of orientation on measured failure strengths of thoracic and lumbar spine segments, J. Mech. Behav. Biomed. Mater., 4, 549–557, 2011.

Castro, A. P. G., Wilson, W., Huyghe, J. M., Ito, K., and Alves, J. L.: Intervertebral disc creep behavior assessment through an open source finite element solver, J. Biomech., 47, 297–301, 2014.

Claus, A., Hides, J., Moseley, G. L., and Hodges, P.: Sitting versus standing: does the intradiscal pressure cause disc degeneration or low back pain?, J. Electromyogr. Kinesiol., 18, 550–558, 2008.

Dennison, C. R., Wild, P. M., Byrnes, P. W. G., Saari, A., Itshayek, E., Wilson, D. C., Zhu, Q. A., Dvorak, M. F. S., Cripton, P. A., and Wilson, D. R.: Ex vivo measurement of lumbar intervertebral disc pressure using fibre-Bragg gratings, J. Biomech., 41, 221–225, 2008.

Dolan, P., Luo, J., Pollintine, P., Landham, P. R., Stefanakis, M., and Adams, M. A.: Intervertebral disc decompression following endplate damage: implications for disc degeneration depend on spinal level and age, Spine, 38, 1473–181, 2013.

Goins, M. L., Wimberley, D. W., Yuan, P. S., Fitzhenry, L. N., and Vaccaro, A. R.: Nucleus pulposus replacement: an emerging technology, Spine J., 5, 317S–324S, 2005.

Iencean, S. M.: Lumbar Intervertebral Disc Herniation Following Experimental Intradiscal Pressure Increase, Acta Neurochir., 142, 669–676, 2000.

Inoue, N. and Espinoza Orías, A. A.: Biomechanics of intervertebral disk degeneration, Orthop. Clin. North Am., 42, 487–99, 2011.

Meakin, J. R. and Hukins, D. W. L.: Effect of removing the nucleus pulposus on the deformation of the annulus fibrosus during compression of the intervertebral disc, J. Biomech., 33, 575–580, 2000.

Menkowitz, M., Stieber, J. R., Wenokor, C., Cohen, J. D., Donald, G. D., and Cresanti-Dakinis, C.: Intradiscal pressure monitoring in the cervical spine, Pain Physician, 8, 163–166, 2005.

Nachemson, A. and Elfström, G.: Intravital dynamic pressure measurements in lumbar discs. A study of common movements, maneuvers and exercises, Scand, J. Rehabil. Med. Suppl., 1, 1–40, 1970.

Sato, K., Kikuchi, S., and Yonezawa, T.: In vivo intradiscal pressure measurement in healthy individuals and in patients with ongoing back problems, Spine, 24, 2468–2474, 1999.

Schechtman, H., Robertson, P. A., and Broom, N. D.: Failure strength of the bovine caudal disc under internal hydrostatic pressure, J. Biomech., 39, 1401–1409, 2006.

Steffen, T., Baramki, H. G., Rubin, R., Antoniou, J., and Aebi, M.: Lumbar intradiscal pressure measured in the anterior and posterolateral annular regions during asymmetrical loading, Clin. Biomech., 13, 495–505, 1998.

Van der Veen, A. J., Mullender, M. G., Kingma, I., van Dieen, J. H., Van, J. H., and Smit, T. H.: Contribution of vertebral [corrected] bodies, endplates, and intervertebral discs to the compression creep of spinal motion segments, J. Biomech., 41, 1260–1268, 2008.

Veres, S. P., Robertson, P. A., and Broom, N. D.: The influence of torsion on disc herniation when combined with flexion, Eur. Spine J., 19, 1468–1478, 2010.

Wilke, H. J., Neef, P., Caimi, M., Hoogland, T., and Claes, L. E.: New in vivo measurements of pressures in the intervertebral disc in daily life, Spine, 24, 755–762, 1999.

8

Shape and force analysis of capillary bridge between two slender structured surfaces

Z. F. Zhu[1], J. Y. Jia[1], H. Z. Fu[2], Y. L. Chen[1], Z. Zeng[1], and D. L. Yu[1]

[1]School of Mechano-Electronic Engineering, Xidian University, Xi'an, China
[2]ZTE Corporation, Shenzhen, China

Correspondence to: Z. F. Zhu (zhaofeizhu@stu.xidian.edu.cn)

Abstract. When a capillary bridge of a constant volume is formed between two surfaces, the shape of the liquid bridge will change as the separation between those surfaces is varied. To investigate the variable forces and Laplace pressure of the capillary bridge, as the shape the bridge evolves, a pseudo-three-dimensional force model of the capillary bridge is developed. Based on the characteristics of the slender structured surface, an efficient method is employed to directly solve the differential equations defining the shape of the capillary bridge. The spacing between the plates satisfying the liquid confined within the hydrophobic region of the structured surface is calculated. The method described in this paper can prevent meshing liquid surfaces such that, compared with Surface Evolver simulations, the computing speed is greatly improved. Finally, by comparing the results of the finite element simulations performed with Surface Evolver with those of the method employed in this paper, the practicality of the method is demonstrated.

1 Introduction

The capillary bridge between two structured surfaces is very important in research into, and application of, micromechanical structures, surface mount technologies, microfluid dynamics, biological bionics, and other technical fields (Stewart et al., 2013; Bowden, 1997; Gau et al., 1999; Mlota et al., 2010; Saad and Neumann, 2014). The forces exerted by the liquid bridge have a great effect on the life and reliability of microelectromechanical systems (MEMS), flip-chip alignment, and the mobility of fluids in micro-channels (Guo et al., 2009; Chen and Soh, 2008; Bush et al., 2010; Dalin et al., 2010; Krammer, 2014). Therefore, research into the mechanical properties of a liquid bridge has great practical significance. By using micro-contact printing, vapor deposition, photolithography, and other techniques, a hydrophobic substrate surface can be patterned with hydrophilic strips to produce a structured surface, so as to control the morphology of the capillary bridge and achieve the desired mechanical properties. This has a high application value in fields such as microelectronics, semiconductors, MEMS, and microchannels (Bowden, 1997; Gau et al., 1999; Bush et al., 2010; Lipowsky, 2001; Broesch and Frechette, 2012; Broesch et al., 2013).

The capillary bridge between two structured surfaces can be used as a simplified model for biological adhesives, microfluidic channels, and self-assembly (Gau et al., 1999; Bush et al., 2010; De Souza et al., 2008; Broesch et al., 2014; Mermoz et al., 2012; Ferraro et al., 2012; Luo et al., 2014). In particular, a structured surface has the shape characteristics of a slender rectangle, which has great potential for application to the field of microchannels (Gau et al., 1999; Valencia et al., 2001; Lipowsky et al., 2000). Some studies have expressed concern about the effect of these slender rectangular structured surfaces. Some of these studies (Broesch and Frechette, 2012; Broesch et al., 2013) have addressed the relationship between the force of the capillary bridge between two slender structured surfaces and the morphological evolution of a liquid–gas interface, but the conclusions drawn could not be applied to situations where the length of the capillary is confined. One study (Swain and Lipowsky, 2000) addressed a slab geometry with a wetting phase confined between two chemically patterned substrates, but the results obtained through 2-D analysis did not correspond to

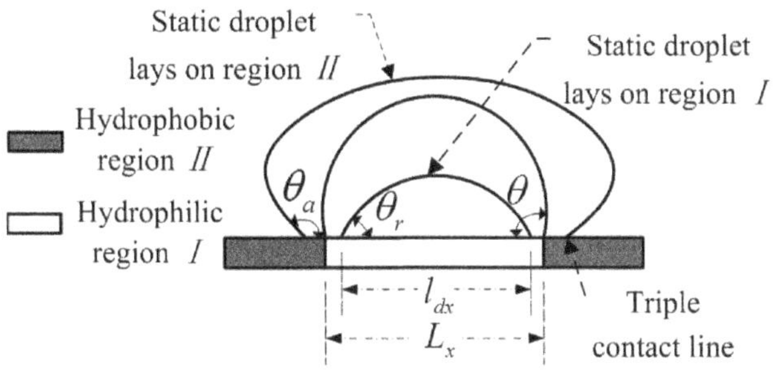

Figure 1. Change in contact angle θ caused by differences in wettability.

the actual situation. Another study (Valencia et al., 2001) addressed the morphology of the liquid phases within chemically structured slit pores, using the finite element method to evolve the surface towards minimal energy, but this suffered from the trade-off between computing speed and accuracy. Surface Evolver uses the gradient descent method to minimize the free energy of the system, and can be applied to a wide range of problems, rather than being optimized to deal with a specific problem (Brakke, 1994). To obtain a precision solution, therefore, it is necessary to progressively refine the surfaces by applying a tessellated mesh and allowing more time for the computing. Surface Evolver has some shortcomings, such as the fact that surfaces are allowed to intersect with each other and convergence to a minimum energy level can be difficult to determine (Brakke, 1994). Therefore, there is a need for a high-efficiency method capable of overcoming these issues.

We adopted the method of equivalent tension to establish a pseudo-three-dimensional force model for the capillary bridge. Considering the marginal effects of the slender capillary bridge profile and the Young–Laplace equation, we set out to solve the differential equations defining the shape of a capillary bridge having a constant volume. To maximize the efficiency with which the initial-value problem can be solved, an efficient mathematical analysis method is used to solve the boundary-value differential equations for the capillary bridge profiles, therefore optimizing the initial value. Finally, the relationship between the forces and the shape parameters of the capillary bridge is explored. For every calculation, we employ the equation defining the equilibrium of the forces to solve the differential equations, then the change processes for the capillary bridge shape parameters are obtained by means of a decline process defining the separation between the plates, which satisfies the physical boundary constraints. As there is no need to consider meshing liquid surfaces, the computing speed is greatly improved relative to the simulations performed with Surface Evolver. Moreover, there are no shortcomings, such as surfaces intersecting with each other or convergence to a minimum energy being difficult to judge.

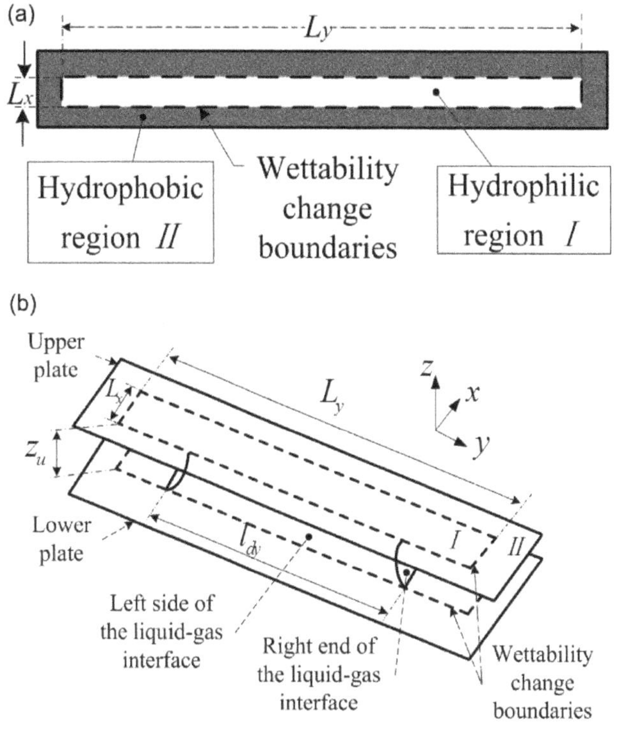

Figure 2. (**a**) Structured surface of the plate and (**b**) capillary bridge.

2 Characteristic parameters and force analysis for the capillary bridge

2.1 Characterizations

On a smooth homogeneous surface, the contact angle between a liquid–gas interface and a liquid–solid interface remains constant. This is known as the "wetting angle" (Langbein, 2002; Bonn et al., 2009). Under a specific gas atmosphere, the size of the wetting angle is affected only by the surface tension between the liquid and the solid (Langbein, 2002; Bonn et al., 2009; Pozrikidis, 2012; Oliver et al., 1977). As shown in Fig. 1, when the triple contact line slips on the structured surface and reaches the point where the wettability changes, then the triple contact line will be pinned and the contact angle will be changed. This is known as "canthotaxis" (Lipowsky et al., 2000; Herminghaus et al., 2008; Liu et al., 2012; Langbein, 2002). If the liquid is static in region I or II, then the contact angles will be θ_r or θ_a, respectively. The variation in the contact angle θ is: $\theta_r \leq \theta \leq \theta_a$ when $l_{dx} = L_x$, $\theta = \theta_r$ when $l_{dx} < L_x$, and $\theta = \theta_a$ when $l_{dx} > L_x$.

As shown in Fig. 2, the slender capillary bridge consists of two plates parallel to each other, both of which have slender rectangular structured surfaces. The structured surface is produced by depositing hydrophilic strips (region I) on a substrate with a hydrophobic surface (region II) through different techniques (Gau et al., 1999; Lipowsky, 2001; Broesch

and Frechette, 2012; Broesch et al., 2013; Valencia et al., 2001) (see Fig. 2a). The length of region I is set to L_y and its width to L_x. Region I has a high aspect ratio ($L_y \gg L_x$). In this paper we only consider a capillary bridge confined within the region I, that is, the case where $l_{dx} = L_x$. Ignoring the roughness of the structured surfaces and evaporation from the capillary bridge, a change in the spacing between the plates causes the two ends of the capillary bridge to advance or recede along the length of the region I, while the contact angles at both ends of the capillary bridge remain constant (Broesch and Frechette, 2012; Broesch et al., 2013; Valencia et al., 2001; Swain and Lipowsky, 2000). This is termed the "slipping" of the triple contact lines on region I. The widths (l_{dx}) of the liquid–solid interfaces are confined by the hydrophobic region II and are constant, but the contact angle and the curvature of the lateral liquid–gas interface along the length of the hydrophilic region I will vary with the decline in the capillary bridge height, termed the "hinge movement" of the bilateral liquid–gas interfaces at the location where the wettability varies (Broesch and Frechette, 2012; Broesch et al., 2013; Valencia et al., 2001; Yaneva et al., 2005). As the spacing between the plates decreases further, the triple contact line at the two ends of the capillary bridge slips and touches region II, at which point the triple contact line is pinned and the all-around liquid–gas interface of the capillary bridge hinges, which inevitably leads to variations such as the forces of the capillary bridge, internal pressure, surface curvature, and contact angles (Lipowsky, 2001; Broesch and Frechette, 2012; Broesch et al., 2013).

2.2 Force analysis for capillary bridge

Based on the three-dimensional shape characteristics of the capillary bridge, a pseudo-three-dimensional force model of the capillary bridge was developed by applying the tension equivalent method. In the Cartesian coordinate system, the upward direction of the force is positive. As shown in Fig. 2, taking advantage of the symmetry of the capillary bridge, the force analysis of the right end and the right side of the liquid–gas interfaces is represented by Fig. 3a and b.

The hydrostatic equilibrium equation of the capillary bridge in the vertical direction can be written as

$$W_z + l_{dx} l_{dy} p_b - \rho g V_0 - 2T\left(l_{dx} \sin\theta_{di} + l_{dy} \sin\theta_{dj}\right) = 0. \quad (1)$$

At the bottom of the capillary bridge, the pressure is

$$p_b = \frac{2T\left(l_{dx} \sin\theta_{di} + l_{dy} \sin\theta_{dj}\right) + \rho g V_0 - W_z}{l_{dx} l_{dy}} \quad (2)$$

where ρ is the density of the liquid and T is the surface tension of the liquid–gas interface. The length of the liquid–solid interface on the lower plate is set to l_{dy}, while its width is set to l_{dx}. θ_{di} and θ_{dj} are the contact angles between the liquid–gas interface and the liquid–solid interface along the width and length of the liquid bridge, respectively. V_0 represents the volume of the liquid; W_z represents the total force

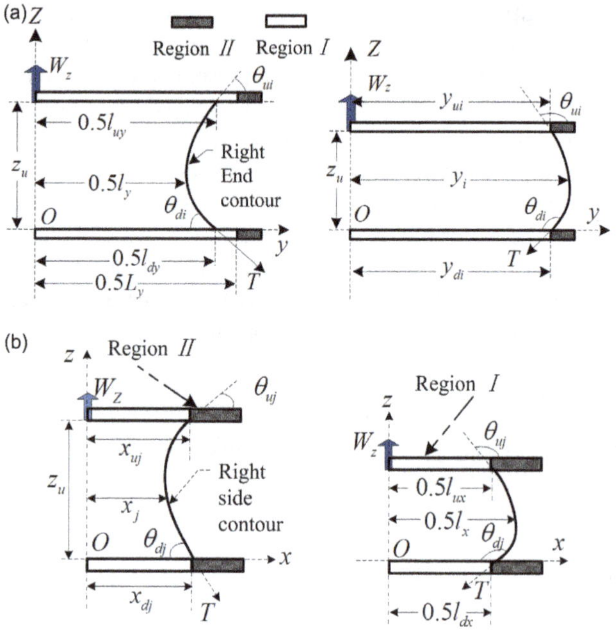

Figure 3. **(a)** Right end of liquid–gas interface force diagram with different spacings and **(b)** lateral liquid–gas interface force diagram with different spacings.

applied to the top of the capillary bridge by the upper plate in the vertical direction, which involves the capillary force and the vertical component of the surface tension.

3 Differential equations defining capillary bridge characterizations

3.1 Transformation of Young–Laplace equation

Unknowns W_z and p_b cannot be solved using Eqs. (1) and (2), but the hydrostatic equilibrium equation shows that the unknowns are closely related to the shape characteristic parameters of the capillary bridge. To solve unknowns W_z and p_b, solving the differential equations defining the capillary bridge shape becomes key to the problem. Assuming that the liquid volume is constant and transforms the Young–Laplace equation (Padday, 1971), we can develop differential equations to define the shape of the capillary bridge. Because of the high aspect ratio of the capillary bridge ($l_y \gg l_x$, where l_y and l_x are the length and width of the capillary bridge, respectively, as shown in Fig. 3), the bilateral curved surfaces of the liquid–gas interface can be regarded as approximating to cylindrical surfaces. Then, one of the major radiuses of curvature of the Young–Laplace equation

$$1/\gamma + 1/\gamma' = \Delta p/T \quad (3)$$

approximates to infinity ($\gamma' \Rightarrow \infty$). The curvature of the bilateral curved surfaces can thus be written as:

$$1/\gamma = \Delta p/T \quad (4)$$

where Δp is the Laplace pressure. We can employ the surrounding atmospheric pressure as a reference ($\Delta p = p_b - p_z$), where $p_z = \rho g z$ represents the variation in the internal pressure due to gravity, and p_b is the pressure at the bottom of the capillary bridge.

The top view at the right end of the slender capillary bridge is shown in Fig. 4. Given the slender shape of the capillary bridge, the major radius of curvature of the two-end curved surfaces can be approximated as being equal to half the capillary bridge width l_x. Based on Eq. (3), the curvature of the two-end curved surfaces can be expressed as:

$$1/\gamma = \Delta p/T - 1/\gamma' \quad (5)$$

where $\gamma' = x_j$, $x_j = 0.5 l_x$ and the capillary bridge width l_x varies with the height coordinate z.

As shown in Fig. 3, the height coordinate z is chosen as an independent variable in order to satisfy the coordinating relationship between the shape characteristic parameters and pressure, such that the differential expressions of the characteristic parameters can be obtained. For different separations between the plates, the volume differential equation for z is obtained by assuming that the capillary bridge volume is constant. The differential equations for the shape characteristic parameters can thus be written as follows:

$$\begin{cases} \frac{d\theta_i}{dz} = \frac{1}{\sin\theta_i}\left(\frac{\Delta p}{T} - \frac{1}{\gamma'}\right) \\ \frac{dy_i}{dz} = 1/\tan\theta_i \\ \frac{d\theta_j}{dz} = \frac{\Delta p}{T\sin\theta_j} \\ \frac{dx_j}{dz} = 1/\tan\theta_j \\ \frac{dV}{dz} = 4x_j y_i - 2\left(2x_j^2 - \frac{\pi x_j^2}{2}\right) \end{cases} \quad (6)$$

where suffixes i and j are the labels of both ends and the bilateral liquid–gas interface profile curves, respectively; $z \in [0, z_u]$, where z_u is the separation between the plates (the height of the capillary bridge in the stable state), θ_i and θ_j are the slope angles of the outline curves in the yoz and xoz planes, respectively, and y_i and x_i are the abscissas of the outline curves in the yoz and xoz planes, respectively (see Fig. 3).

3.2 Boundary conditions of capillary bridge between slender plates

Based on the constraints and the characterization of the capillary bridge, the boundary values for the bottom and top of both ends and the bilateral liquid–gas interface profile curves are related as follows:

1. when $z = z_u$,

$$x_j(z_u) = 0.5 l_{ux} = 0.5 L_x;\ \theta_j(z_u) = \theta_{uj};$$
$$\theta_i(z_u) = \theta_{ui};\ y_i(z_u) = 0.5 l_{uy};\ V(z_u) = V_0, \quad (7)$$

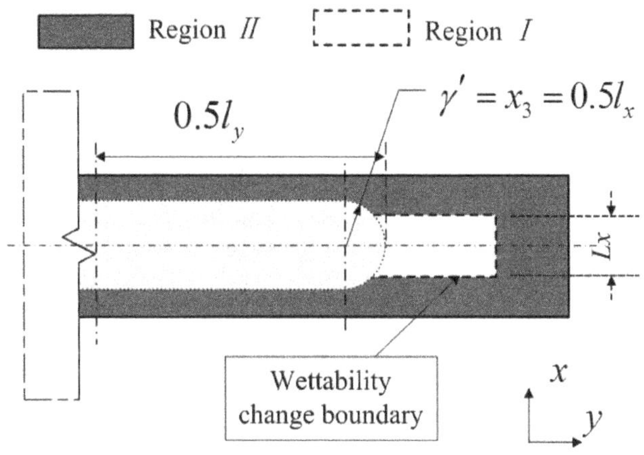

Figure 4. Top view of right end of capillary bridge.

2. when $z = 0$,

$$x_j(0) = 0.5 l_{dx} = 0.5 L_x;\ \theta_j(0) = \theta_{dj};\ \theta_i(0) = \theta_{di};$$
$$y_i(0) = 0.5 l_{dy};\ V(0) = 0, \quad (8)$$

where $\theta_{dj} \in [\theta_r, \theta_a]$. The length of the liquid–solid interface on the capillary bridge top is set to l_{uy}, and its width to l_{ux}. When the triple contact lines at the two ends of the capillary bridge have not yet touched region II, $\theta_i(z_u) = \theta_{ui} = \theta_r$, $\theta_i(0) = \theta_{di} = \theta_r$. When the triple contact lines touch the ends of the region I, $y_i(z_u) = 0.5 l_{uy} = 0.5 L_y$, $y_i(0) = 0.5 l_{dy} = 0.5 L_y$.

The solution to differential Eq. (6) is equivalent to solving for the boundary values of first-order nonlinear differential equations, obtaining five unknowns with only one independent variable z ($z \in [0, z_u]$). Because of the mutual coupling between the contact angle θ_{di} and θ_{dj} at the bottom of the capillary bridge, differential Eq. (6) can thus be solved.

4 Methods for solving differential equations

4.1 Making the boundary-value problem equal to an initial-value problem

To maximize the efficiency with which an initial-value problem can be solved, the boundary-value conditions for Eq. (8) are assumed to be equivalent to the initial-value conditions, but the solution to the initial-value problem cannot satisfy the boundary-value of Eq. (7), which becomes a problem of objective function minimization. Obtaining the boundary-value conditions for Eq. (7) is equivalent to minimizing relative error objective functions (Eqs. 9 and 10).

When $l_{dy} < L_y, \theta_i(z_u) = \theta_i(0) = \theta_r$. The objective function can then be written as

$$\mathrm{Min}:F\left(\theta_{dj}^{*},l_{dy}^{*},W_{z}^{*}\right)=\frac{(\theta_i(z_u)-\theta_r)^2}{(\theta_r)^2}+\frac{\left(x_j(z_u)-0.5L_x\right)^2}{(0.5l_{dx})^2}+\frac{(V(z_u)-V_0)^2}{V_0^2} \tag{9}$$

which includes the implied variables such as bilateral contact angles θ_{dj}^{*}, the length l_{dy}^{*} at the bottom of the liquid–solid interface and the force W_z^{*} on the top of the capillary bridge to be optimized. $[z_{u2}^{*}, z_{u1}^{*}]$ are the spacing ranges between the plates (capillary bridge height z_u) satisfying boundary-value conditions for Eqs. (7) and (8). If the spacing between the plates is greater than z_{u1}^{*}, there is a tendency for the bilateral contact angles $\theta_{dj}^{*}>\theta_a$. Thus, the triple contact lines constituted by the bilateral liquid–gas interfaces and plates will be free from restraint and slide into region II. When the triple contact lines at the end of the capillary bridge touch region II, the separation between the plates is $z_{u2}^{*}, l_{dy}=L_y$ and $\theta_i(z_u)=\theta_i(0)=\theta_r$. With a further reduction in the height z_u, the liquid–gas surfaces of both sides and two ends will hinge at the same time.

When $l_{dy}=L_y$, $y_i(0)=0.5l_{dy}=0.5L_y$. The objective function is as follows:

$$\mathrm{Min}:F\left(\theta_{di}^{*},\theta_{dj}^{*},W_{z}^{*}\right)=\frac{\left(y_i(z_u)-0.5L_y\right)^2}{\left(0.5l_{dy}\right)^2}+\frac{\left(x_j(z_u)-0.5L_x\right)^2}{(0.5l_{dx})^2}+\frac{(V(z_u)-V_0)^2}{V_0^2} \tag{10}$$

which includes the implied variables, such as the contact angles of the two ends θ_{di}^{*}, the bilateral contact angles θ_{dj}^{*} and the force W_z^{*} on the top of the capillary bridge, which has to be optimized. Here, $[z_{u3}^{*}, z_{u2}^{*}]$ are the capillary bridge heights z_u satisfying boundary-value conditions for Eqs. (7) and (8). If the spacing between the plates is less than z_{u3}^{*}, the triple contact lines will be free from restraint and slide into region II.

For a height range $z_u\in[z_{u3}^{*}, z_{u1}^{*}]$, as the height z_u is reduced, there are m capillary bridge heights ($z_u=z_u(0)$, $z_u(1),\cdots,z_u(m)$). Under those constraint conditions in which the capillary bridge could degrade, there are m values of bilateral contact angles. When $z=z_u$, the implicit variables are obtained by optimizing the implicit variables (if $l_{dy}<L_y$, the implicit variables are θ_{dj}^{*}, l_{dy}^{*} and W_z^{*}; if $l_{dy}=L_y$, the implicit variables are θ_{di}^{*}, θ_{dj}^{*} and W_z^{*}) in order to attain the permissible errors of the objective function (Eqs. 9 or 10), and to make the boundary-values at the top end of the capillary bridge satisfy Eq. (7).

The flowchart for solving differential Eq. (6) is shown in Fig. 5.

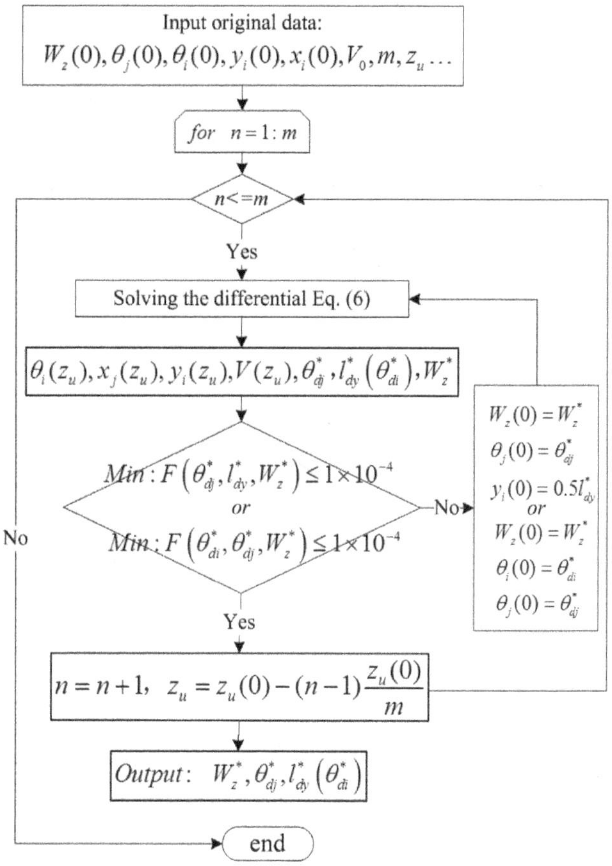

Figure 5. Flowchart for solving differential equations.

4.2 Range of capillary bridge height

If the boundary values at the top and bottom of the outline curves satisfy Eqs. (7) and (8), the method introduced in Sect. 4.1 can be employed to calculate the spacing ranges between the plates, provided the liquid does not overflow into hydrophobic region II. The height z_{u1}^{*} which satisfies the constraint conditions can be obtained by optimizing objective function (Eq. 11), when $\theta_j(0)=\theta_{dj}=\theta_a$ and $l_{dy}<L_y$. The height z_{u2}^{*} can be obtained by optimizing objective function (Eq. 12), when $\theta_i(z_u)=\theta_i(0)=\theta_r$ and $y_i(z_u)=0.5l_{uy}=0.5L_y$. The height z_{u3}^{*} which satisfies the constraint conditions can be obtained by optimizing objective function (Eq. 13), when $\theta_j(0)=\theta_{dj}=\theta_a$ and $y_i(z_u)=0.5l_{uy}=0.5L_y$.

$$\mathrm{Min}:F\left(z_{u1}^{*},l_{dy}^{*},W_{z}^{*}\right)=\frac{\left(\theta_i\left(z_{u1}^{*}\right)-\theta_r\right)^2}{\theta_r^2}+\frac{\left(x_j\left(z_{u1}^{*}\right)-0.5l_{dx}\right)^2}{(0.5l_{dx})^2}+\frac{\left(V\left(z_{u1}^{*}\right)-V_0\right)^2}{V_0^2} \tag{11}$$

$$\mathrm{Min}:F\left(z_{u2}^{*},\theta_{dj}^{*},W_{z}^{*}\right)=\frac{\left(\theta_i\left(z_{u2}^{*}\right)-\theta_r\right)^2}{\theta_r^2}+\frac{\left(x_j\left(z_{u2}^{*}\right)-0.5l_{dx}\right)^2}{(0.5l_{dx})^2}+\frac{\left(V\left(z_{u2}^{*}\right)-V_0\right)^2}{V_0^2} \tag{12}$$

Table 1. Parameters for capillary bridge and flat plate.

Parameters	L_x (mm)	V_0 (μL)	L_y (mm)	Wetting angles θ_r	Wetting angle θ_a	Density ρ (kg mm^{-3})	Surface tension T (N mm^{-1})
Group a	1	5	10	$30, 45, 60°$	135°	1×10^{-6}	7.2×10^{-5}
Group b	1	7.5	10	$30, 45, 60°$	135°	1×10^{-6}	7.2×10^{-5}
Group c	0.5	5	10	$30, 45, 60°$	135°	1×10^{-6}	7.2×10^{-5}

$$\text{Min}: F\left(z_{u3}^*, \theta_{di}^*, W_z^*\right) = \frac{\left(y_i\left(z_{u3}^*\right) - 0.5L_y\right)^2}{\left(0.5L_y\right)^2} + \frac{\left(x_j\left(z_{u3}^*\right) - 0.5l_{dx}\right)^2}{(0.5l_{dx})^2} + \frac{\left(V\left(z_{u3}^*\right) - V_0\right)^2}{V_0^2} \quad (13)$$

Objective functions (Eqs. 11 to 13) are implied optimization variables: $z_{u1}^*, l_{dy}^*, W_z^*$; $z_{u2}^*, \theta_{dj}^*, W_z^*$ and $z_{u3}^*, \theta_{di}^*, W_z^*$.

5 Results and discussion

Three groups of liquid bridge systems are listed in Table 1. The hydrophilic regions I for different wetting angles, the curves of the morphological parameters, the forces, and internal pressures for a range of heights are analyzed. Moreover, the results are compared with the Surface Evolver simulations.

5.1 Relationships between morphological parameters

The outline curves satisfying boundary-value conditions for Eqs. (7) and (8) are obtained by programming for solving differential Eq. (6). Figures 6 and 7 show the outline curves for $V_0 = 5\,\mu$L, $L_x = 1$ mm, and $\theta_r = 30°$.

When the range of the spacing between the two plates is $z_u \in [z_{u2}^*, z_{u1}^*]$, the triple contact lines of the two ends do not touch hydrophobic region II ($l_{dy} < L_y$). In Fig. 6a, both sides of the liquid–gas interface gradually change from convex to concave and the contact angles θ_{dj} decrease with the height z_u. However, the width at the bottom of the capillary bridge is constant ($l_{dx} = L_x$), because the bilateral triple contact lines are pinned by hydrophobic region II along the length of region I, and the bilateral liquid–gas interfaces are in a "hinge movement" state. In Fig. 6b, the contact angles θ_{di} of the capillary bridge's two ends are constant ($\theta_{di} = \theta_r$), but the spacing between the two plates can vary. The length of the capillary bridge can change freely, the contour surfaces of its two ends are always concave, and the triple contact lines of the two ends "slip" in region I. As shown in Fig. 6b, it can be found that gravity has a significant effect on the morphology of the capillary bridge (there is a large difference between abscissas $y(0)$ and $y(z_u)$ of the upper and lower endpoints a_n and b_n of the contour curves), when the height of the capillary bridge is greater ($z_u > 1.25$ mm). When $z_u < 1.25$ mm, $|y(0) - y(z_u)| \Rightarrow 0$, gravity has an increasingly smaller influence on the morphology of the capillary bridge. The contour curves of the capillary bridge will approximate to a circular arc.

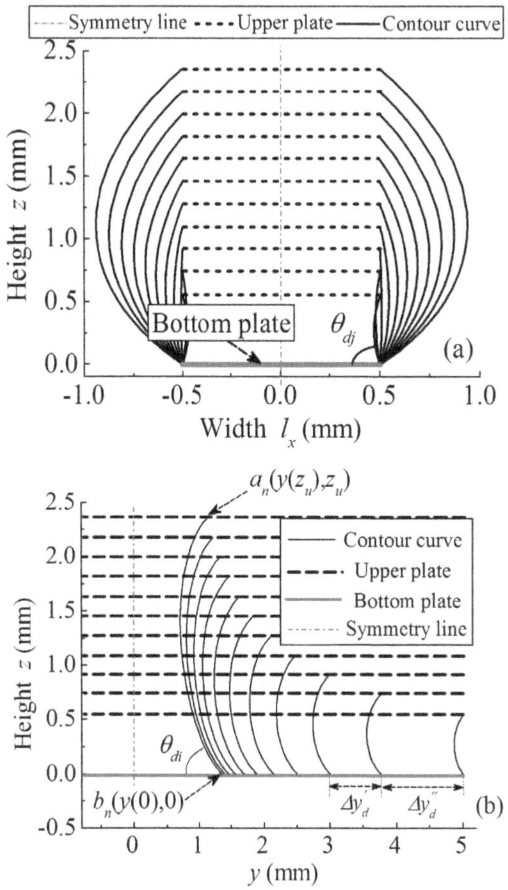

Figure 6. Profiles of the capillary bridge when the two ends' triple contact lines are unconstrained. **(a)** Bilateral side contour curves and **(b)** right end contour curves.

When the height $z_u \in [z_{u3}^*, z_{u2}^*]$, the triple contact lines are also pinned by hydrophobic region II ($l_{dy} = L_y$) at the two ends of region I. In Fig. 7, the liquid–gas interfaces around the capillary bridge are in a "hinge movement" state and gradually bulge outward.

For different volumes V_0 or widths L_x, the bottom lengths of the liquid–solid interfaces are as shown in Fig. 8, when

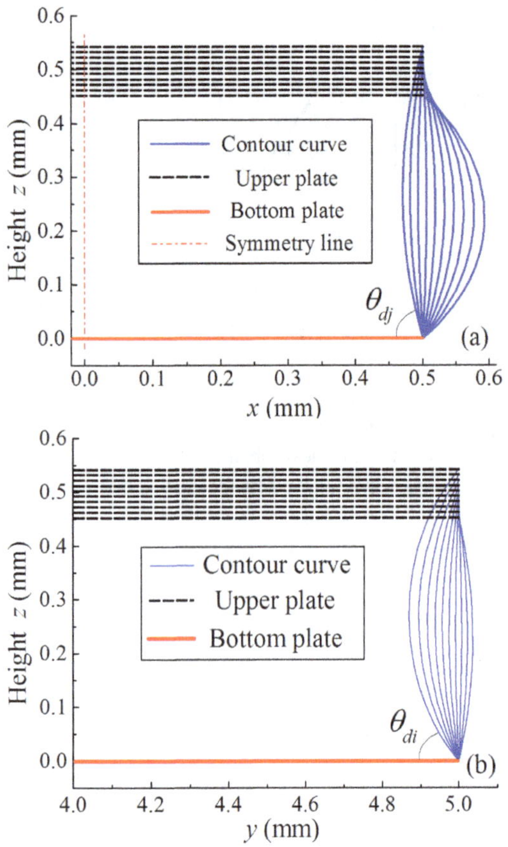

Figure 7. Profiles of capillary bridge when triple contact lines are pinned. **(a)** Right-side contour curves and **(b)** right-end contour curves.

the triple contact lines of the two ends "slip" in region I as the height z_u falls. It can be found that the lengths of the liquid–solid interfaces are in quantitative agreement with the result of the Surface Evolver simulations when the height z_u is low, but as the height increases, we find that the results from our theory underestimate the simulation results. When the height of the capillary bridge is greater ($z_u/L_x > 1.5$), we assume that the deviation observed at higher values of z_u is caused by a decrease in the aspect ratio (l_y/l_x) of the bridge, at which point one of the major curvature radiuses of Eq. (3) approximates to infinity ($\gamma' \Rightarrow \infty$). By integrating Fig. 8a and c it can be found that, for the same volume V_0, a smaller value of L_x will greatly reduce the variation in the height z_u (z^*_{u1} becomes smaller, z^*_{u3} becomes larger).

For variable V_0 or region I width L_x, Fig. 9 indicates the variations in the bilateral contact angles θ_{dj} with the reduction in the height z_u. Based on the changes in the wettability of region I, the ranges of the contact angles are [θ_r, θ_a], (θ_r is 30, 45 or 60°; θ_a is 135°). When the capillary bridge can freely advance along the length of region I, θ_{dj} decreases with the height, which is in agreement with the results of experiments described in the literature (Broesch and Frechette, 2012) as well as those of the Surface Evolver simulations. Once the triple contact lines at the left and right ends of the capillary bridge have been pinned by region II ($l_{dy} = L_y$), θ_{dj} will increase as the height z_u decreases. As the height of the capillary bridge increases ($z_u/L_x > 1.5$), the deviation also increases relative to the results of the Surface Evolver simulations. The variation in the bilateral contact angles θ_{dj} is in agreement with the results of the Surface Evolver simulations, except when the wetting angle $\theta_r = 60°$ and $L_x = 0.5$ mm (see the inset in Fig. 9c), because the results of the Surface Evolver simulations showed that the ends of the capillary bridge did not touch region II in the spacing variable range determined using our theory.

By combining Figs. 9 and 10, it can be seen that θ_{dj} is always greater than θ_{di} ($\theta_{dj} > \theta_{di}$), corresponding to the same height z_u, which implies that a bulge appears on the side of the liquid–gas interfaces first, and then the bilateral triple contact lines depart from pinning and overflow into hydrophobic region II when the height z_u is greater than z^*_{u1} or less than z^*_{u3}. The contact angle θ_{di} remains constant when both ends of the liquid bridge are able to freely "slip". When the two ends of the capillary bridge are constrained by the region II ($l_{dy} = L_y$), θ_{di} will increase. Based on Figs. 9 to 10, the ranges of variable z_u are influenced by the changes in the wetting angle θ_r, volume V_0, and the width L_x of the region I. For the same group of parameters in Table 1, θ_r only influences z^*_{u1}, the upper limit on the range of the height z_u. If θ_r were smaller, z^*_{u1} could attain a greater value. V_0 mainly influences z^*_{u3}, the lower limit on the range of the height z_u. If V_0 were smaller, z^*_{u3} could attain a smaller value. The variations in the bilateral contact angles θ_{di} are in agreement with the results of the Surface Evolver simulations, except when the wetting angle $\theta_r = 60°$ and $L_x = 0.5$ mm (see the inset in Fig. 10c), because the Surface Evolver simulations indicate that the ends of the capillary bridge do not touch region II, such that $\theta_{di} = 60°$ remains constant.

5.2 Stiffness characteristics curves of capillary bridge

The force W_z acting on the capillary bridge upper end is related to z_u. The height of the capillary bridge as determined by the curves known as the "stiffness characteristics curve" of W_z is shown in Fig. 11. As z_u decreases, the force W_z continues to increase, to the point where the triple contact lines at the ends of the capillary bridge move into the "slipping" stage. According to Eq. (1), the change in W_z depends on variables l_{dy}, p_b and θ_{dj}. As $\sin\theta_{dj}$ and l_{dy} increase, p_b decreases (see Fig. 12), thus demonstrating that the variation in W_z is mainly a result of the surface tension. As the width of region I decreases, the change in the wettability of region I exerts a greater influence on the change in W_z (see Fig. 11c). As z_u further decreases, the triple contact lines touch region II, which are pinned. At this point, the all-around liquid–gas interfaces are in the "hinge movement" stage. Here, the pressure p_b is rapidly increasing, but

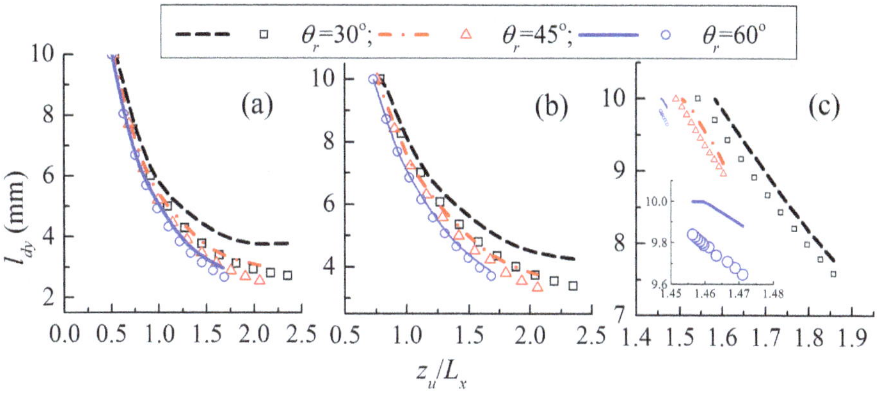

Figure 8. Variations in bottom length of liquid–solid interfaces. **(a)**, **(b)**, and **(c)** refer to parameters groups a, b, and c listed in Table 1 for a range of wetting angles. The data points are as obtained with our theory and the lines are those obtained from the Surface Evolver simulations. The inset shows the change in the bottom length of the liquid–solid interfaces when the wetting angle $\theta_r = 60°$.

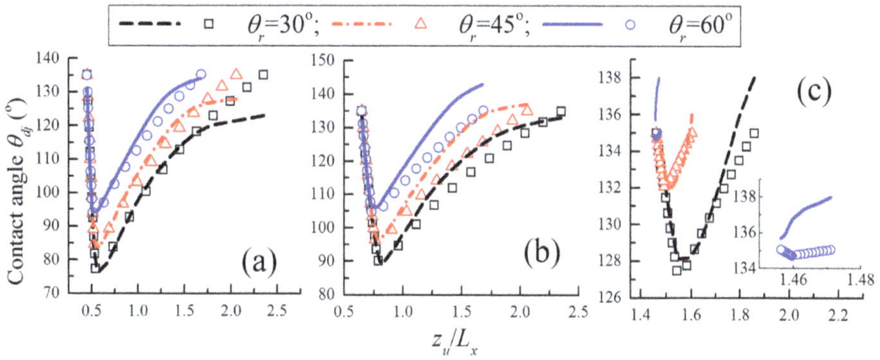

Figure 9. Changes in contact angle θ_{dj} at both sides. Here, **(a)**, **(b)**, and **(c)** correspond to parameter groups a, b, and c listed in in Table 1 for different wetting angles. The data points are as obtained with our theory and the lines are those obtained from the Surface Evolver simulations.

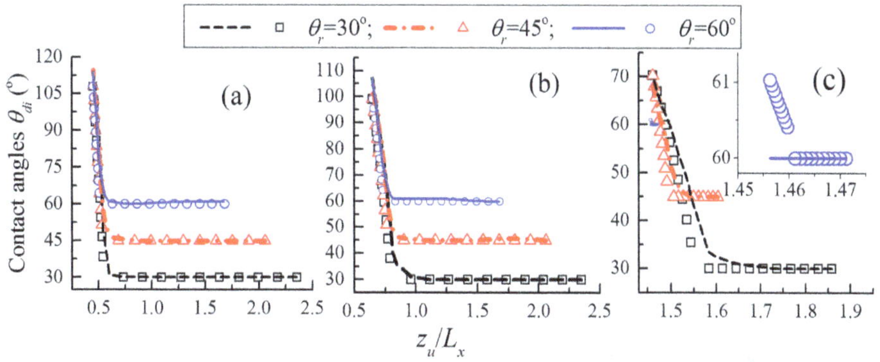

Figure 10. Changes in contact angles θ_{di} at both ends. Here, **(a)**, **(b)**, and **(c)** refer to parameter groups a, b, and c that are listed in Table 1 for different wetting angles. The data points are as obtained with our theory and the lines are those obtained from the Surface Evolver simulations.

the force W_z exerted on the capillary bridge by the upper plate is rapidly declining and it changes from stress to pressure as the height decreases, which is mainly affected by the capillary force on the capillary bridge. When we compare the predictions obtained from the Surface Evolver simulations with the results obtained with our method, we find that the two sets of results are in quantitative agreement, but as the region I widths $L_x = 0.5$ mm and wetting angle $\theta_r = 60°$ (see the inset in Fig. 11c), we can see that the results of the Sur-

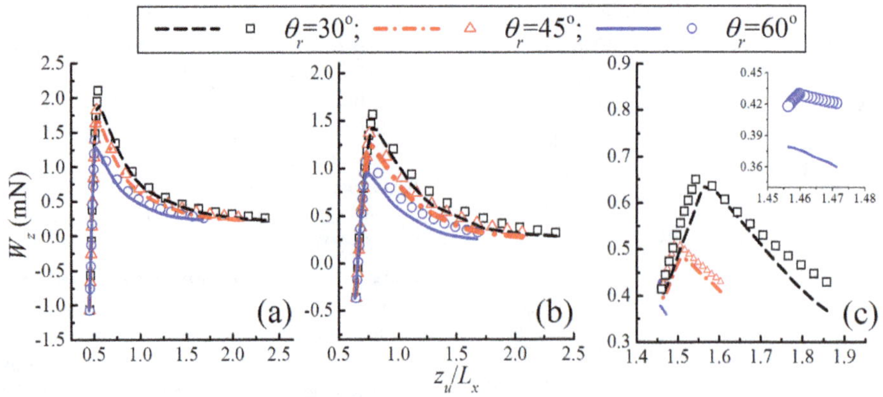

Figure 11. Stiffness characteristic curves of W_z. Here, **(a)**, **(b)**, and **(c)** correspond to parameter groups a, b, and c, listed in Table 1 for different wetting angles. The data points are as obtained with our theory and the lines are those obtained from the Surface Evolver simulations.

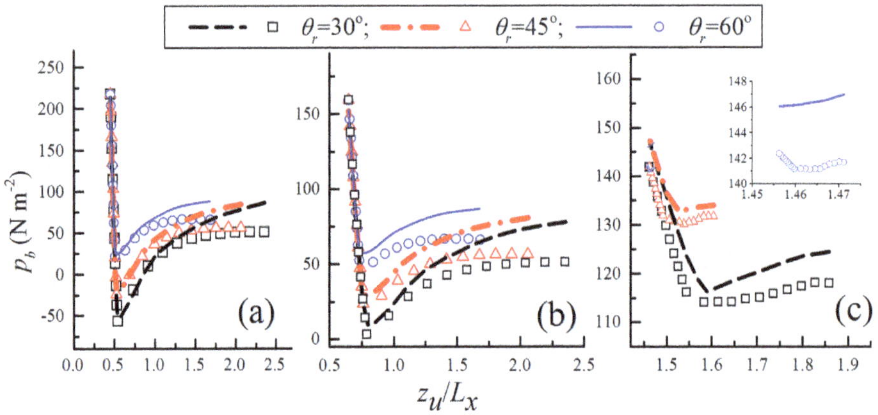

Figure 12. Variations in bottom pressure of capillary bridge. Here, **(a)**, **(b)**, and **(c)** refer to groups a, b, and c listed in Table 1 for different wetting angles. The data points are as obtained with our theory and the lines are those obtained from the Surface Evolver simulations.

face Evolver simulations are about 15 % less than the results obtained with our theory.

Figure 12 shows the changes in the pressure p_b. When the triple contact lines are "slipping", the decline in the upper plate causes the capillary bridge to elongate. When the value of p_b is falling slowly, a transition from negative to positive values at around $z_u/L_x = 1$ can be seen for parameter group a and the wetting angle $\theta_r = 30°$. This corresponds directly to the transition in the contact angles θ_{dj} from greater than 90° to less than 90°, as shown in Fig. 9a. When the all-around liquid–gas interfaces are in the "hinge movement" stage, the capillary bridge does not elongate any further. The descent of the upper plate causes the internal pressure to increase rapidly. This corresponds to a variation in the contact angles θ_{dj} and θ_{dj}, shown in Figs. 9 and 10. When the height of the capillary bridge is greater ($z_u/L_x > 1.5$), we can see that the results of our theory underestimate the results obtained with the Surface Evolver simulations, due to the breakdown of the Eqs. (4) and (5) approximation, as discussed in the previous sections.

6 Conclusions

A pseudo-three-dimensional force model of a capillary bridge was developed by the application of the tension equivalent method based on the three-dimensional shape characteristic of the capillary bridge. First, the rules governing the capillary bridge's characteristic parameters were identified by integrating the initial values and final values of differential equations for the capillary bridge characteristic parameters and by optimizing the initial values for obtaining the boundary values of differential equations. Then the rules governing the changes of the contact angles were analyzed for the "slipping" and "hinge movement" stages. Further, the stiffness characteristic curves of the capillary bridge were obtained by analyzing the relationship between the forces and the separations between the plates. Finally, the changes in the internal pressure of the capillary bridge were explored. It is found that the rules governing the forces and the capillary bridge's characteristic parameters are in better agreement with the Surface Evolver simulations, and agree especially well for low val-

ues of z_u/L_x. The theory highlights the fact that the computing speed is greatly improved, relative to the Surface Evolver simulations. Moreover, there are no shortcomings, such as surfaces intersecting with each other or convergence to the minimum energy level being difficult to judge. This is particularly useful for the study of these issues. This would reduce the computing time needed to determine the equilibrium positions when a package with a large number of solder joints is assembled onto a PCB.

Acknowledgements. The authors would like to thank the supports by the National Natural Science Foundation of China (No. 61201021, No. 51306134).

References

Bonn, D., Eggers, J., Indekeu, J., Meunier, J., and Rolley, E.: Wetting and spreading, Rev. Mod. Phys., 81, 739–805, 2009.

Bowden, N.: Self-Assembly of Mesoscale Objects into Ordered Two-Dimensional Arrays, Science, 276, 233–235, 1997.

Brakke, K. A.: Surface evolver manual. Mathematics Department, Susquehanna Univerisity, Selinsgrove, PA, 1994.

Broesch, D. J. and Frechette, J.: From Concave to Convex: Capillary Bridges in Slit Pore Geometry, Langmuir, 28, 15548–15554, 2012.

Broesch, D. J., Dutka, F., and Frechette, J.: Curvature of Capillary Bridges as a Competition between Wetting and Confinement, Langmuir, 29, 15558–15564, 2013.

Broesch, D. J., Shiang, E., and Frechette, J.: Role of substrate aspect ratio on the robustness of capillary alignment, Appl. Phys. Lett., 104, 1–5, 2014.

Bush, J. W. M., Peaudecerf, F., Prakash, M., and Quéré, D.: On a tweezer for droplets, Adv. Colloid Interf. Sci., 161, 10–14, 2010.

Chen, S. H. and Soh, A. K.: The capillary force in micro- and nano-indentation with different indenter shapes, Int. J. Solids Struct., 45, 3122–3137, 2008.

Dalin, J., Wilde, J., Zulfiqar, A., Lazarou, P., Synodinos, A., and Aspragathos, N.: Electrostatic attraction and surface-tension-driven forces for accurate self-assembly of microparts, Microelect. Eng., 87, 159–162, 2010.

De Souza, E. J., Brinkmann, M., Mohrdieck, C., and Arzt, E.: Enhancement of Capillary Forces by Multiple Liquid Bridges, Langmuir, 24, 8813–8820, 2008.

Ferraro, D., Semprebon, C., Tóth, T., Locatelli, E., Pierno, M., Mistura, G., and Brinkmann, M.: Morphological Transitions of Droplets Wetting Rectangular Domains, Langmuir, 28, 13919–13923, 2012.

Gau, H., Herminghaus, S., Lenz, P., and Lipowsky, R.: Liquid morphologies on structured surfaces: from microchannels to microchips, Science, 283, 46–49, 1999.

Guo, J.-G., Zhou, L.-J., and Zhao, Y.-P.: Instability analysis of torsional MEMS/NEMS actuators under capillary force, J. Colloid Interf. Sci., 331, 458–462, 2009.

Herminghaus, S., Brinkmann, M., and Seemann, R.: Wetting and dewetting of complex surface geometries, Annu. Rev. Mater. Res., 38, 101–121, 2008.

Krammer, O.: Modelling the self-alignment of passive chip components during reflow soldering, Microelect. Reliabil., 54, 457–463, 2014.

Langbein, D.: Canthotaxis/Wetting Barriers/Pinning Lines, in: Capillary Surfaces, Springer Tracts in Modern Physics, Springer, Berlin, Heidelberg, 149–177, 2002.

Lipowsky, R.: Morphological wetting transitions at chemically structured surfaces, Curr. Opin. Colloid Interf. Sci., 6, 40–48, 2001.

Lipowsky, R., Lenz, P., and Swain, P. S.: Wetting and dewetting of structured and imprinted surfaces, Colloids Surf. A, 161, 3–22, 2000.

Liu, J., Xia, R., and Zhou, X.: A new look on wetting models: continuum analysis, Sci. China Phys. Mech. Astron., 55, 2158–2166, 2012.

Luo, C., Heng, X., and Xiang, M.: Behavior of a Liquid Drop between Two Nonparallel Plates, Langmuir, 30, 8373–8380, 2014.

Mermoz, S., Sanchez, L., Di Cioccio, L., Berthier, J., Deloffre, E., and Fretigny, C.: Impact of containment and deposition method on sub-micron chip-to-wafer self-assembly yield, IEEE International 3D Systems Integration Conference (3DIC), Osaka, USA, 1–5, 2012.

Mlota, N. J., Toveyb, C. A., and Hua, D. L.: Fire ants self-assemble into waterproof rafts to survive floods, P. Natl. Acad. Sci., 108, 7669–7673, 2010.

Oliver, J. F., Huh, C., and Mason, S. G.: Resistance to spreading of liquids by sharp edges, J. Colloid Interf. Sci., 59, 568–581, 1977.

Padday, J. F.: The Profiles of Axially Symmetric Menisci, Philos. T. Roy. Soc. Lond. A, 269, 265–293, 1971.

Pozrikidis, C.: Stability of sessile and pendant liquid drops, J. Eng. Math., 72, 1–20, doi:10.1007/s10665-011-9459-3, 2012.

Saad, S. M. I. and Neumann, A. W.: Total Gaussian curvature, drop shapes and the range of applicability of drop shape techniques, Adv. Colloid Interf. Sci., 204, 1–14, doi:10.1016/j.cis.2013.12.001, 2014.

Stewart, M. P., Hodel, A. W., Spielhofer, A., Cattin, C. J., Müller, D. J., and Helenius, J.: Wedged AFM-cantilevers for parallel plate cell mechanics, Methods, 60, 186–194, 2013.

Swain, P. S. and Lipowsky, R.: Wetting between structured surfaces: Liquid bridges and induced forces, Europhys. Lett., 49, 203–209, 2000.

Valencia, A., Brinkmann, M., and Lipowsky, R.: Liquid Bridges in Chemically Structured Slit Pores, Langmuir, 17, 3390–3399, 2001.

Yaneva, J., Milchev, A., and Binder, K.: Polymer droplets on substrates with striped surface domains: molecular dynamics simulations of equilibrium structure and liquid bridge rupture, J. Phys. Condens. Matter, 17, S4199–S4211, doi:10.1088/0953-8984/17/49/014, 2005.

9

Towards developing product applications of thick origami using the offset panel technique

Michael R. Morgan[1], **Robert J. Lang**[2], **Spencer P. Magleby**[1], **and Larry L. Howell**[1]

[1]Department of Mechanical Engineering, Brigham Young University, Provo, UT 84602, USA
[2]Lang Origami, Alamo, CA 94507, USA

Correspondence to: Spencer P. Magleby (magleby@byu.edu)

Abstract. Several methods have been developed to accommodate for the use of thick materials in origami models which preserve either the model's full range of motion or its kinematics. The offset panel technique (OPT) preserves both the range of motion and the kinematics while allowing for a great deal of flexibility in design. This work explores new possibilities for origami-based product applications presented by the OPT. Examples are included to illustrate fundamental capabilities that can be realized with thick materials such as accommodation of various materials in a design and manipulation of panel geometry resulting in an increased stiffness and strength. These capabilities demonstrate the potential of techniques such as the OPT to further inspire origami-based solutions to engineering problems.

1 Introduction

In recent years, origami's potential for innovative solutions facilitated by its complex behaviors, yet simple fabrication methods has caught the attention of scientists and engineers. Some of the proposed applications of origami include deployable space applications (Schenk and Guest, 2011; Zirbel et al., 2013; Wu and You, 2010), nano-structure fabrication (Arora et al., 2006), robotics (Felton et al., 2014), and medical equipment (Kuribayashi et al., 2006; Francis et al., 2014). While origami's potential can be seen and even explored using traditional paper origami, many engineering applications would generally require materials with stiffness and strength. With increased stiffness and strength, however, comes some degree of thickness, and as thickness increases folding becomes less and less feasible.

Several formal methods have been developed for accommodating thickness in origami-based design. These include methods that shift rotational axes to edges of panels (Tachi, 2011; Hoberman, 2010), mount trimmed panels onto membranes (Zirbel et al., 2013), employ spatial mechanisms at vertices (Chen et al., 2015), taper the edges of the panels (Tachi, 2011), and replace creases with a rigid link and two axes to allow folding of adjacent panels (Ku and Demaine, 2015). The method used in this paper, the offset panel technique (OPT) developed by Edmondson et al. (2014), preserves the kinematics of an origami model and allows a full range of motion. While the benefits of these two traits can come at the cost of self-intersection and some complexity, the technique does allow for a variety of shapes, materials, and movements to be employed in a thick model based on a given origami pattern.

Previously, a number of basic configuration-related capabilities of the OPT have been shown by Edmondson et al. (2015). For example, the technique can provide designers with flexibility as it accommodates uniform and varying panel thickness, gaps between panels, and freedom in joint plane placement. In this paper, the intent of the authors is to move beyond describing the basic capabilities of the OPT (to accommodate thick panels) to illustrating how the technique can be used to facilitate the development and engineering of origami-based devices and products. The examples visually demonstrate that employing the OPT with various origami patterns can facilitate the design of mechanisms that use a variety of materials, have links (panels) that are non-planar shapes, and that can resist loads and/or transfer forces. Engineers and designers can use the illustrated examples as inspiration for further designs that use origami patterns.

Table 1. A comparison of thickness accommodation methods.

Method	Kinematics preserved	ROM preserved	Single DOF	Unfolds flat	Application considerations
Axis-shift Tachi (2011)	No	Yes	Yes	Yes	Limited to selected fold patterns
Offset joint Hoberman (1991) Hoberman (2010)	No	Yes	Yes	Yes, varied thickness	Limited to selected fold patterns
Membrane folds Zirbel et al. (2013)	No	Yes, if gaps between panels $\geq 2\times$ thickness	No, gaps allow movement	Yes	Deployed system requires tension at edges to keep membranes stretched
Tapered panels Tachi (2011)	Yes	No	Yes	Yes	Required tapering of panels limits possible geometry and materials
Offset crease Abel et al. (2015)	No	Yes	No	Yes	Panels required to be trimmed to avoid self-interference at vertices
Spatial linkages Chen et al. (2015)	No	Yes	Yes	No	Fold angles and panel thicknesses limited by the spatial mechanism
Offset panel Edmondson et al. (2014)	Yes	Yes	Yes	No	Cutouts in panels may be required to avoid self-intersection

2 Background

As preparation for extending the application of the OPT, we briefly review three areas. These areas combined lead to an understanding of new ways to create mechanical products by adapting origami models and motions for thick materials.

2.1 Modeling origami

To take advantage of an origami mechanism's potential for application, it is useful to have mathematical models of its motion and behaviors. Origami can be modeled as a web of coupled spherical mechanisms where each vertex is the center of a spherical mechanism, each panel is a link, and each fold is a joint (e.g. Greenberg et al., 2011; Bowen et al., 2013; Lang and Hull, 2005; Evans et al., 2015; Abel et al., 2015; Tachi, 2009a; Miura, 1989). Spherical kinematic mechanisms belong to a subset of three-dimensional kinematics in which any point on the mechanism is constrained to be coincident with a spherical surface whose center is the point of intersection of all joints within the mechanism. The behavior of a spherical mechanism is defined by the location of the rotational axes. In other words, if a link's shape or size changes, as long as the rotational axes have not been altered, the motion will be the same (Chiang, 2000; Bowen et al., 2014).

2.2 Thickness accommodation

Mathematical models have been developed which have been used to predict the behavior of a given fold pattern (Tachi, 2009b) and which can generate crease patterns based on a desired form (Tachi, 2009a; Lang, 1996). These mathematical models usually assume zero-thickness materials. Due to paper's relatively thin nature, zero-thickness is a suitable approximation for traditional origami. However, this approximation breaks down for thicker origami materials (Francis et al., 2013).

Several techniques have been developed to accommodate thick materials. Each of the thickness accommodation techniques has its own set of strengths, weaknesses, and limitations. Figures 1 and 2 illustrate seven methods, including the OPT, and Table 1 lists the methods and compares them against four characteristics that will be important to discussing applications for the OPT. Following is a description of these characteristics.

Kinematics preserved indicates if the kinematics of the base origami model (a zero-thickness model) is preserved with the method. While matching kinematics may not be important for some applications, in many it will be essential to achieve the same degrees of freedom, the consistency, or the predictability of a motion identical to the origami model.

ROM preserved indicates whether or not the range of motion, from fully folded to fully deployed, is preserved. Many methods do not allow for full range of motion due to clashes of panels/edges. In some applications full

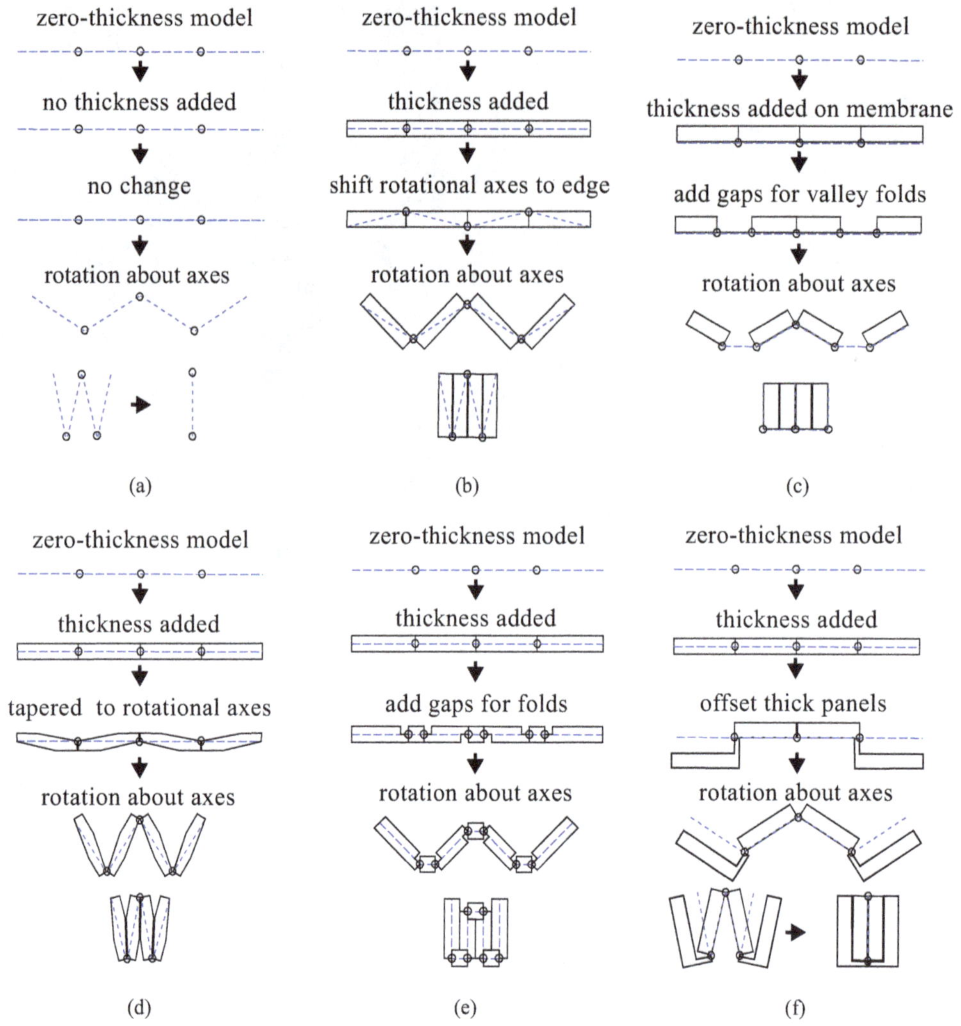

Figure 1. An illustration of the concepts of different thickness accommodation methods. All images, except **(e)**, are from Edmondson et al. (2014). **(a)** The zero-thickness model describes the basic kinematic behavior of the model. **(b)** The axis-shift method as demonstrated by Tachi (2011) shifts each rotational axis to either the top or bottom of the thick material. While slightly different conceptually, the method described by Hoberman (2010) can be illustrated identically. **(c)** The membrane folds method by Zirbel et al. (2013) mounts thick-material facets to a flexible membrane. **(d)** The tapered panels method from Tachi (2011) trims material from the panel edges to maintain the kinematics. **(e)** The offset crease technique, described by Abel et al. (2015), is similar to the membrane folds method, but calls for rigid material in the gaps between panels. This method was inspired by work done by Hoberman (1991) **(f)** The offset panel technique shown by Edmondson et al. (2014) offsets each panel from a selected joint plane and extends the rotational axes back to the joint plane.

motion may not be required, but in many there is a need for the folds to move through the full 180°.

Single DOF indicates if the method will result in a single degree-of-freedom (DOF) system (assuming that the pattern itself has a single DOF). Many rigid-foldable origami models have one DOF. For many applications, especially those implementing a deployable application, a single DOF is desirable.

Unfolds flat indicates whether or not the thick origami system resulting from the method will be flat in its fully unfolded shape. Flat is defined in this case as resulting in a system where the entire upper face of each panel lies in the same plane. Traditional paper origami unfolds flat by this definition. Most of the methods do not produce fully flat configurations as the panels are offset or tapered. The OPT does not result in flat configurations, but the offset panels (or panel substitutes) can be parallel to the joint plane.

Application considerations presents notes on key aspects which should be considered when creating practical applications with the method. Depending on the application, the considerations may be impediments or used as advantages.

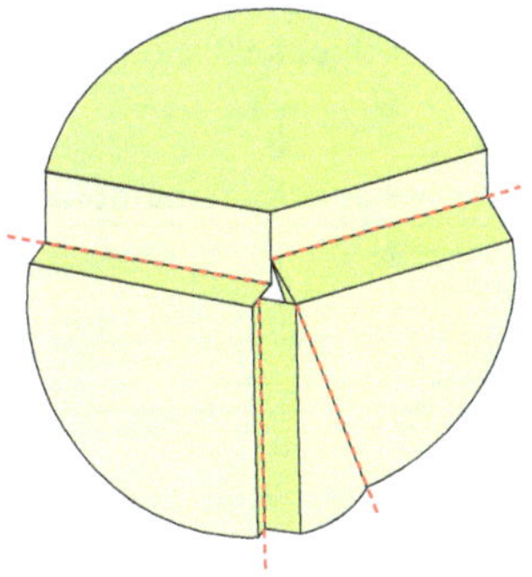

Figure 2. Chen et al. (2015) demonstrates a thickness accommodation method which uses spatial linkages at the vertices of the origami pattern to achieve its motion. A single origami vertex is shown that employs such a mechanism. This illustration is separate because the method cannot be portrayed in the same two-dimensional form as the other methods. The red dotted lines indicate the rotational axes. This method was shown to work with degree-4, degree-5, and degree-6 vertices.

In this paper, the applications that are presented take advantage of the OPT's capabilities to preserve kinematics, allow for full range of motion, and open with a single DOF. OPT-based systems generally do not fold flat and have a tiered appearance in the open position. This is used as an advantage in some of the following examples.

2.3 Offset panel technique

One of the major advantages of the OPT is that it maintains both the kinematic behavior and the full 180° of motion as demonstrated by Edmondson et al. (2014). The former permits designers to take advantage of the mathematical models already developed while the latter allows for fully opened and closed models. The OPT also allows for flexibility in a design. Since an origami-inspired design is constrained by only the preservation of the location of the axes and self-intersection, attributes such as varied panel thickness, spacing between panels, and selectable joint plane placement are all possible with the OPT (see Edmondson et al., 2015).

3 Fundamental capabilities of the offset panel technique

A major benefit of traditional origami is the simplicity that it offers – the desired model is a product of a single monolithic piece of paper which undergoes no process other than folding. While some of this simplicity is lost in "thick origami", there are at least two fundamental capabilities possessed by origami models implemented by the OPT which are not exhibited by traditional origami. These are the variation of panel geometry and the material(s) used in a design. We refer to these as fundamental capabilities because others are either subsets of them or results of combining them.

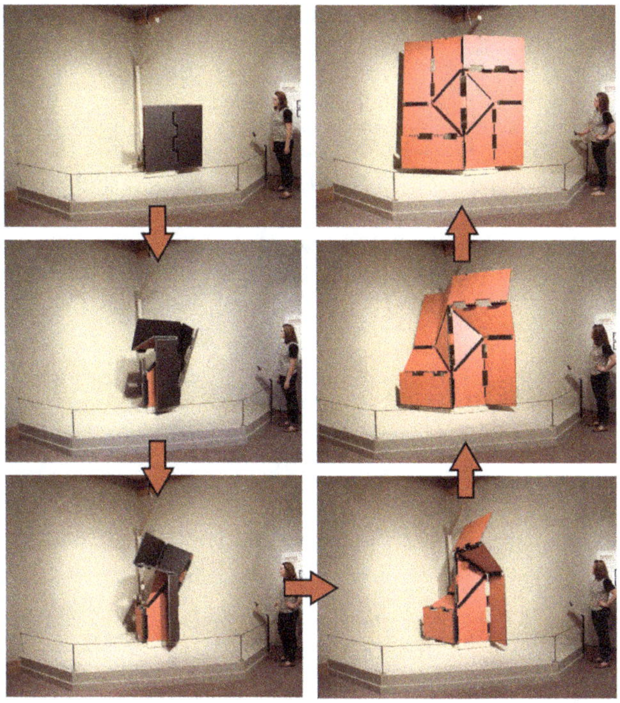

Figure 3. The unfolding of an origami-based kinetic sculpture which employs the OPT. The mechanism is made of panels with a thickness of one inch. This was part of an exhibit in BYU's Museum of Art.

3.1 Panel and joint materials

With the option of using non-paper materials, it becomes clear that virtually any solid material can be utilized in designs that incorporate folding. Designers of origami-inspired products now have the ability to accommodate materials that are, for example, transparent/opaque, conductive/insulative, lubricative/abrasive, adhesive, stiff, modifiable, expansive, electrically charged, absorbent, or reflective.

Additionally, not only can a model be made of virtually any solid material, since thick origami panels are fabricated individually in most cases, the individual panels of a model can be assigned different materials. For example, Fig. 3 shows a kinetic sculpture based on an origami fold pattern. In the design of this model, a lightweight material was desired for the moving panels, while strength was needed in the supporting ground panel. A lightweight foam board was chosen to be used for the moving panels and MDF was used for the ground panel. The different sculpture materials can best be seen from the back, as shown in Fig. 4.

This kinetic sculpture is based on the square twist fold pattern shown in Fig. 5a. Figure 5b illustrates the closed stack of panels (looking down at the sculpture from the top). In this design, the panels all have an equal thickness of one inch and, while interference has been accounted for, the panel profiles have not been altered. Figure 5c shows the same panel stack but includes the offsets which have the same color as the pan-

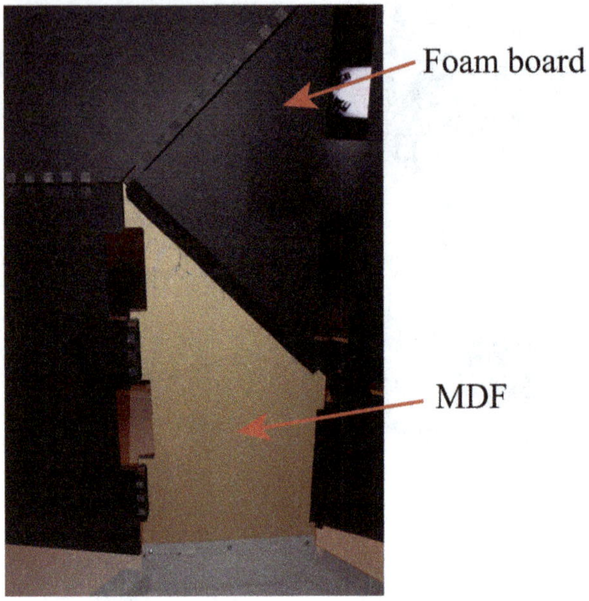

Figure 4. In the kinetic sculpture, the grounded panel is made of MDF while all others are made of a foam board.

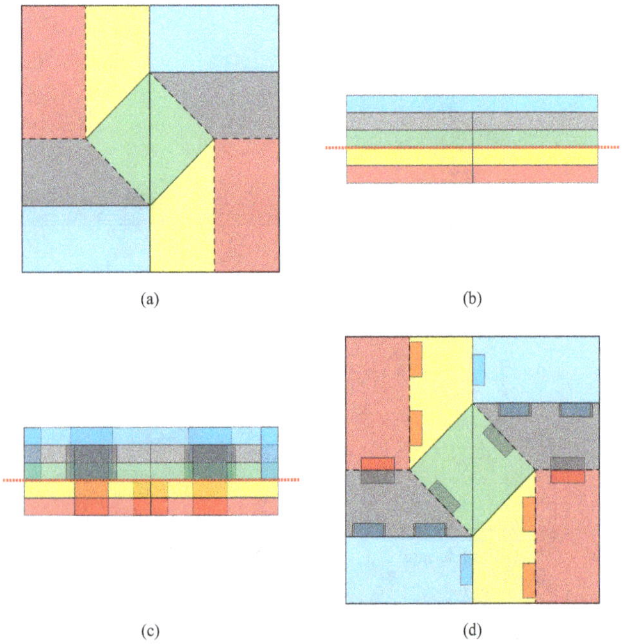

Figure 5. A schematic of the pattern used in the design of the kinetic sculpture. **(a)** The square twist crease pattern. **(b)** A side view of the stacked panels and joint plane (represented by the red line). **(c)** The panel stack with offsets. **(d)** The unfolded panels including the offsets for the OPT model.

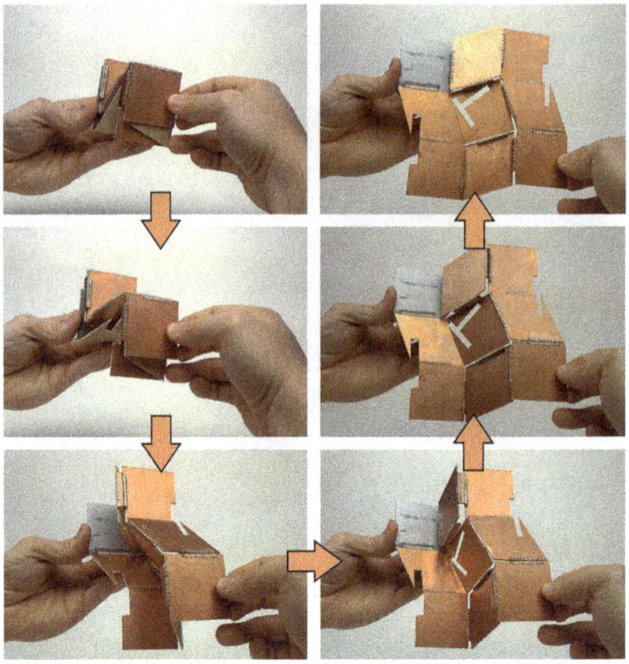

Figure 6. A foldable circuit board with one panel constructed from metal. The thickness of the panel is approximately 0.063 inches.

els to which they are rigidly attached. The offsets are shown to demonstrate the preservation of the axes' locations which in the closed configuration all lie on a single plane (referred to as the joint plane) indicated by the dotted red line.

The next example shows the possibilty of a foldable circuit board which uses the OPT. This is shown in Fig. 6. All but one panel is made from a PCB substrate, while the exception is made of a metal plate intended to act as the ground layer. While simpler crease patterns (i.e., a tesselated tri-fold pattern) could offer the similar stowed-to-deployed area ratios, this pattern facilitates connections between each panel and its neighbors and a single DOF.

A square twist pattern, as illustrated in Fig. 7a, was used as a base for this circuit board. The illustration in Fig. 7b shows the stacked panels from the side and Fig. 7c shows the same panel stack with the offsets.

3.2 Panel geometry

One of the most applicable and inspiring capabilities of thick origami (the OPT in particular) is the freedom a designer has to alter the geometry of the panels. Once it is determined that an origami-inspired design will employ the OPT, the only limitations are the preservation of the rotational axes and the prevention of self-intersection. Panel material can be added and/or removed to add structure, give form, achieve motion, etc. Combining this idea with that of using various materials, not only is it possible to alter the shape of a model's panel material, but panels can be replaced entirely by mechanical parts, electrical parts, displays, wheels, optical devices, solar

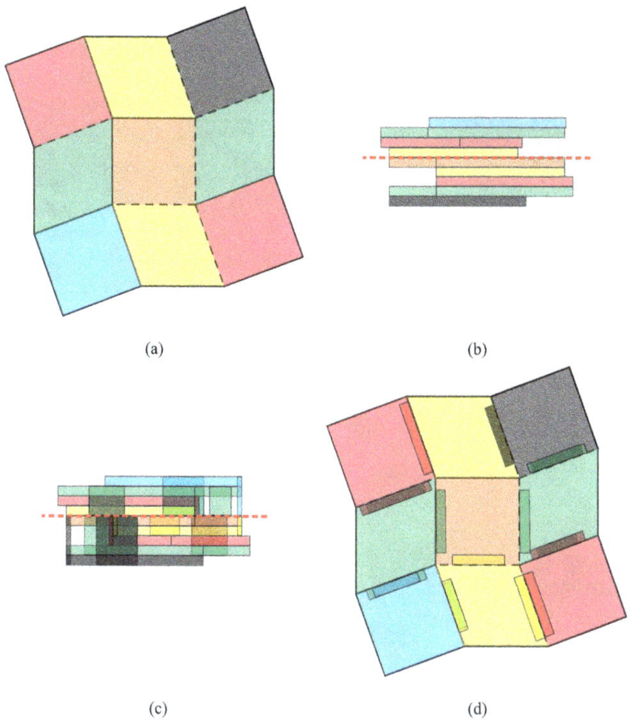

Figure 7. A schematic of **(a)** the pattern used in the design of the foldable circuit board. **(b)** A side view of the panel stack and the joint plane. **(c)** The stacked panels with offsets. **(d)** The unfolded pattern with the offsets. **(b)** and **(c)** are 2X scale.

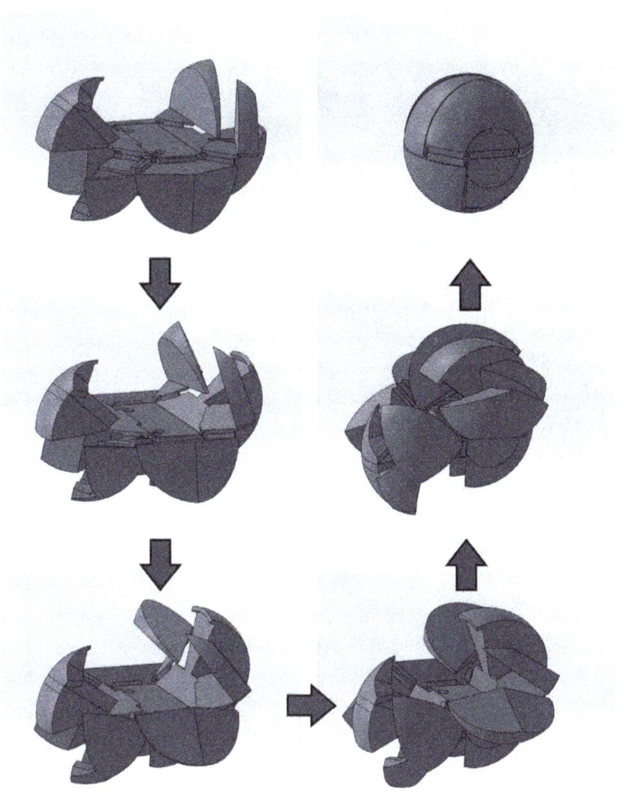

Figure 8. A kinematic simulation model of how panel geometry can vary, in this case to form a sphere.

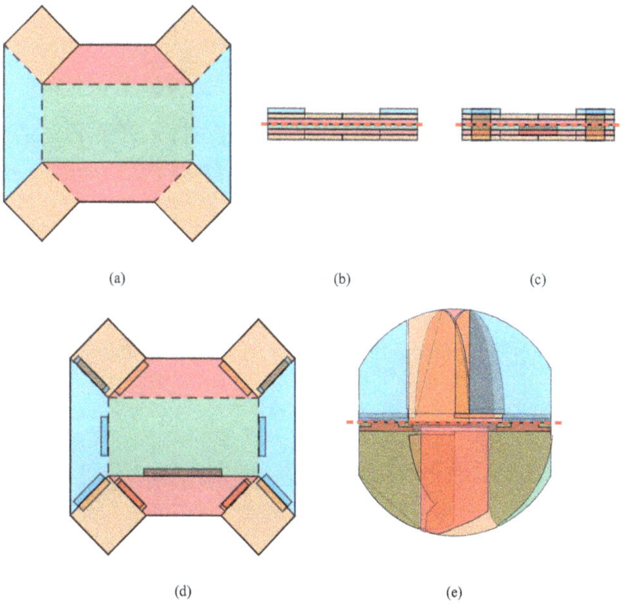

Figure 9. A schematic of the pattern used in the design of the foldable sphere. **(a)** The crease pattern. **(b)** The panel stack and joint plane. **(c)** The panel stack with offsets. **(d)** The offsets shown with the open model. **(e)** The panel stack with offsets and modified panel geometry.

panels, molds, etc. As long as axes are not moved and self-intersection is avoided, the model will maintain the kinematic behavior and the range of motion of the source model.

To demonstrate the manipulation of panel geometry, a model of a folding sphere is shown in Fig. 8. From the closed configuration of a basic OPT model with uniform thickness, material is added to each panel in such a way that the model takes on the shape of a sphere. This is done without affecting the rotational axes and with care taken to ensure no self-intersection.

A version of the square twist as shown in Fig. 9a formed the origami base of this foldable sphere. Figure 9b and 9c show the panel stack alone and with the offsets, respectively, and Fig. 9d shows the panel stack with offsets and the material added to form a sphere.

As another example of non-constant panel geometry, see the engineer's toolbox in Fig. 10. This model explores the utility of adding and removing material by carving out material from the panels to form compartments for supplies and adding material to panels to form the walls of the box. The box's opening motion is shown in Fig. 11.

The same square twist pattern as used with the kinetic sculpture was used for the toolbox, but with a different crease assignment as shown in Fig. 12a. To facilitate the closing of the box, a small space was added between the panels as demonstrated by Edmondson et al. (2015). This spacing is

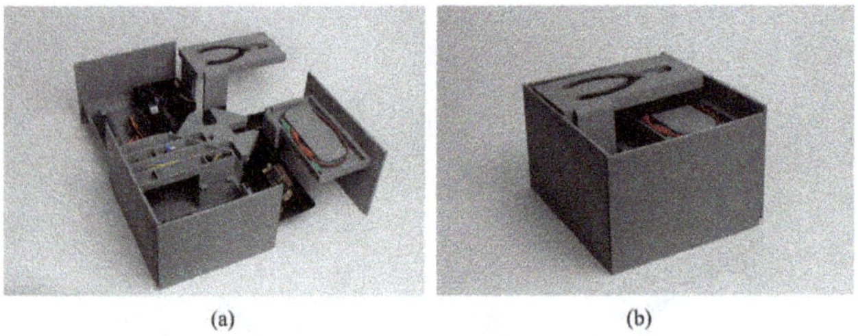

Figure 10. An electrical engineer's toolbox created from an origami crease pattern and the OPT in **(a)** the unfolded postion and **(b)** the folded position. The panel thickness in this model is 0.5 inches.

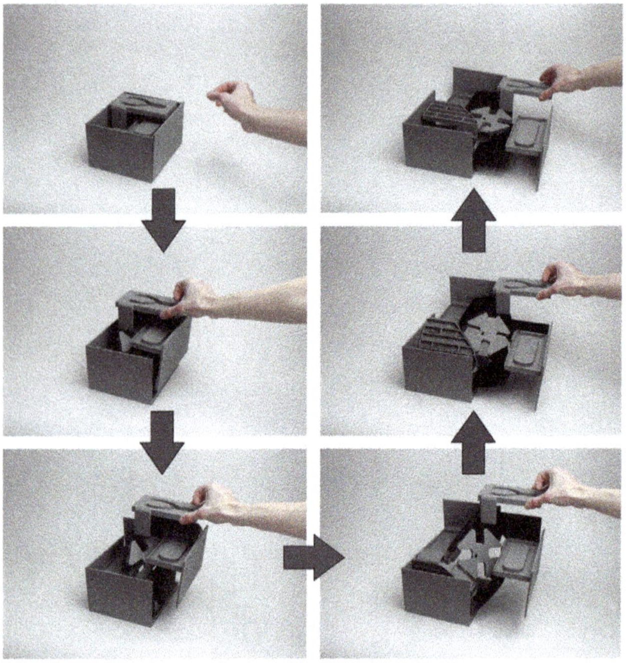

Figure 11. A demonstration of the motion of the toolbox.

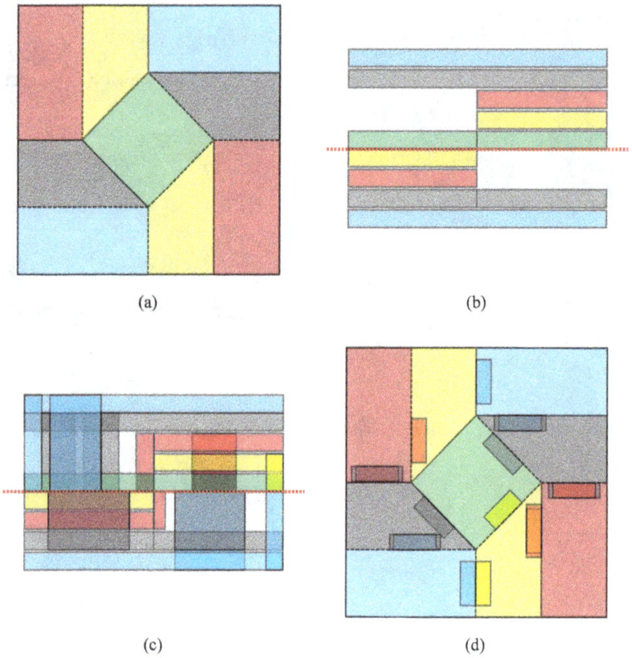

Figure 12. A schematic of the pattern used in the design of the toolbox. **(a)** The crease pattern. **(b)** The stacked panels and the joint plane (in red). **(c)** The stacked panels with added offsets. **(d)** The unfolded pattern including the offsets.

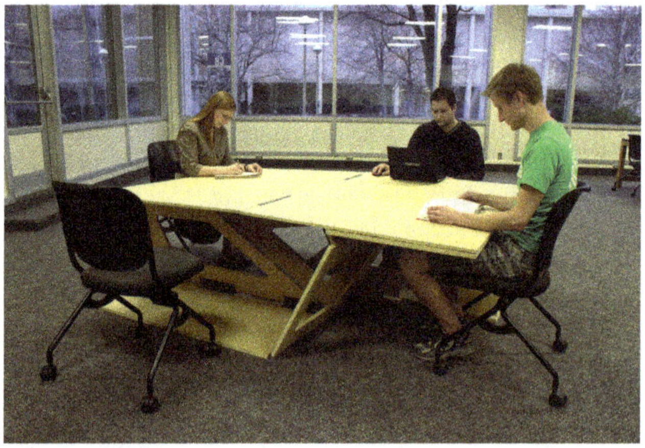

Figure 13. This origami-based table supports a significant amount of weight. Unlike the other examples in this paper, the table is designed to unfold to an intermediate position that does not correspond to the zero-thickness model's fully unfolded position. The panels in this design are 0.75 inches.

shown in Fig. 12b and the same panel stack is shown in Fig. 12c, but with offsets.

4 Resultant capabilities – stiffness and strength

Having discussed the use of various materials and geometry, it is now befitting to examine an example of a capability that is a result of simultaneously varying materials and geometry. We will explore the use of materials commonly used in engineering combined with the necessary geometry to develop designs which can support loads and apply forces. Based on the design specifications of an origami-inspired product, a fold pattern can be selected which would give the desired behavior – a particular motion, mechanical advantage, shape, or unfolded/folded size ratio. After the desired pattern has been determined, materials with sufficient stiffness and strength to perform the desired task can be selected and any geometrical modifications necessary can be made.

For example, an origami-inspired table is shown in Fig. 13. In the folded configuration, the table is compact. In the unfolded configuration, the table is much larger and, based on the kinematics and the materials of the model, can support a significant amount of weight. The table's unfolding motion

Figure 14. An origami-inspired table is shown through its opening motion.

Figure 15. The origami-inspired lift mechanism shown through its motion as it lifts the black weight. The panels of the lift are 0.75 inches.

is demonstrated in Fig. 14. The table uses the same square twist pattern as the kinetic sculpture and also has the same panel stacking configuration (see Fig. 5).

In the next example, we consider a lift. By using a single reverse fold, a substantial mechanical advantage is possible. In this case, the mechanical advantage is approximately 20 at the open state (top-left in Fig. 15) and gradually decreases to 5 at its highest point (top-right in Fig. 15). To withstand the stresses that accompany the loads and large mechanical advantage of this example, MDF was the selected material.

The lift employs a reverse fold which is a single vertex origami pattern as shown in Fig. 16a. Figure 16b shows the stack of the panels and Fig. 16c shows that same stack with the offsets and altered geometry. To create the offsets for this model, the entire inner faces of the long green panels were extended inward to the joint plane while leaving clearance for the small blue panels. This is equivalent to assigning different thicknesses to the panels as shown possible by Edmondson et al. (2015).

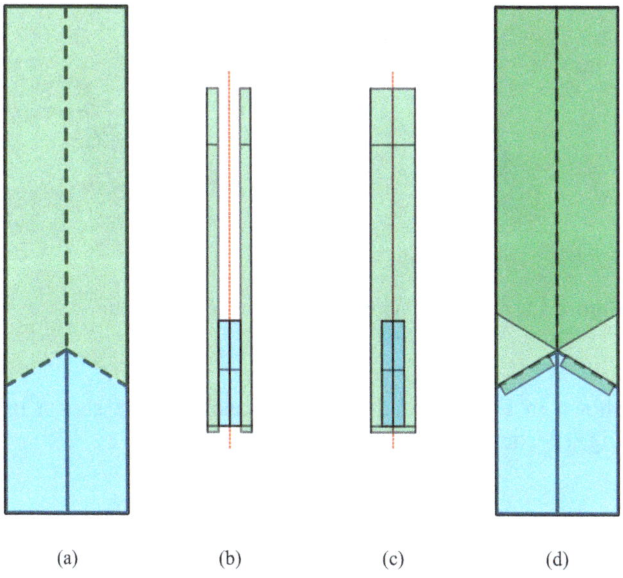

Figure 16. The schematic of the reverse fold pattern used in the design of the lift. The dotted red line represents the joint plane. **(a)** The fold pattern. **(b)** A side view of the panel stack. **(c)** The panel stack with offsets which in this design are large and cover the majority of the inside face of each green panel. **(d)** The open pattern with the offsets.

5 Conclusions

Origami's simple fabrication methods, infinite possibilities, and predictability provide it potential to emerge as a source of inspiration for many innovative designs. While paper origami models are useful for quickly visualizing and prototyping origami-inspired products, paper is often insufficient as a material for a finished product. However, the use of other materials in such origami-inspired designs often presents a handful of major difficulties including the folding of these materials and the interference issues that accompany it. This paper has illustrated how designs based on origami patterns

can be realized using non-paper materials in a variety of shapes and configurations. The examples show that common engineering-related product objectives can be achieved with origami-based mechanisms. The techniques shown in the examples can be applied to a wide variety of origami patterns.

Several methods for accommodating thickness have been developed, each with its own capabilities and limitations. While the examples in this paper use the offset panel technique, the capabilities of accommodation of various materials, manipulation of panel geometry, and strength and stiffness can be applied to other thickness accommodation methods. These ideas add new and exciting possibilities for origami-inspired design and facilitate the development of creative solutions to real-world problems.

Acknowledgement. This material is based on work supported by the National Science Foundation and Air Force Office of Scientific Research under NSF Grant EFRI-ODISSEI-1240417.

References

Abel, Z., Cantarella, J., Demaine, E. D., Eppstein, D., Hull, T. C., Ku, J. S., Lang, R. J., and Tachi, T.: Rigid Origami Vertices: Conditions and Forcing Sets, arXiv.org, math.MG, available at: http://arxiv.org/abs/1507.01644v1 (last access: 22 July 2015), 2015.

Arora, W. J., In, H. J., Buchner, T., Yang, S., Smith, H. I., and Barbastathis, G.: Nanostructured Origami™ 3D Fabrication and Self Assembly Process for Soldier Combat Systems, Sel. Top. Electr. Syst., 42, 473–477, 2006.

Bowen, L. A., Grames, C. L., Magleby, S. P., Lang, R. J., and Howell, L. L.: A Classification of Action Origami as Systems of Spherical Mechanisms, J. Mech. Design, 135, 111008, doi:10.1115/1.4025379, 2013.

Bowen, L. A., Baxter, W., Magleby, S. P., and Howell, L. L.: A Position Analysis of Coupled Spherical Mechanisms Found in Action Origami, Mech. Mach. Theory, 77, 13–44, 2014.

Chen, Y., Peng, R., and You, Z.: Origami of thick panels, Science, 349, 396–400, doi:10.1126/science.aab2870, 2015.

Chiang, C. H.: Kinematics of Spherical Mechanisms, Krieger Publishing Company, Malabar, FL, 2000.

Edmondson, B. J., Lang, R. J., Magleby, S. P., and Howell, L. L.: An Offset Panel Technique for Thick Rigidly Foldable Origami, Proceedings of the ASME International Design Engineering Technical Conferences, Buffalo, NY, 18–20 August 2014, DETC2014-35606, 2014.

Edmondson, B. J., Lang, R. J., Morgan, M. R., Magleby, S. P., and Howell, L. L.: Thick Rigidly Foldable Structures Realized by an Offset Panel Technique, in: Origami 6, American Mathematical Society, 1, 149–161, 2015.

Evans, T. A., Lang, R. J., Magleby, S. P., and Howell, L. L.: Rigidly Foldable Origami Twists, in: Origami 6, AMS, 1, 119–130, 2015.

Felton, S., Tolley, M., Demaine, E., Rus, D., and Wood, R.: A method for building self-folding machines, Science, 345, 644–646, doi:10.1126/science.1252610, 2014.

Francis, K. C., Blanch, J. E., Magleby, S. P., and Howell, L. L.: Origami-like creases in sheet materials for compliant mechanism design, Mech. Sci., 4, 371–380, doi:10.5194/ms-4-371-2013, 2013.

Francis, K. C., Rupert, L. T., Lang, R. J., Morgan, D. C., Magleby, S. P., and Howell, L. L.: From crease pattern to product: Considerations to engineering origami-adapted designs, in: Proceedings of ASME 2014 International Design Engineering Technical Conferences & Computers and Information in Engineering Conference, 17–20 August 2014, Buffalo, NY, 2014.

Greenberg, H. C., Gong, M. L., Magleby, S. P., and Howell, L. L.: Identifying links between origami and compliant mechanisms, Mech. Sci., 2, 217–225, doi:10.5194/ms-2-217-2011, 2011.

Hoberman, C.: Reversibly expandable structures, US Patent 4,981,732, United States Patent and Trademark Office, 1991.

Hoberman, C.: Folding Structures Made of Thick Hinged Panels, US 7794019, United States Patent and Trademark Office, 2010.

Ku, J. S. and Demaine, E. D.: Folding Flat Crease Patterns With Thick Materials, Proceedings of the ASME International Design Engineering Technical Conferences, 2–5 August 2015, Boston, MA, DETC2015-48039, 2015.

Kuribayashi, K., Tsuchiya, K., You, Z., Tomus, D., Umemoto, M., Ito, K., and Sasaki, M.: Self-deployable origami stent grafts as a biomedical application of Ni-rich TiNi shape memory alloy foil, Mat. Sci. Eng. A-Struct., 419, 131–137, 2006.

Lang, R.: A computational algorithm for origami design, in: Proceedings of the twelfth annual symposium on Computational geometry, ACM, 98–105, 1996.

Lang, R. J. and Hull, T. C.: Origami design secrets: mathematical methods for an ancient art, The Mathematical Intelligencer, 27, 92–95, 2005.

Miura, K.: A note on intrinsic geometry of origami, in: Proceedings of the First International Meeting of Origami Science and Technology, Ferrara, Italy, edited by: Huzita, H., 239–249, 1989.

Schenk, M. and Guest, S.: Origami Folding: A Structural Engineering Approach, in: Origami 5: Fifth International Meeting of Origami Science, Mathematics, and Education, edited by: Wang-Iverson, P., Lang, R., and Yim, M., CRC Press, 291–304, 2011.

Tachi, T.: Origamizing Polyhedral Surfaces, IEEE T. Vis. Comput. Gr., 16, 298–311, doi:10.1109/tvcg.2009.67, 2009a.

Tachi, T.: Simulation of Rigid Origami, in: Origami 4: Fourth International Meeting of Origami Science, Mathematics, and Education, edited by Lang, R., A K Peters, Ltd., 175–188, 2009b.

Tachi, T.: Rigid-Foldable Thick Origami, in: Origami 5: Fifth International Meeting of Origami Science, Mathematics, and Education, edited by: Wang-Iverson, P., Lang, R., and Yim, M., CRC Press, 253–264, 2011.

Wu, W. and You, Z.: Modelling rigid origami with quaternions and dual quaternions, P. Roy. Soc. A-Math. Phy., 466, 2155–2174, 2010.

Zirbel, S. A., Lang, R. J., Thomson, M. W., Sigel, D. A., Walkemeyer, P. E., Trease, B. P., Magleby, S. P., and Howell, L. L.: Accommodating Thickness in Origami-Based Deployable Arrays 1, J. Mech. Design, 135, 111005, doi:10.1115/1.4025372, 2013.

10

Experimental study on transient behavior of embedded spiral-coil heat exchanger

E. I. Jassim

Department of Mechanical Engineering, Prince Mohammad Bin Fahd University, PMU, Al-Khobar, Saudi Arabia

Correspondence to: E. I. Jassim (ejassim@pmu.edu.sa)

Abstract. Spiral coil offers a substantial amount of heat transfer area at a considerably low cost as it does not only have a lower wall resistance but it also achieves a better heat transfer rate in comparison to conventional U-tube arrangement. The general aim of the study is to assess different configurations of spiral coil heat exchangers that can eventually operate in a highly efficient manner.

The paper documents the transient behavior of spiral-shaped tubes when the coil is embedded in a rectangular conducting slab. Different arrangements and number of turns per unit length, with fixed volumes, are considered in order to figure out the optimal configuration that maximizes the performance of the heat transfer. The implementation presented in the study is conducted to demonstrate the viability of the use of a large conducting body as supplemental heat storage.

The system uses flowing water in the coil and stagnant water in the container. The copper-made coils situated in the center of the slab carries the cold fluid while the container fluid acts as a storage-medium. The water temperature at several depths of the container was measured to ensure uniformity in the temperature distribution of the container medium.

Results have shown that the coil orientation, the number of loops, and the Reynolds number, substantially influence the rate of the heat transfer. The vertically-embedded spiral coil has a better performance than the horizontally-embedded spiral coil. Doubling the number of loops is shown to enhance the performance of the coil. Increasing Reynolds Number leads to better coil performance.

1 Introduction

For almost a century after the introduction of the curved tubes heat exchanger, much attention have recently been drawn towards the enhancement techniques in energy and the mass transfer of spiral coil heat exchanger. The popularity of the spiral coil heat exchanger is coming from its wide applications, due to usefulness and suitability, as it proves to be compact in design, unique in geometrical structure, and feasible in manufacturing. However, these benefits are not the only reasons behind the attraction of this type of heat exchanger. The eminence of such system over the straight tube type, in terms of heat and mass transfer, gives them domination over the conventional type thermal system. Today, with the advancements in science and technology, spiral tube heat exchangers have become prevalent in numerous thermal applications globally.

The primary reason of such superiority is attributed to the motion induced by the coil curvature, as it relies on the dimensionless number named as the Dean Number, defined as the ratio of centrifugal force to inertial force. It can be expressed as:

$$De = Re\sqrt{\frac{d_\mathrm{p}}{2R}} \tag{1}$$

where d_p is the pipe diameter, and R is the radius of curvature of the coil.

W. R. Dean first induced Dean Vortices during his study on how heat transfer can be enhanced in spiral piping systems

(Dean, 1927a, b). The Dean vortex considers several important factors, such as curvature ratio, viscosity, and Reynolds number. The Dean Vortices method has rarely been examined in regards to the study of the heat transfer enhancement of spiral piping. Nevertheless, recent research on enhancement of heat transfer in the spiral piping has been conducted using the Dean Vortices method. Recent studies and investigations have proved that the thermal coupling between spiral pipes and heat transfer depends, to a great extent, on Dean Vortex's motion (Elazm et al., 2013).

Dean Vortices also influence the flow's critical Reynolds number. The critical Reynolds Number that describes the transition of the flow inside coiled tubes from laminar to turbulent is much higher than that for conventional straight tubes heat exchanger. Dean Vortex tends to restrict turbulence inception, delaying the flow transition to the turbulent. Hence, it should be considered together with the Reynolds Number in describing the flow regime. Srinivasan et al. (1970) came up with a correlation of the critical Reynolds number as a function of the coil geometry:

$$Re_{\mathrm{c}} = 2100\left(1 + 12\sqrt{\frac{d_{\mathrm{p}}}{2R}}\right). \tag{2}$$

The design criteria of the spiral tube heat exchanger vary with the intended functionality (Sanner et al., 2003). For instance, in geothermal application, heat exchangers are available in two types: open, and closed type heat exchangers. The former is a direct system between the ground and the system. The latter, however, is an indirect type of system in which the pipes are buried in vertical or horizontal patterns. The condition of the nearby fluid determines the suitability and appropriateness of vertical versus horizontal ground heat exchangers (Alalaimi et al., 2014).

Temperature complexities that are caused by curvature-based torsion arise a unique advantage to the process. It can only be observed in the case of spiral pipes (Sasmito et al., 2011). Consequently, much research study and implementation of spiral pipe heat transfer exchangers are nowadays demanded for various applications.

The spiral piping heat-coupling mechanism is attributed to many factors, such as geometrical configurations, compact size, bigger thermal conduction space, number of loops, etc. Numerous studies have been conducted to investigate the configuration and geometry of spiral pipes that could enhance the thermal coupling between them and the conducting volume. Heo and Chung (2012) have described thermal coupling using spiral piping with various shapes and sizes, such as the length of the coil, the number of turns, the coil radius, and the helix diameter. The natural transfer of thermal energy was also studied under different conditions, including the spiral pipe dimensions, rate of flow of the fluid, the bath temperature, and the inlet temperature (Heo and Chung, 2012). He proved that the spiral shape is more effective than the U-shape in terms of the rate of heat transfer.

Other studies conducted by different researchers have emphasized the fact that spiral tubes are indeed better than any other form of tubes, as they effectively act as heat transfer enhancers. Therefore, despite the differences of studies on spiral pipe usage in industrial applications, the outcome results of both experimental and analytical studies are the same.

Jamshidi et al. (2013) experimentally analyzed the enhancement of heat transfer for the coiled tube heat exchanger. The parameters of interest set in their study were: tube diameter, coil curvature, number of loops, coil pitch, and shell and tube side flow rate. They utilized the Wilson plot and the Taguchi technique during the tests. The shell side Reynolds number was found to be the primary factor for the reduction in thermal resistance. However, Zhu et al. (2014) reported that the Nusselt number of the flow inside the spiral tubes increase with the Reynolds number.

Heat transfer by natural convection has been experimentally studied by Ali (2006) using vertical coil at a range of Prandtl number 250–400. Different configurations of the helix-to-tube diameter ratio 30, 20.83, 17.5, 13.33, and 10 were employed. Three empirical correlations based on characteristics length were developed for: $250 < Pr < 400$.

$$\begin{aligned} Nu_{\mathrm{L}} &= 0.619 Ra_{\mathrm{L}}^{0.3} && \text{valid for } 4.37\times10^{10} \le Ra_{\mathrm{L}} \le 5.5\times10^{14} \\ Nu_{\mathrm{L}} &= 0.555 Gr_{\mathrm{L}}^{0.301} Pr^{0.314} && \text{valid for } 1\times10^{8} \le Gr_{\mathrm{L}} \le 5.0\times10^{14} \text{ and } 4.4 \le Pr \le 345 \\ Nu_{\mathrm{L}} &= 0.714 Ra_{\mathrm{L}}^{0.294} && \text{valid for } 4.35\times10^{10} \le Ra_{\mathrm{L}} \le 8.0\times10^{14} \end{aligned} \tag{3}$$

The impact of curvature ratio was also addressed by Beigzadeh and Rahimi (2012) using artificial neural networks (ANNs). The characteristics of the flow, including heat transfer, were of interest.

Experimental tests on spiral coil heat exchanger for boiling purposes has been reported by Elsayed et al. (2012). Effects of tube and helix diameters on the heat transfer coefficients were investigated using different sizes. It was observed that an enhancement of up to 63 % was obtained in the heat transfer coefficient, when the smaller tube diameter was employed. A similar conclusion was reported by Gupta et al. (2011). Observing from the experiments on different tube sizes, they found that a smaller size tube enhances the performance of the coil by increasing the rate of heat transfer. Decreasing the helix diameter of the coil could also improve the heat transfer coefficient up to a level of 150 % (Elsayed et al., 2012).

Freng et al. (2012) numerically studied the thermo-fluid features of the spiral coil heat exchanger at a different Dean Number and pitch size. Previous experimental works validated the CFD methodology. The Nusselt number increases as the pitch size increases. However, as long as the curvature ratio remains constant, no alteration in the behavior of the Nusselt Number was observed when the torsion ratio is changed (Freng et al., 2012).

Recently Seara et al. (2014) developed a numerical model to illustrate the thermal and hydraulic characteristics of the vertical spiral heat exchanger. They validated the numerical

outcomes with experimental data obtained from lab facilities under various operating conditions. Among several different correlations, the following expression has showed good agreement with the experimental results.

$$Nu_{\mathrm{do}} = 0.499 Ra_{\mathrm{do}}^{0.2633} \text{ for } 4.67 \times 10^6 \leq Ra_{\mathrm{do}} \leq 3.54 \times 10^7 \quad (4)$$

Spiral coils buried in conducting volumes have scarcely been addressed in the literatures. Applications of such a type heat exchanger include: geothermal heat pumps, extracting oil and gas, and other solid-fluid interaction applications. Despite several numerical, analytical, and experimental papers that have been published to address the heat transfer mode of embedded coil, limited studies have been conducted to optimize the configuration and orientation of the coil, as well as the heat transfer mechanism during the transient process. Parameters such as number of loops per unit tube length, mass flow rate, coil orientation, and coil pitch are some of the many factors that influence the heat transfer efficiency. Yet, further emphasis is required to address their effects on coil performance and heat transfer rate. The main objectives of this paper include studying and experimentally analyzing the behavior of spiral coil embedded in a conducting container, highlighting the impact of orientation and number of loops on the coil performance, as well as presenting the spatial and temporal mechanisms of the heat transfer process.

2 Theories and assumptions

2.1 The 1-D unsteady state energy analysis

Consider a conducting volume that contains a medium (water) of mass M, volume V, density ρ, and specific heat C_{p}, at a uniform initial temperature T_{wi}. At time $t = 0$, a coiled-tube pipe with surface area A_{s}, carrying a fluid of mass flow $\dot{m}_{\mathrm{l}}$, is sunk into the medium. Energy transfer occurs between the medium in the container and the fluid inside the coil by natural convection (h). For simplicity, we assume that the transient temperature of the medium $T(t)$ is less than its initial temperature T_{wi}. However, the energy analysis could also be used for the opposite case. We also assume that the medium inside the container has a uniform instantaneous temperature that only changes with time, $T = T(t)$. The analysis ignores the coil thermal resistance since it is comprised of a high conductive material (copper). Copper thermal resistance is of order 10^{-2} of convective heat transfer. The error in the convective heat transfer coefficient of the coil side was not exceeding 1 %.

The temperature of the medium drops by a differential value $\mathrm{d}T$ for a differential time interval $\mathrm{d}t$.

The energy balance of the system for the time interval $\mathrm{d}t$ can be expressed as:

$$\begin{aligned} M \times C_{\mathrm{p}} \times \frac{\mathrm{d}T}{\mathrm{d}t} &= h \times A_{\mathrm{s}} \times (T - T_{\mathrm{s}}) \frac{\mathrm{d}(T - T_{\mathrm{s}})}{T - T_{\mathrm{s}}} \\ &= \frac{h \times A_{\mathrm{s}}}{\rho \times V \times C_{\mathrm{p}}} \times \mathrm{d}t \text{ or } \frac{T(t) - T_{\mathrm{s}}}{T_{\mathrm{wi}} - T_{\mathrm{s}}} \\ &= e^{-B \times t} \text{ where } B = \frac{h A_{\mathrm{s}}}{\rho V C_{\mathrm{p}}}. \end{aligned} \quad (5)$$

The surface area of the coil will be compared using two scenarios. The first scenario ignores the effect of the curvature and assumes the 1-D conventional tube surface area is applied; i.e. ($A_{\mathrm{s}} = \pi\, d_{\mathrm{p}}\, l$), where l is the coil length. In contrast, the second scenario considers the coil geometry by introducing the total outside heat transfer area (A_{coil}) for spiral coil. The area can be obtained from the following expression (Wijeysundera et al., 1996):

$$A_{\mathrm{coil}} = \pi D A_{\mathrm{o}} n (1 + 2\pi \alpha n / D) \quad (6)$$

where A_{o} stands for the heat transfer area per unit length ($\pi\, d_{\mathrm{p}}$); n is the number of turns; D is the helix diameter; and α is the rate of increase of the helix radius, in ($\mathrm{m\,rad}^{-1}$).

The external natural convective heat transfer coefficient (h) for each time step can be obtained from the following expressions (Yunus and Afshin, 2010):

$$Nu = \left\{ 0.825 + \frac{0.387 Ra_{\mathrm{L}}^{1/6}}{\left[1 + (0.492/Pr)^{9/16}\right]^{8/27}} \right\}^2 \text{ for verticle coil}$$

$$Nu = \left\{ 0.6 + \frac{0.387 Ra_{\mathrm{D}}^{1/6}}{\left[1 + (0.599/)^{9/16}\right]^{8/27}} \right\}^2 \text{ for horizontal coil} \quad (7a)$$

where $Nu = \frac{h\, L_{\mathrm{c}}}{k}$; for vertical position and $Nu = \frac{h\, D}{k}$; for horizontal position

$$Ra = Gr_{\mathrm{Lc}} Pr = \frac{g\beta (T(t) - T_{\mathrm{s}}) L_{\mathrm{c}}^3}{\nu^2} Pr L_{\mathrm{c}} = \begin{cases} H_{\mathrm{t}} & \text{height of the coil for veritcal position} \\ D & \text{Helix diameter for horizontal position} \end{cases}$$

Helix diameter is obtained from the following expression (Ali, 2006):

$$D = \frac{\sum\limits_{i=1}^{N} \sqrt{D_{\mathrm{si}}^2 - \left(\frac{H_{\mathrm{i}}}{\pi}\right)^2}}{N} \quad (8)$$

where D_{si} is the slanted outer turn diameter for each loop (Fig. 1b).

The transient surface temperature of the spiral coil can be obtained by employing Eq. (5).

2.2 The coil inlet and exit temperatures

The temperature sensors at inlet and outlet of the pipe are mounted outside the conducting volume, as seen in Fig. 2. Since the coil has an extended straight portion at its ends, it is necessary to predict the coil inlet and exit temperatures, (T_{ci}) and (T_{ce}), from the measured inlet and exit temperatures, (T_{i}) and (T_{e}), respectively. The following expressions

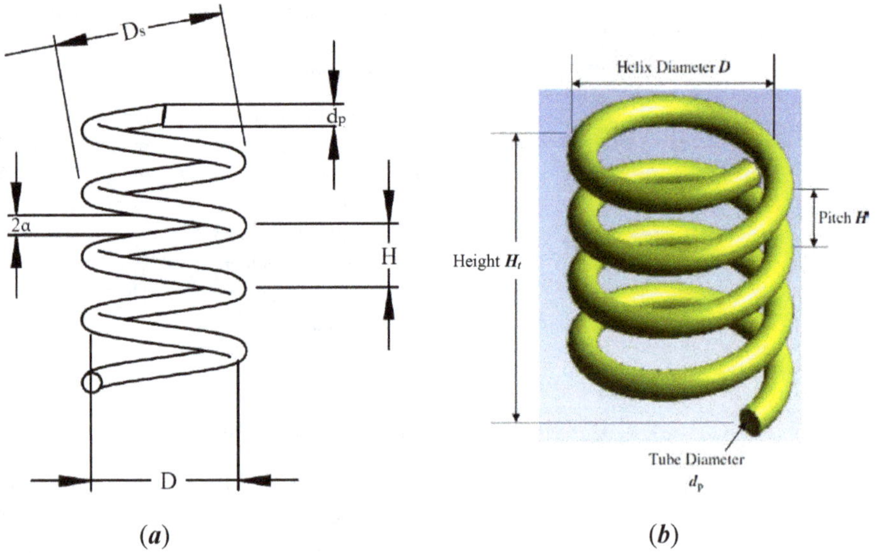

Figure 1. Typical geometry of spiral coil: **(a)** plane view and **(b)** isometric.

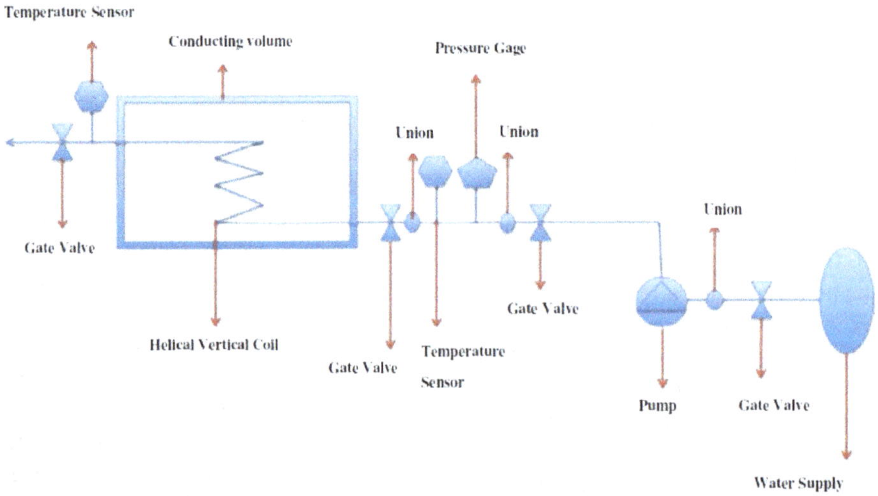

Figure 2. Experiment apparatus.

have been derived (see Appendix A) and employed to determine the fluid temperatures at the coil's ends.

$$T_{ci} = T_i\left(\frac{1-0.5\mathrm{NTU}}{1+0.5\mathrm{NTU}}\right) + T(t) \times \left(\frac{\mathrm{NTU}}{1+0.5\mathrm{NTU}}\right) \quad (9a)$$

$$T_{ce} = T_e\left(\frac{1+0.5\mathrm{NTU}}{1-0.5\mathrm{NTU}}\right) - T(t) \times \left(\frac{\mathrm{NTU}}{1-0.5\mathrm{NTU}}\right) \quad (9b)$$

For convenience, the temperature difference across the straight portion of the coil was calculated at the entrance and exit regions (see Appendix A). It has concluded that not more than 1 % of the increment in temperature has been gained across the uncoiled pipe while 99 % of the temperature rise occurs across the spiral coil. Hence, it is reasonable to assume that the measured inlet and exit temperatures are equal to their corresponding inlet and exit temperatures of the helical part.

3 Experimental set-up

For the purpose of the experiment, copper-tube coils with a diameter of 19 mm and a length of 10.6 m are used. Only 5 % of the length is straight before and after the helical portion. The tubes are carefully furnished from the inner surface to reduce flow resistance. External pipes are connected to the coil with straight joints at the two ends. Other than the container, insulation was applied to the external pipes to avoid any heat gain/loss from the surroundings. Thermal insulation material, wool blanket ($k = 0.04\ \mathrm{W\,m^{-1}\,K^{-1}}$), covers the conducting volume (container) to minimize the heat

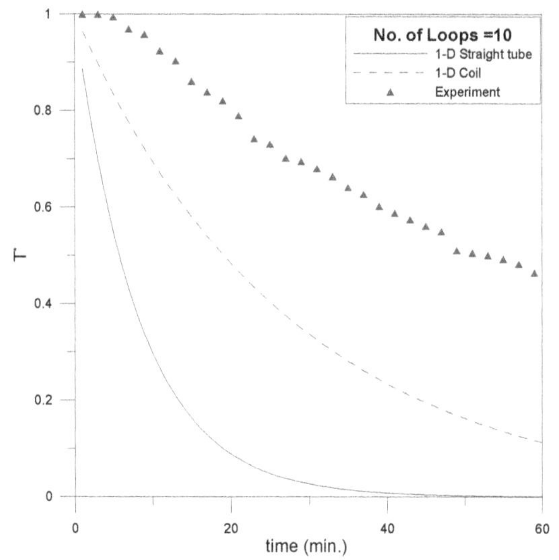

Figure 3. Impact of curvature on transient behavior of the coil.

losses to the surroundings. The heat lost by the container to the surrounding does not exceed 0.5 % of the heat exchange between the coil and the water in the container. For such small amount of heat loss, the assumption of perfectly insulated system for the container is fair and reasonable.

Four electric heaters, each holding a capacity of 2500 W, are mounted at the bottom of the container to provide heat energy to the water of the container. A thermal regulator is installed to stabilize the water pool temperature within a ±0.1 °C precision. The pump and the ball valve regulate the flow of fluids inside the spiral pipes. The ball valve is specifically designed to maintain a particular flow rate of fluid flowing within the coil when the pump is under operation. A centrifugal pump is connected to pump fluid through the spiral pipes at varying Reynolds numbers.

During the experiment, the fluid flowing inside the coil absorbs heat energy from the water of the container while passing through the spiral tube. The inlet/outlet temperatures of the spiral pipes are measured using the installed digital thermometers. Similarly, the inlet/outlet pressures were also measured using the installed pressure transducers. Important data were captured after every moment and recorded until a steady state occurs.

4 Results and discussion

Experiments are carried out for different boundary conditions and coil parameters, such as: coil pitch, number of loops, Reynolds number, and coil orientation. The constraints of the experiments are illustrated in Table 1.

Table 1. Thermal properties and system dimensions.

Property	Unit	Value
Pipe diameter	mm	19
Coil length	m	10.6
Container ($l \times w \times h$)	m	$1.14 \times 0.65 \times 0.48$
Density	$kg\,m^{-3}$	994
Dynamic viscosity	$kg\,m^{-1}\,s^{-1}$	0.653×10^{-3}
Thermal conductivity	$W\,m^{-1}\,K^{-1}$	0.648
Specific heat	$J\,kg^{-1}\,K^{-1}$	4190
Prandtl number		4.23

4.1 Coil geometry

The coil selected for studying the influence of its geometry has a helix diameter of ($D = 15$ cm), a pitch of ($H = 2$ cm), and 10 loops. The coil is firmly mounted vertically and sunk completely inside the pool. The coil inlet/exit temperatures and pressures, container medium temperature, and flow rate are recorded each minute.

A dimensionless temperature of the medium is plotted against time for a time span of an hour to ensure the occurrence of the steady state condition. Three curves are plotted: 1-D tube (ignoring the curvature of the coil), 1-D coil (including the curvature of the coil), and the experiment results. As seen in Fig. 3, the temperature of the medium in the container drops exponentially for all cases. However, the steep of the curves are not identical. In fact, the reduction in the medium temperature for the 1-D coil is not as sharp as the reduction for the 1-D tube. This could be attributed to the impact of coil geometry, which is excluded in the analysis of the 1-D tube, hence strengthening the essential demand of considering the coil configuration during the spiral coil design process. Although the results of the 1-D coiled-type analysis are somewhat acceptable, the experiment outcome is shown to be notably different since the Dean vortex is excluded in 1-D theory.

Figure 4 depicts the performance of the coil, Eq. (10), with time only for the case of experiment measurements. The coil efficiency asymptotically reaches its steady value (24 %) quickly, despite the process still being in the transient region.

$$\eta = \frac{T_{ce}(t) - T_{ci}(t)}{T_{wi}(t) - T_{ci}(t)} \tag{10}$$

4.2 Number of loops

Another coil of 5 loops with similar length and pitch, but a different loop diameter (30 cm), is vertically investigated under similar conditions that are illustrated in Table 1. The behavior of the coil, in correspondence with the change in the number of loops, is presented in Figs. 5–7. The comparison of the normalized temperature of the medium clearly portrays that increasing the number of loops also increases

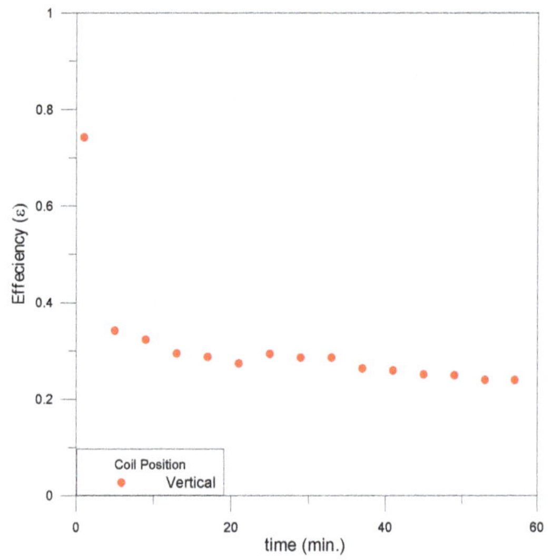

Figure 4. Efficiency variation of the coil with time.

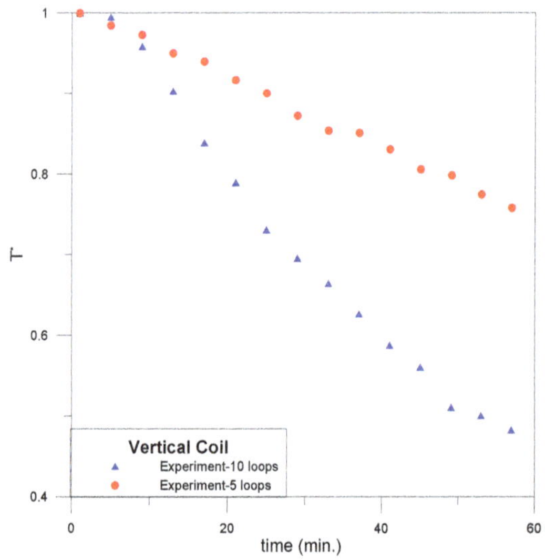

Figure 5. Influence of number of loops.

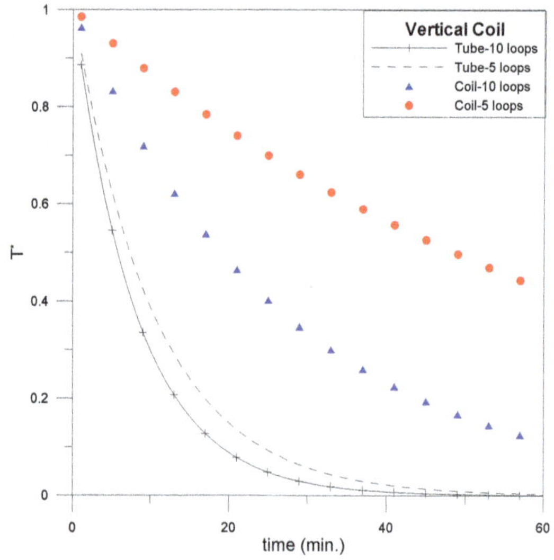

Figure 6. Comparison of 1-D analysis for straight tub vs. coil.

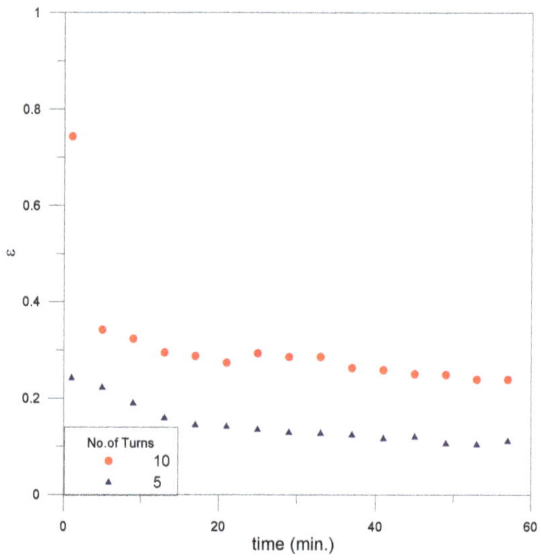

Figure 7. Effect of coil loops on the performance.

the rate of heat transfer, despite the length of the two coils being identical.

Figure 6 shows that disregarding the curvature of the coil leads to an insignificant impact of the loops on the coil performance. Hence, one would expect a remarkable error in the designed parameters during the process of analysis of the spiral coil heat exchanger.

Coil efficiency is expected to improve following the enhancement that occurred in the rate of the heat transfer, as depicted in Fig. 7.

4.3 Coil orientation

To investigate the impact of coil orientation, the 10-looped vertical coil is fabricated to sink horizontally in the container.

Figure 8 presents the normalized temperature variation with a time step for vertical and horizontal positions. Based on the experiment results, in terms of the rate of heat transfer, placing the coil vertically is shown to be better than placing the coil horizontally. However, no such outcome was obtained when 1-D analysis was applied, despite the coil curvature being considered. This important conclusion manifests the drawbacks of relying on the 1-D analysis of the coil and enforces the demand of experimental studies for obtaining reliable results.

The performance of the vertically-positioned coil is shown to be more enhanced than the horizontal position, as presented in Fig. 9.

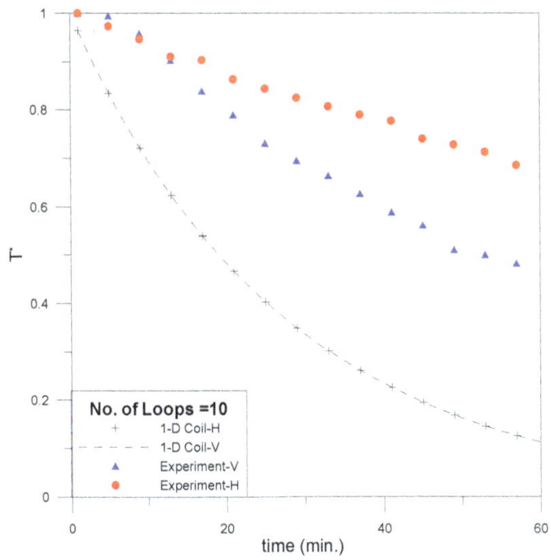

Figure 8. Temperature variation predicted by 1-D coiled-type and experiment.

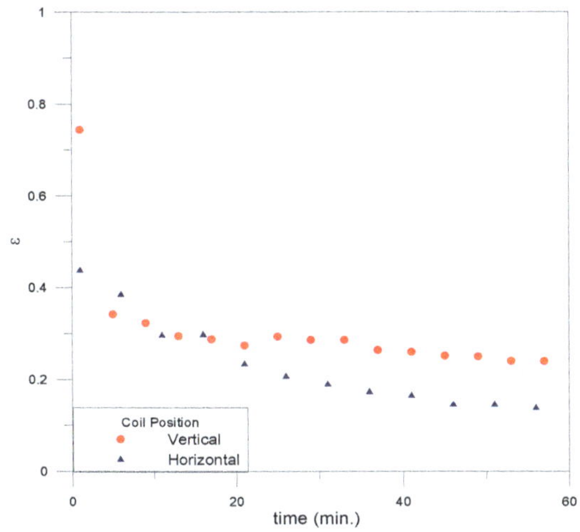

Figure 9. Comparison of efficiency between vertical vs. horizontal coil.

4.4 Effect of mass flow rate

The 10-looped vertical coil is then tested for two different flow rates to illustrate the influence of Reynolds Number on the performance of the coil. Figure 10 shows the theoretical 1-D analysis for the tube verses coil heat exchangers. It is clearly seen that the performance is slightly enhanced when Reynolds number increases, though more tests should be conducted at different *Re* in order to support such conclusion. Figure 11, which illustrates the experimental results, also shows similar behavior of Re. This could be attributed to the fact that the Nusselt Number of the fluid carried by the coil improves as *Re* increases; hence, enhancing the convective heat transfer with the medium in the container.

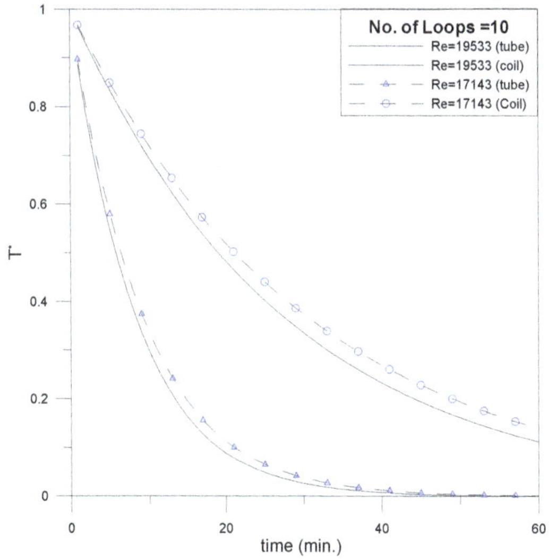

Figure 10. Effect of Reynolds number on the rate of heat transfer.

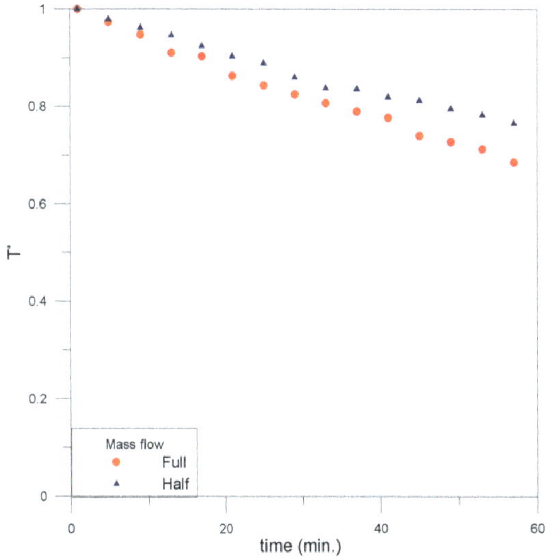

Figure 11. Coil performance for different mass flow.

5 Uncertainty analysis

Inconsistencies exit in experimental engineering analyses due to precision and accuracy measurements. The purpose of this section is to describe the uncertainty analyses of the sensors that were used during the implementation of the experimental tests.

The total outside heat transfer area for spiral coil is listed earlier in Eq. (5):

$$A_{coil} = \pi D A_o n(1 + 2\pi \alpha n/D) \tag{11}$$

where A_o stands for the heat transfer area per unit length ($\pi\, d_p$); n is number of turns; D is helix diameter; and α is the rate of increase of radius, in ($\mathrm{m\,rad^{-1}}$).

Spiral coil heat exchanger effectiveness can be written in the following form:

$$\varepsilon = \frac{\rho \dot{Q} c_p \Delta T}{h A_t \mathrm{LMTD}}; \text{ where LMTD} = \frac{2T_w + \Delta T}{\ln \frac{T_w - T_o}{T_w - T_i}}. \quad (12a)$$

For simplicity, we assume that the effectiveness does not have sensitivity to fluid properties such as density, viscosity, specific heat, and conductivity. However, it is sensitive to the convective heat transfer coefficient (h), which in turn is influenced by the Reynolds number.

Hence, Eq. (11a) becomes:

$$\varepsilon = \frac{\rho \dot{Q} c_p \Delta T}{0.023 Re^{0.8} Pr^{0.4} \times A_t \times \mathrm{LMTD}} \quad (12b)$$

or in terms of parameters' sensitivity:

$$\varepsilon = \varepsilon \left(Q^{0.2}, D, d_p^{0.8}, \Delta T, T_w \right). \quad (13)$$

The independent variables, or the measured parameters, are: inlet and outlet flow temperatures and pressures (T, p), helix diameter (D), fluid flow rate (Q), pipe diameter (d_p), and the conducting volume temperature (T_w). Hence, the uncertainty equation of the coil performance (ε) is:

$$U_\varepsilon = \left[\left(\frac{\partial U_\varepsilon}{\partial Q} \delta Q \right)^2 + \left(\frac{\partial U_\varepsilon}{\partial D} \delta D \right)^2 + \left(\frac{\partial U_\varepsilon}{\partial d_p} \delta d_p \right)^2 + \left(\frac{\partial U_\varepsilon}{\partial (\Delta T)} \delta (\Delta T) \right)^2 + \left(\frac{\partial U_\varepsilon}{\partial T_w} \delta T_w \right)^2 \right]^{1/2}. \quad (14)$$

White (2003) elaborates the approach that was used in the analysis.

The precision of the flow-meter device used in the experiment is 1.5 %, the tolerance in the size of the pipe is 0.5 %, the deviation in the temperature sensors are within 0.1 K, and that of the helix diameter is 2 %.

$$U_\varepsilon = \left[(0.2 \times 1.5)^2 + (1 \times 2)^2 + (0.8 \times 0.5)^2 + (1 \times 0.2)^2 + (1 \times 0.1)^2 \right]^{0.5}$$

$$U_\varepsilon = 2.16\%$$

The dominant parameter that caused considerable impact was the helix diameter.

6 Conclusions

Experimental studies on transient natural convection heat transfer from spiral coil are conducted. Different coils are embedded in a conducting volume container to study the transient behavior of heat transfer. Impact of coil geometry and orientation, number of loops, and Reynolds number are documented. Results have shown that coil geometry significantly influences the performance and efficiency of the heat transfer. The following summarize the outcomes:

1. Vertically-positioned coil was shown to be more effective than horizontally-positioned coil in terms of transferring heat. The enhancement was found to be doubled.
2. Although including the coil curvature during the theoretical analysis somewhat interprets the actual behavior of the coil, such analysis fails to predict the effect of coil orientation. This supports the demand of experimental investigation as an alternative robust method.
3. Doubling the number of loops per meter length of spiral coil enhances the rate of heat transfer by ~130 %.
4. Increasing the Reynolds number also improves the performance of the coil. However, further experiments at different values should be conducted to support such conclusion.

Appendix A

Consider a straight pipe with a length (L) and diameter (d) carrying a fluid with density (ρ), specific heat (C_p), viscosity (μ), and thermal conductivity (k). The pipe is surrounded by another fluid with different properties. The fluid inlet and exit temperatures are respectively (T_i) and (T_e) and the surrounding temperature is ($T\infty$).

Using the energy balance law for the 1-D case:

$$\dot{Q} = \dot{m} \times C_p \times (T_e - T_i) = U \times A_s \times \Delta T_{ave} \quad \text{(A1)}$$

where U is the overall heat transfer coefficient, and could be determined from:

$$\frac{1}{UA_s} = \frac{1}{h_i A_i} + \frac{\ln(d_{po}/d_{pi})}{2\pi k L_{str.}} + \frac{1}{h_o A_o}. \quad \text{(A2)}$$

h_i is the forced convective heat transfer coefficient of the fluid inside the pipe, that was obtained from the Dittus–Boelter equation for straight pipe (Yunus and Afshin, 2010):

$$Nu = 0.023 Re^{4/5} Pr^n; \quad n = 0.4 \text{ for heating};$$
$$n = 0.3 \text{ for cooling}.$$

h_o is the natural convective heat transfer coefficient of the fluid surrounding the pipe.

ΔT_{ave} is the average temperature difference, evaluated for simplicity, from the arithmetic mean.

$$\Delta T_{ave} \approx \frac{(T_\infty - T_i) + (T_\infty - T_e)}{2} \quad \text{(A3)}$$

The error in using the arithmetic mean instead of the logarithmic mean is less than 1 % (Yunus and Afshin, 2010).

Combine A1 and A3 then solve for T_e (exit temperature of the straight portion which is the inlet coil temperature T_{ci}):

$$T_{ci} = T_i \left(\frac{1 - 0.5\text{NTU}}{1 + 0.5\text{NTU}}\right) + T(t) \times \left(\frac{\text{NTU}}{1 + 0.5\text{NTU}}\right).$$

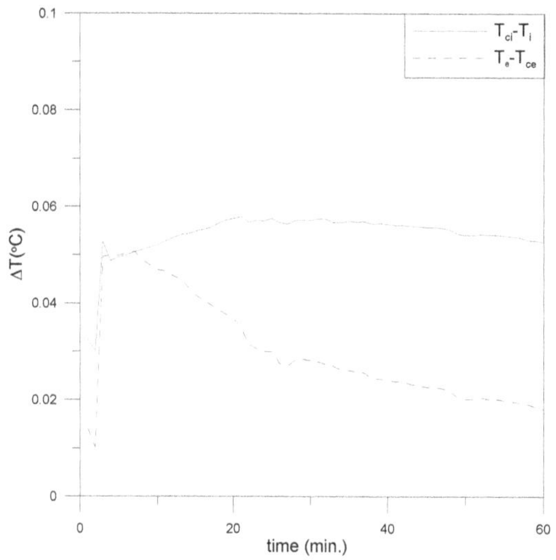

Figure A1. Temperature difference across the uncoiled portion.

Solve again for T_i (inlet temperature of the straight portion which is the outlet coil temperature T_{ce}):

$$T_{ce} = T_e \left(\frac{1 + 0.5\text{NTU}}{1 - 0.5\text{NTU}}\right) - T(t) \times \left(\frac{\text{NTU}}{1 - 0.5\text{NTU}}\right). \quad \text{(A4)}$$

NTU in Eq. (A4) stands for number of transfer unit: $\text{NTU} = \frac{UA_s}{\dot{m}C_p}$; and $T(t)$ is the transient temperature of the surrounding fluid.

Figure A1 illustrates the difference in temperature of the fluid inside the spiral coil, with time across the uncoiled portion at entrance and exit. The assumption of using measured temperatures as coil inlet and outlet temperatures can be easily justified by looking at the maximum temperature difference in the graph. The graph shows that the temperature difference did not even reach the precision of the temperature sensors fixed in then apparatus.

References

Alalaimi, M., Lorente, S., and Bejan, A.: Thermal coupling between a spiral pipe and a conducting volume, Int. J. Heat Mass Trans., 77, 202–207, 2014.

Ali, M.: Natural convection heat transfer from vertical helical coils in oil, Heat Trans. Eng., 27, 79–85, 2006.

Beigzadeh, R. and Rahimi, M.: Prediction of heat transfer and flow characteristics in helically coiled tubes using artificial neural network, Int. Commun. Heat Mass, 39, 1279–1285, 2012.

Dean, W. R.: Note on the motion of fluid in a curved pipe, Philos. Mag., 4, 208–223, 1927a.

Dean, W. R.: The stream line motion of fluid in a curved pipe, Philos. Mag., 5, 208–223, 1927b.

Elazm, M., Ragheb, A., Elsafty, A., and Teamah, M.: Numerical Investigation for the Heat Transfer Enhancement in Helical Cone Coils over Ordinary Helical Coils, J. Eng. Sci. Tech., 8.1, 1–15, 2013.

Elsayed, A., Dadah, R. A., Mahmoud, S., and Rezk, A.: Investigation of flow boiling heat transfer inside small diameter helically coiled tubes, Int. J. Refrig., 35, 2179–2187, 2012.

Freng, Y. M., Lin, W. C., and Chieng, C. C.: Numerically investigated effects of different Dean number and pitch size on flow and heat transfer characteristics in a helically coiled tube heat exchanger, Appl. Therm. Eng., 36, 378–385, 2012.

Gupta, R., Wanchoo, R. K., and Ali, T.: Laminar flow in helical coils a parametric study, Industrial and Engineering Chemistry Research, 50, 1150–1157, 2011.

Heo, J. and Chung, B.: Influence of helical tube dimensions on open channel natural convection heat transfer, Int. J. Heat Mass Trans., 55, 2829–2834, 2012.

Jamshidi, N., Farhadi, M., Ganji, D., and Sedighi, K.: Experimental analysis of heat transfer enhancement in shell and tube helical tube heat exchangers, Appl. Therm. Eng., 51, 644–652, 2013.

Sanner, B., Karytsas, C., Mendrinos, D., and Rybach, L.: Current status of ground source heat pumps and underground thermal energy storage in Europe, Geothermics, 32, 579–588, 2003.

Sasmito, A. P., Kurnia, J. C., and Mujumdar, A. S.: Numerical evaluation of laminar heat transfer enhancement in nanofluid flow in coiled square tubes, Nanoscale Res. Lett., 6.1, 1–14, 2011.

Seara, J. F., Pontevedra, C. P., and Dopazo, J. A.: On the performance of a vertical helical coil heat exchanger numerical model and experimental validation, Appl. Therm. Eng., 62, 680–689, 2014.

Srinivasan, P. S., Nandapurkar, S. S., and Holland, F. A.: Friction factor for coils, Trans. Inst. Chem. Eng., 48, 156–161, 1970.

White, F. M.: Fluid Mechanics, 5th Edn., McGraw-Hill Inc., New York, 2003.

Wijeysundera, N., Ho, J., and Rajasekar, S.: The Effectiveness of a Spiral Coil Heat Exchanger, Int. Commun. Heat Mass, 23, 623–631, 1996.

Yunus, A. C. and Afshin, J. G.: Heat and Mass Transfer, 4th Edn., McGraw Hill, New York, USA, 2010.

Zhu, H., Hanqing, W., and Guangxiao, K.: Experimental study on the heat transfer enhancement by Dean Vortices in spiral tubes, Int. J. Energy Environ., 5, 317–326, 2014.

11

Representation of the kinematic topology of mechanisms for kinematic analysis

A. Müller

Johannes Kepler University, Altenbergerstr. 69, 4040 Linz, Austria

Correspondence to: A. Müller (a.mueller@jku.at)

Abstract. The kinematic modeling of multi-loop mechanisms requires a systematic representation of the kinematic topology, i.e. the arrangement of links and joints. A linear graph, called the topological graph, is used to this end. Various forms of this graph have been introduced for application in mechanism kinematics and multibody dynamics aiming at matrix formulations of the governing equations. For the (higher-order) kinematic analysis of mechanisms a simple yet stringent representation of the topological information is often sufficient. This paper proposes a simple concept and notation for use in kinematic analysis. Upon a topological graph, an order relation of links and joints is introduced allowing for recursive computation of the mechanism configuration. An ordering is also introduced on the topologically independent fundamental cycles. The latter is indispensable for formulating generically independent loop closure constraints. These are presented for linkages with only lower pairs, as well as for mechanisms with one higher kinematic pair per fundamental cycle. The corresponding formulation is known as cut-body and cut-joint approach, respectively.

1 Introduction

The *kinematic topology* of a mechanism refers to the existence and the arrangement of links and joints, i.e. kinematic constraints. It is hence an adjacency relation that can be represented by an undirected linear graph, referred to as the *topological graph* denoted with Γ. This graph provides the basis for a systematic treatment of the kinematics of mechanisms and multibody systems (MBS).

Various types of topological graphs have been proposed in the literature. They have been an important aspect for modeling of complex MBS. Wittenburg (1977, 2008) introduced a linear graph to represent general interconnections of rigid bodies within a MBS. This linear graph includes not only kinematic interconnections but also physical interactions like springs and dampers. This concept was taken up by Arczewski (1992a, b, c). An exhaustive overview of graph representations in MBS dynamics was presented by Jain (2011a, b). All these formulations can be used to derive compact matrix formulations of the MBS motion equations.

For kinematical investigations of mechanisms a graph representation of the kinematic topology has been proposed by Davis (1981, 2015). In particular the velocity constraints were considered as a variant of Kirchoff's law for electric circuits, and it was concluded that the principle concepts and results available for electric networks can be adopted to the velocity analysis of mechanisms. This approach addresses the velocity analysis leading to systems of linear equations, and thus adjacency matrices, incidence matrices, etc. could be used to manipulate the governing equations. Wohlhart (2004), for example, used the topological graph to derive the first-order constraints for topologically independent loops, and to deduced the connectivity of links.

Topological graphs have further interesting features related to generic properties of the mechanism. The essential kinematic properties (of generic realizations) were investigated by Simoni et al. (2013) using topological graphs. It was shown that properties like mobility and connectivity are preserved by any automorphism of the graph. This may be important for topological synthesis.

Still, topological information are rarely exploited for kinematic analysis. One consequence often observed is that redundant loop constraints are imposed for multi-loop mechanisms. This becomes critical in particular if higher-order analyses are pursued. Furthermore, despite the vast literature

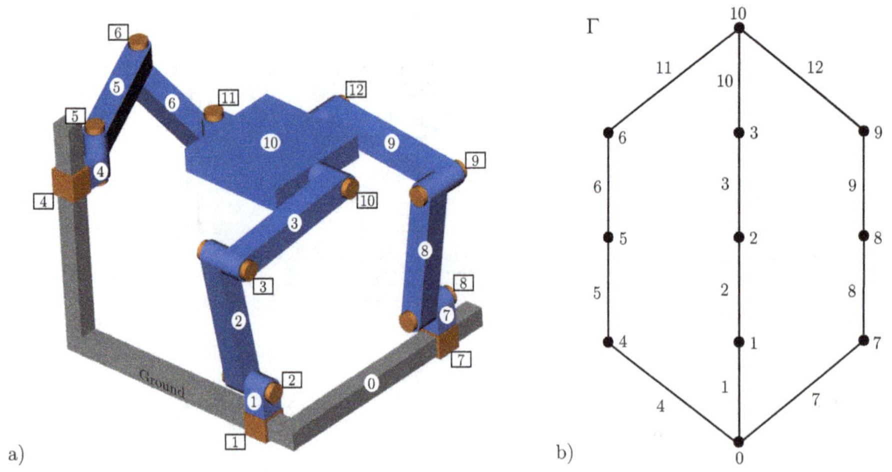

Figure 1. **(a)** 3 DOF parallel manipulator consisting of 10 moving bodies (labeled with circles) connected by 12 joints (labeled with rectangles). **(b)** Topological graph Γ for the manipulator.

on graph modeling of mechanisms and MBS topology, there is no established approach and notation used in mechanism theory. The aim of this paper is to summarize the basic concept of graph representation and of formulations of loop constraints for multi-loop mechanisms in a way appropriate for the higher-order kinematic analysis.

In this paper, the graph representation of the kinematic topology is recalled, and the essential relations necessary for introducing loop constraints are derived. Throughout the paper relative coordinates (joint coordinates) are used to parameterize the configuration allowing for a recursive evaluation of the mechanism kinematics. The essential topological relations are:

1. an order relation to define predecessors of bodies and joints,
2. an indicator of the direction in which a relative joint motion is defined, and
3. an order relation defining predecessors of bodies and joints within topologically independent loops.

The first and second allow for recursive determination of configurations of bodies, and the third for a recursive formulation of topologically independent loop constraints. To this end, (1) an oriented spanning tree $\vec{G}$, (2) an oriented topological graph $\vec{\Gamma}$, and (3) an oriented system of fundamental cycles (FC) are introduced.

These topological relations provide the basis for kinematic investigations, in particular the higher-order kinematic analysis. Matrix representations of topological relations (incidence, adjacency, etc.) are omitted as they are of little help for higher-order constraints.

2 Graph representation of mechanism topology

2.1 Topological graphs

The constituent structural elements of a mechanism are the bodies (links, members) and the joints between them. The topological graph is an undirected graph $\Gamma = (B, J)$, where B is the set of vertices (representing bodies) and J is the set of edges (representing joints). The graph Γ is simple, i.e. two vertices are connected by no more than one edge. The number of joints and bodies is denoted with $N := |J|$ and $M := |B|$, respectively. Bodies are indexed with greek letters $\alpha = 0, \ldots, M-1$. The index 0 is often used to refer to a "fixed member" or to the "ground". Joints are indexed with Latin characters $i = 1, \ldots, N$.

An edge is an unordered pair of vertices denoted $J_i = (B_\alpha, B_\beta) \in J$. This indicates that body B_α and B_β are connected by joint J_i. For the sake of simplicity, the shorthand notations $J_i = (\alpha, \beta) \in J$ and $(\alpha, \beta) \in J$ are used.

Figure 1b shows the topological graph for the parallel mechanism in Fig. 1a that was reported by Carricato and Parenti-Castelli (2002), Kim and Tsai (2002) and Kong and Gosselin (2002). The mechanism contains $M = 11$ bodies and $N = 12$ joints. Notice that for sake of simplicity, in all figures throughout the paper, edges are denoted simply with the index α instead of B_α.

2.2 Joint orientations and oriented topological graphs

The joint $J_i = (B_\alpha, B_\beta)$ constrains the relative motion of the two bodies B_β and B_α. That is, it can be considered to determine the relative configuration of B_α w.r.t. B_β or that of B_β w.r.t. B_α. The definition of this "joint direction" is an indispensable step within the kinematics modeling that further includes introduction of certain joint variables (angles, displacements) to describe the joint motion. This is

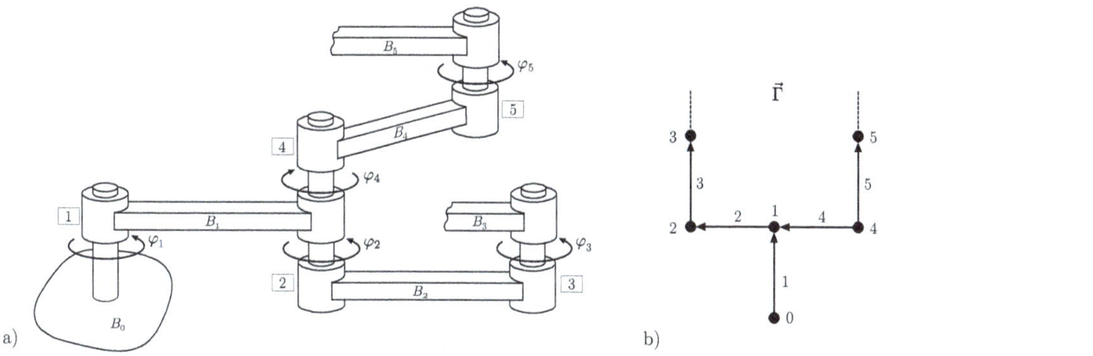

Figure 2. **(a)** A linkage with 5 revolute joints. The direction of positive joint angles is shown. **(b)** Edges in the oriented topological graph $\vec{\Gamma}$ are directed according to the positive joint angles.

not revealed by the undirected edge $J_i \in \Gamma$. For example in Fig. 2a, the joint angle of the revolute joint J_2 measures the rotation of B_2 relative to B_1, whereas the angle of joint J_4 measures the rotation of B_1 w.r.t. B_4, so that the joint angle has the opposite meaning. In order to represent the directions in which the joint motions are to be interpreted, an oriented graph $\vec{\Gamma}$ is introduced. This is obtained by considering the edges of Γ as ordered pairs of vertices. That is, if $(B_\alpha, B_\beta) \in \vec{\Gamma}$, then $(B_\beta, B_\alpha) \notin \vec{\Gamma}$. The vertex B_α is the source (or tail) and B_β is the target (or head) of the edge. This is graphically indicated by an arrow. Then $J_i = (B_\beta, B_\alpha) \in \vec{\Gamma}$ means that joint J_i is assumed to define the motion of B_α w.r.t. B_β. Edges of $\vec{\Gamma}$ are called *arcs*. Figure 2b shows the oriented topological graph for the mechanism in Fig. 2a.

Remark 1 *Frequently the joint orientations are introduced upon algorithmic considerations. In particular, directions are often assigned so to form a root-directed spanning (Jain, 2011a; Wittenburg, 2008). This limits the generality, however. Moreover, in an MBS modeling environment, joints are introduced with well-defined and prescribed orientations. These orientations are in general different from those introduced for an oriented spanning tree, which is merely an algorithm construct (see next section and Sect. 3.3).*

2.3 Spanning trees and predecessor relations

A kinematic chain can be evaluated recursively by starting from an initial body. For mechanisms with kinematic loops there is a priori no unique chain between two bodies. Such can be introduced with help of a spanning tree of Γ. A *spanning tree*, denoted $\mathcal{G} = (B, J^{\mathcal{G}})$, is an acyclic subgraph of Γ, i.e. there is exactly one path between any two vertices. The spanning tree of a graph is not unique. Figure 3b shows a spanning tree for the manipulator in Fig. 1.

The recursive evaluation of the kinematics further requires an order relation that assigns to each body and joint its direct predecessor. Such a relation is induced by directing the spanning tree.

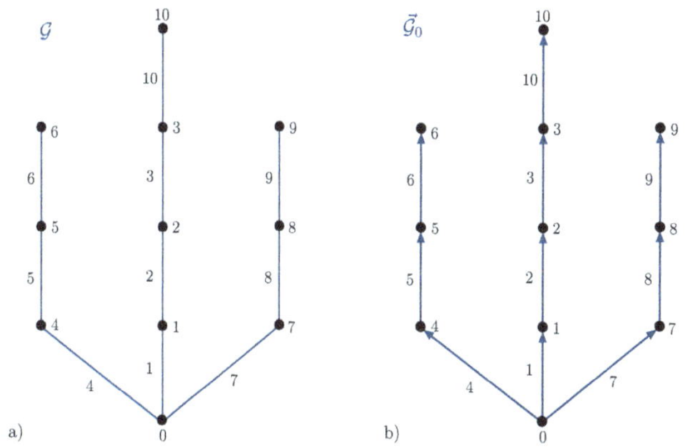

Figure 3. **(a)** A spanning tree $\mathcal{G}$ for the topological graph in Fig. 1. **(b)** The corresponding unique root-directed tree $\vec{\mathcal{G}}_0$.

A *root-directed tree*, denoted $\vec{\mathcal{G}}_0$, is introduced such that there is a directed path in $\vec{\mathcal{G}}_0$ from B_0 (the ground) to every vertex of $\vec{\Gamma}$. B_0 is the *root* of the tree. Edges of $\vec{\mathcal{G}}_0$ are called arcs. Figure 3b shows the root-directed spanning tree for the PKM example in Fig. 1.

Remark 2 Jain (2011a) *called* $\vec{\mathcal{G}}_0$ *the "standard digraph" associated to a mechanism. Moreover, this has been used to actually define directions of joints. Consequently, the original allocation of joint directions would be changed according to the particular (but not unique) directed spanning tree (see remark 1).*

The predecessor of a body can now be defined relative to the root-directed spanning tree. Body B_β is the *direct predecessor* of B_α, if they are connected by an arc, i.e. $(B_\beta, B_\alpha) \in \vec{\mathcal{G}}_0$. This is denoted as $B_\beta = B_\alpha - 1$ (for short $\beta = \alpha - 1$).

Joint J_j is the *direct predecessor* of J_i, if the target of J_j is the source of J_i, i.e. $J_j = (\cdot, \alpha) \in \vec{\mathcal{G}}_0$ and $J_j = (\alpha, \cdot) \in \vec{\mathcal{G}}_0$. This is denoted as $J_j = J_i - 1$ (for short $j = i - 1$). Joint J_j is a *predecessor* of J_i, if there is a finite k, such that $J_j = J_i - 1 - 1 \ \ldots \ -1$ (k times). This is expressed as $J_j = J_i - k$. Being a predecessor is indicated by $J_j < J_i$. Because of the tree topology it is possible that two bodies have the same predecessor, i.e. $\alpha - 1 = \beta - 1$, and analogously for joints.

Denote with $J(B_\alpha)$ the tree-joint that connects B_α with its predecessor, i.e. $J(B_\alpha) = (\beta, \alpha) \in \mathcal{G}$, and with $J_{\mathrm{root}}(B_\alpha)$ the joint that connects to the root B_0 in the path from B_α within $\mathcal{G}$.

The tree in Fig. 3b, induces the following predecessor relations

$J_1 = J_2 - 1$	$J_2 = J_3 - 1$	$J_3 = J_{10} - 1$	$J_4 = J_5 - 1$	
$J_5 = J_6 - 1$	$J_7 = J_8 - 1$	$J_8 = J_9 - 1$		
$B_0 = B_1 - 1$	$B_0 = B_4 - 1$	$B_0 = B_7 - 1$	$B_1 = B_2 - 1$	$B_2 = B_3 - 1$
$B_3 = B_{10} - 1$	$B_4 = B_5 - 1$	$B_5 = B_6 - 1$	$B_7 = B_8 - 1$	$B_8 = B_9 - 1$

and the following assignment of tree-joints connecting the bodies

$$J_{\mathrm{root}}(B_\alpha) = J_1, \alpha = 1,2,3,10 \quad J_{\mathrm{root}}(B_\alpha) = J_4, \alpha = 4,5,6$$
$$J_{\mathrm{root}}(B_\alpha) = J_7, \alpha = 7,8,9 \quad J(B_\alpha) = J_i, \alpha = i.$$

2.4 Cotrees and fundamental cycles

The edges of Γ that are not in $\mathcal{G}$ constitute the *cotree*, denoted $\mathcal{H} = (B, J^{\mathcal{H}})$, with $J^{\mathcal{H}} := J / J^{\mathcal{G}}$. The cotree edges (or cut-edges) are the
textitchords of the spanning tree. The topological graph of a mechanism, where all restrains are due to kinematic couplings, consists of exactly one connected component. Such a graph possesses $\gamma = N - M + 1$ independent *fundamental cycles* (FC), also called fundamental loops. The integer γ is called the cyclomatic number (or Euler number) of Γ. A FC of Γ is a closed path (i.e. a sequence of edges) without repeated edges or vertices that contains exactly one cotree-edge. Thus $\mathcal{H}$ possesses γ edges. The FCs are denoted with Λ_l, where l is the index of the cotree-edge in the FC. The FCs are not unique. The manipulator in Fig. 1a has $12 - 11 + 1 = 2$ FCs, thus $\mathcal{H}$ comprises 2 edges. Figure 4 shows the cotree $\mathcal{H}$ to the spanning tree in Fig. 3a and the corresponding two FCs.

3 Recursive determination of mechanism configurations

3.1 Rigid body configurations

A rigid body is kinematically represented by a body-fixed reference frame. The configuration of B_α w.r.t. to a global reference frame can be represented by a homogenous transformation matrix (Murray et al., 1994; Selig, 2005)

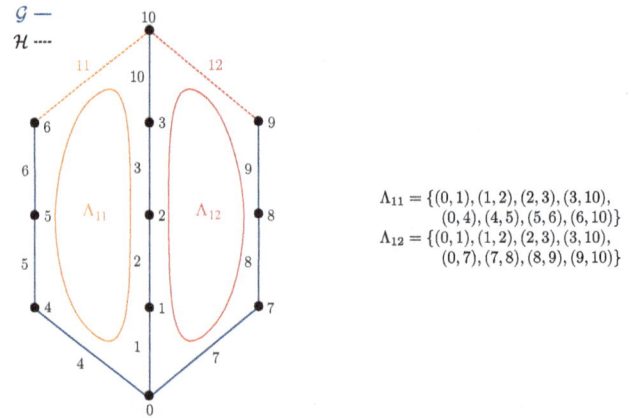

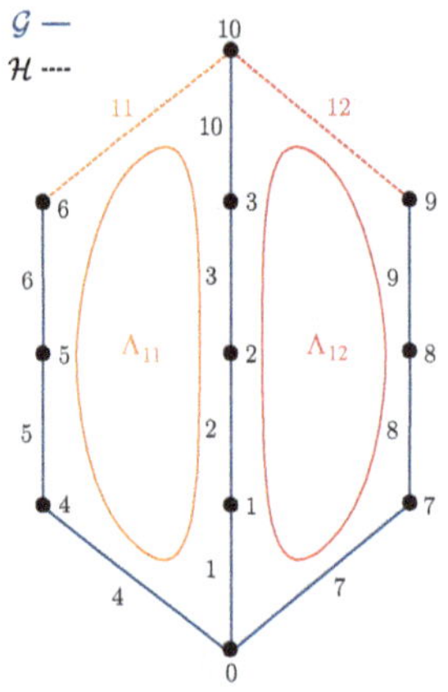

$$\Lambda_{11} = \{(0,1),(1,2),(2,3),(3,10),(0,4),(4,5),(5,6),(6,10)\}$$
$$\Lambda_{12} = \{(0,1),(1,2),(2,3),(3,10),(0,7),(7,8),(8,9),(9,10)\}$$

Figure 4. Cotree $\mathcal{H}$ to the spanning tree $\mathcal{G}$ in fig. 2a). Λ_{11} and Λ_{12} are two FCs according to $\mathcal{H}$. Here the notation (α, β) is used instead of (B_α, B_β).

$$\mathbf{C}_\alpha = \begin{pmatrix} \mathbf{R}_\alpha & \boldsymbol{r}_\alpha \\ \mathbf{0} & 1 \end{pmatrix} \in SE(3) \qquad (1)$$

where $\mathbf{R}_\alpha \in SO(3)$ is the rotation matrix transforming coordinates from the body-fixed reference frame to the global frame, and $\boldsymbol{r}_\alpha \in \mathbb{R}^3$ is the position vector to the origin of the body-fixed reference frame expressed in the world frame.

3.2 Relative joint motions – lower pair joints

The joint motion is interpreted according to the direction of the joint. Let the tree-joint J_i be connecting body B_α and B_β. According to its direction $J_i = (\beta, \alpha) \in \vec{\Gamma}$ it determines the relative configuration of B_α w.r.t. B_β, which is given as

$$\mathbf{D}_i := \mathbf{C}_\beta^{-1} \mathbf{C}_\alpha \in SE(3). \qquad (2)$$

This relation follows immediately, since $\mathbf{C}_\alpha$ and $\mathbf{C}_\beta$ is the configuration of body α and β w.r.t. the global frame. That is, $\mathbf{C}_\alpha$ transforms from body-fixed reference frame on B_α to the global frame, and $\mathbf{C}_\beta^{-1}$ transforms from global frame to the body-fixed frame on B_β. Hence, $\mathbf{D}_i$ transforms from body-fixed frame on B_α to body-fixed frame on B_β.

The majority of technical joints can be modeled as combination of lower kinematic pairs (Uicker et al., 2013). Moreover, their motion can be expressed as combination of 1-DOF screw motions, with pure rotations and translation as special cases, which is a traditional approach in mechanisms and MBS modeling. The relative configuration of joint J_i is then (with appropriate choice of reference frames) determined with the exponential mapping on SE(3) as

$$\mathbf{D}_i = \exp(q_i \mathbf{Y}_i) \qquad (3)$$

where $\mathbf{Y}_i \in \mathbb{R}^6$ is the screw coordinate vector and q_i the joint variable (Selig, 2005). This is a basic result in space kinematics (Angeles, 2003; McCarthy, 1990). Details are omitted here as this is beyond the scope of this paper.

3.3 Mechanism configuration

Successive combination of the relative configurations of tree-joints in the spanning tree allows to determine the configuration of all bodies in the mechanism. This requires taking into account the assigned directions of the tree-joints. To this end, an indicator function is introduced as

$$\sigma(J_i) = \begin{cases} 1, & B_\beta \text{ is direct predecessor of } B_\alpha \\ -1, & B_\alpha \text{ is direct predecessor of } B_\beta \\ 0, & J_i \text{ is not a tree-joint,} \end{cases} \quad \text{for } J_i = (\beta, \alpha).$$

The short hand notation $\sigma(i)$ is used for simplicity. More precisely, $\sigma(i) = 1$, if $(B_\beta, B_\alpha) \in \vec{\mathcal{G}}_0 \wedge (B_\beta, B_\alpha) \in \vec{\Gamma}$; it is $\sigma(i) = -1$, if $(B_\beta, B_\alpha) \in \vec{\mathcal{G}}_0 \wedge (B_\alpha, B_\beta) \in \vec{\Gamma}$; and $\sigma(i) = 0$, if $(B_\beta, B_\alpha) \notin \mathcal{G}$.

For the manipulator example with Γ in Fig. 1b, a joint orientation is chosen according to $\vec{\Gamma}$ in Fig. 5, which also shows the indicator function.

The relative configuration of body B_α w.r.t. its predecessor $B_\beta = B_\alpha - 1$ due to the tree-joint $J_i = (\beta, \alpha)$ is then $\mathbf{D}_i^{\sigma(i)}$, and with Eq. (3) this is expressed as $\mathbf{D}_i^{\sigma(i)} = \exp(\sigma(i) q_i \mathbf{Y}_i)$. That is, the transformation is reversed, if J_i is directed (i.e. measured) opposite to the root-directed tree.

The joint variable q_i is interpreted according to the joint direction. For a revolute joint, Eq. (3) is the transformation matrix (Eq. 1) in terms of the rotation angle q_i. The meaning of the joint direction is apparent. If the joint angle is measured from body B_α to B_β, i.e. if it is directed opposite to the root-directed tree, then $\sigma(i) = -1$, so that its meaning is reversed, and $\mathbf{D}_i^{\sigma(i)} = \exp(-q_i \mathbf{Y}_i)$. For instance, the rotation angle of the revolute joint J_4 as defined in Fig. 2a is reversed in order to determine the motion of B_4 w.r.t. B_1.

With the relative configuration Eq. (2) of the tree-joints, the configuration of an arbitrary body B_α is given as

$$\mathbf{C}_\alpha = \mathbf{D}_r^{\sigma(r)} \cdot \ldots \cdot \mathbf{D}_{i-2}^{\sigma(i-2)} \cdot \mathbf{D}_{i-1}^{\sigma(i-1)} \cdot \mathbf{D}_i^{\sigma(i)}, \quad r = J_{\text{root}}(\alpha), \quad i = J(\alpha) \tag{4}$$

and using the expression Eq. (3) for lower pair joints yields

$$\mathbf{C}_\alpha = \exp(\sigma(r) q_r \mathbf{Y}_r) \cdot \ldots \cdot \exp(\sigma(i-1) q_{i-1} \mathbf{Y}_{i-1}) \cdot \exp(\sigma(i) q_i \mathbf{Y}_i), \text{with } r = J_{\text{root}}(\alpha), \quad i = J(\alpha). \tag{5}$$

For example, the configuration of B_{10} of the mechanism in Fig. 1, according to the oriented tree in Fig. 3b and the oriented topological graph in Fig. 5, is

$$\mathbf{C}_{10} = \mathbf{D}_1 \cdot \mathbf{D}_2^{-1} \cdot \mathbf{D}_3 \cdot \mathbf{D}_{10}^{-1} = \exp(q_1 \mathbf{Y}_1) \cdot \exp(-q_2 \mathbf{Y}_2) \cdot \exp(q_3 \mathbf{Y}_3) \cdot \exp(-q_{10} \mathbf{Y}_{10}).$$

For the mechanism in Fig. 2a the configuration of B_5 is (with $\sigma(1) = \sigma(5) = 1, \sigma(4) = -1$)

$$\mathbf{C}_5 = \mathbf{D}_1 \cdot \mathbf{D}_4^{-1} \cdot \mathbf{D}_5 = \exp(q_1 \mathbf{Y}_1) \cdot \exp(-q_4 \mathbf{Y}_4) \cdot \exp(q_5 \mathbf{Y}_5).$$

The interpretation of the recursive relations Eqs. (4) and (5) is straightforward. The configuration of B_α is the combination of the joint configurations when traversing the kinematic chain in $\vec{\mathcal{G}}_0$ starting from B_0 to B_α while noting the joint directions encoded in $\vec{\Gamma}$.

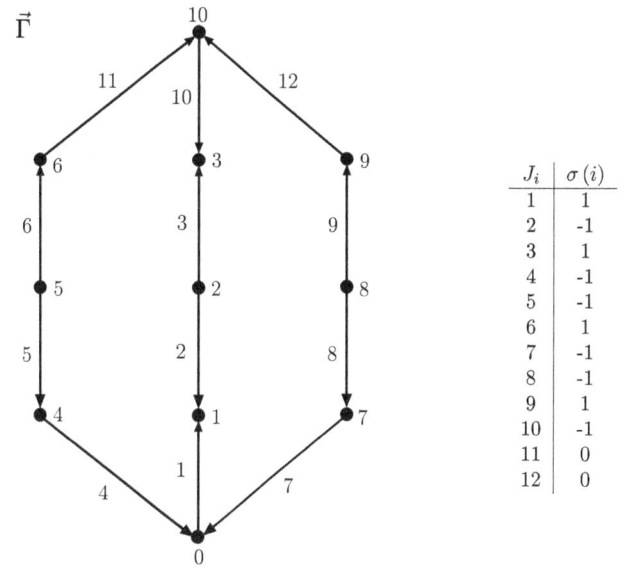

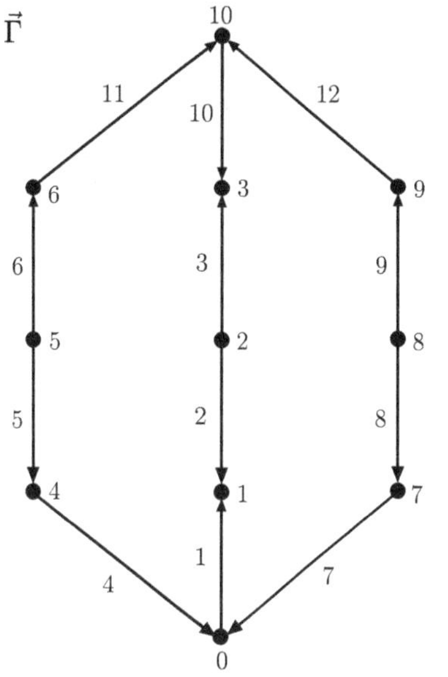

J_i	$\sigma(i)$
1	1
2	-1
3	1
4	-1
5	-1
6	1
7	-1
8	-1
9	1
10	-1
11	0
12	0

Figure 5. **(a)** Oriented topological graph $\vec{\Gamma}$ for Γ in Fig. 1b. **(b)** The function $\sigma(i)$ indicates the direction of joint J_i relative to the root-directed tree $\vec{G}$ in Fig. 3b.

Remark 3 *The formulation Eq. (5) is referred to as the product of exponentials (POE) formula (Brockett, 1984) that gave rise to very compact formulations for the mechanism kinematics and algorithms for the dynamics of MBS (Ploen and Park, 1997; Park, 1994) employing the Lie group SE(3). An important aspect is that it leads to simple explicit algebraic relations for velocities, accelerations, and higher-order derivatives of any order (Müller, 2014b, a). Furthermore, this provides the basis for higher-order kinematic analysis of mechanisms. This has been pursued in Rico et al. (1999, 2008) and Müller and Rico (2008) for single-loop mechanisms. The extension to multi-loop mechanisms requires a systematic yet simple description of the mechanism topology. This is the aim of the present paper.*

In summary, starting from the basic topological information encoded in Γ, the determination of the configurations of the bodies in a mechanism requires introduction of

- the oriented topological graph $\vec{\Gamma}$ in order to represent the joint orientations as assigned in the kinematics modeling, and
- the root-directed tree $\vec{\mathcal{G}}_0$ in order to define an ordering that determines a unique predecessor for each body.

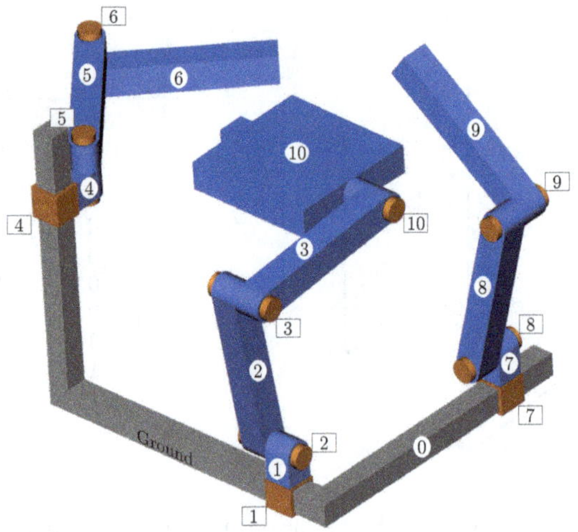

Figure 6. The tree-topology mechanism obtained after removing the cut-joints J_{11} and J_{12}.

4 Kinematic loop constraints

For each of the γ FCs (and only for these) a system of kinematic constraints is introduced. Such loop constraints can be formulated in two basically different ways: the *cut-joint* and the *cut-body* formulation.

For the *cut-body* formulation, a body within the loop, called the "cut-body", is virtually cut open so to obtain an open kinematic chain comprising all joints in the loop. The loop closure constraints require the cut-body to be (re)connected.

For the *cut-joint* approach, a joint within the loop, called the "cut-joint", is eliminated (cut open) from the kinematic model leading to an open kinematic chain comprising all joints in the loop except the cut-joint. The loop closure constraints restrict the relative motion of the two bodies connected by the cut-joint according to its mobility.

4.1 Cut-joint approach

This method is used for kinematics modeling in computational MBS dynamics. Consider the FC Λ_l with cotree-edge $J_l = (B_\beta, B_\alpha)$. The joint J_l is used as cut-joint, and removed from the FC. This leaves two open kinematic chains with the respective terminal bodies B_β and B_α of the spanning tree. Their configurations are determined by the tree-joint configurations via Eq. (4). A system of cut-joint constraints is then formulated for J_l of the form

$$h_l\left(\mathbf{C}_\alpha, \mathbf{C}_\beta\right) = 0. \qquad (6)$$

The constraints Eq. (6) restrict the relative displacement and orientation, i.e. the configuration, of B_β and B_α according to the mobility of joint J_l. These are well-known for various joint types (Uicker et al., 2013; Wittenburg, 2008). Cotree edges are the chords of the spanning tree, which bears an obvious kinematic meaning: a cut-joint reconnects the two terminal vertices of the spanning tree that are linked by a chord so to close the FC.

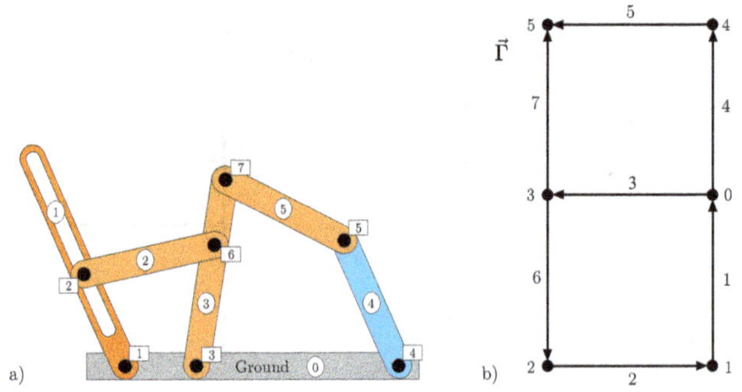

Figure 7. **(a)** Linkage with a higher kinematic pair J_2 (pin-in-slot joint). **(b)** An oriented topological graph $\vec{\Gamma}$ for this linkage.

Remark 4 *The constraints Eq. (6) only involve the joint variables of the tree-joints since the cut-joint is removed from the kinematic model. The cut-joint approach is used in recursive MBS dynamics algorithms. The main reason is that the dynamic motion equations of the tree topology system (the mechanism defined by $\mathcal{G}$) can be derived and evaluated, possibly with low-order algorithms, and the constraints be imposed. As example, the tree-topology system for the manipulator in Fig. 1 according to the tree in Fig. 3 is shown in Fig. 6. The cut-joints J_{11} and J_{12} are removed.*

Remark 5 *The overall system of loop constraints is possibly redundant due to the particular mechanism geometry. The computational treatment of such situations is a topic of ongoing research (Arabyan and Wu, 1998; Meijaard, 1993; Wojtyra and Fraczek, 2013). It should be remarked that the POE formulation and its underlying Lie group concept allows to deduce redundant loop constraints and eventually to determine a reduced non-redundant set of constraints (Müller, 2011, 2014c).*

The cut-joint formulation is also advantageous for the kinematic analysis when only one higher kinematic pair is present in a FC. Then the configuration, velocity, and acceleration, etc. of the two open chains with terminal B_β and B_α can be expressed by the POE Eq. (5) since all other tree-joints are lower pairs. For instance, the mechanism in Fig. 7a comprises a higher kinematic pair: the pin-in-slot joint J_2. An oriented topological graph is shown in Fig. 7b. The spanning tree in Fig. 8a is introduced so that J_2 is in the cotree. Figure 8b shows the corresponding FCs.

The higher pair $J_2 = (B_2, B_1)$ is the cut-joint in Λ_2. It imposes two rotational constraints and two translational constraints on the relative motion of body B_2 and B_1 (Uicker et al., 2013) that are summarized as $h_2(\mathbf{C}_1, \mathbf{C}_2) = \mathbf{0}$. The configurations of B_2 and B_1 are determined by the lower pairs of

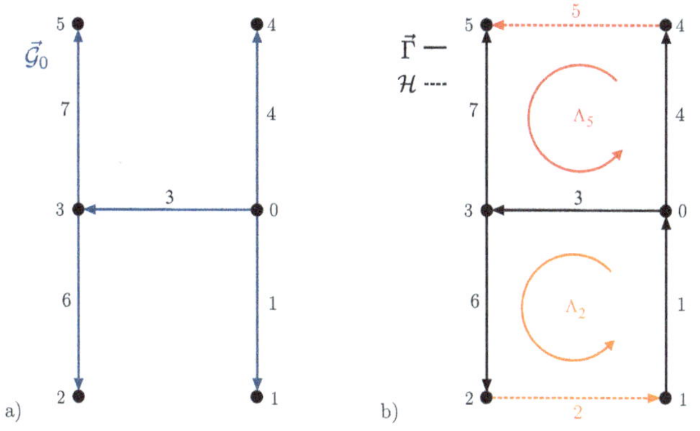

Figure 8. **(a)** A root-directed spanning tree $\vec{\mathcal{G}}_0$ for the linkage in Fig. 5. **(b)** Oriented FCs Λ_2 and Λ_5.

the tree topology system. With the orientation of tree-joints defined by $\vec{\Gamma}$ and $\vec{\mathcal{G}}_0$, these are

$$\mathbf{C}_1 = \mathbf{D}_1^{-1} = \exp(-q_1 \mathbf{Y}_1)$$
$$\mathbf{C}_2 = \mathbf{D}_3 \mathbf{D}_6 = \exp(q_3 \mathbf{Y}_3)\exp(q_6 \mathbf{Y}_6).$$

The loop constraints $h_5(\mathbf{C}_4, \mathbf{C}_5) = \mathbf{0}$ for Λ_5 due to the revolute (lower pair) joint J_5 are derived with

$$\mathbf{C}_4 = \mathbf{D}_4 = \exp(q_4 \mathbf{Y}_4)$$
$$\mathbf{C}_5 = \mathbf{D}_3 \mathbf{D}_7 = \exp(q_3 \mathbf{Y}_3)\exp(q_7 \mathbf{Y}_7).$$

In summary, the cut-joint formulation requires introduction of

- the oriented topological graph $\vec{\Gamma}$ representing the joint orientations,
- the root-directed tree $\vec{\mathcal{G}}_0$ in order to determine the configuration of the bodies connected by the cut-joint, and
- the FCs Λ_l defining the kinematic loops for which closure constraints are introduced.

4.2 Cut-body approach

This method is used for kinematic analysis of linkages, i.e. closed kinematic chains comprising only lower pairs. Instead of eliminating the cotree-joint $J_l = (B_\beta, B_\alpha)$, it is regarded as part of the closed kinematic chain defined by the FC Λ_l. This requires taking into account the orientation of the joints within the FC. To this end, an orientation of the FC Λ_l is introduced such that it is aligned with the cotree-edge J_l. The orientations of edges relative to Λ_l are indicated by the cycle incidence function

$$\sigma_l(J_i) = \begin{cases} 1, & (B_\beta, B_\alpha) \in \vec{\Gamma} \text{ is aligned with } \Lambda_l \\ -1, & (B_\beta, B_\alpha) \in \vec{\Gamma} \text{ is directed opposite to } \Lambda_l \\ 0, & (B_\beta, B_\alpha) \notin \Lambda_l, \end{cases} \qquad \text{for } J_i = (\beta, \alpha).$$

Notice that $\sigma_l(l) = 1$. These numbers are commonly arranged in the cycle incidence matrix of the oriented graph.

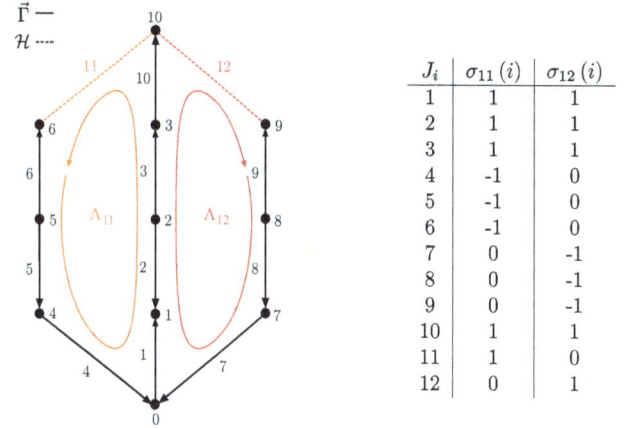

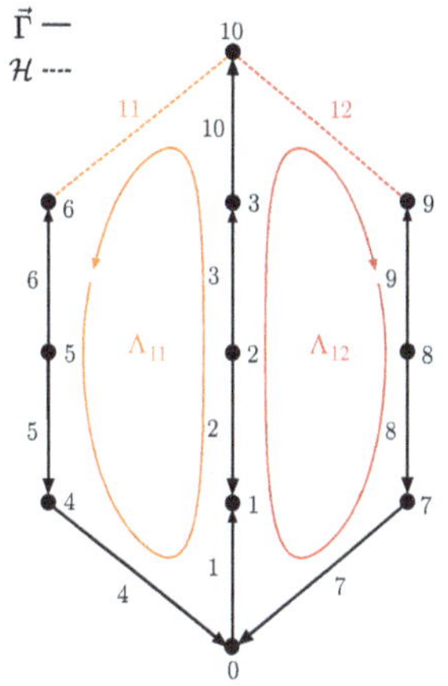

J_i	$\sigma_{11}(i)$	$\sigma_{12}(i)$
1	1	1
2	1	1
3	1	1
4	-1	0
5	-1	0
6	-1	0
7	0	-1
8	0	-1
9	0	-1
10	1	1
11	1	0
12	0	1

Figure 9. Two oriented FC for the oriented topological graph $\vec{\Gamma}$ in Fig. 4). The cycle incidence function $\sigma_l(J_i)$ indicates whether joint J_i is oriented along or oposite to the FC Λ_l.

The orientation of Λ_l induces an order relation in the FC. J_j is considered as predecessor of J_i in Λ_l, if it is met after J_i when traversing the FC Λ_l starting from J_l. This is denoted with $J_j <_l J_i$. Clearly $J_i <_l J_l$ for all $i \neq l$. Joint J_l and the last joint in the FC connect to the same body B_α, i.e. $J_i = (\cdot, B_\alpha)$ and $J_l = (B_\alpha, \cdot)$.

In the manipulator example, the two FCs in Fig. 4 can be oriented as in Fig. 9 that also shows the cycle incidence function. The ordering in Λ_{11}, for instance, is such that $10 <_{11} 3 <_{11} 2 <_{11} 1 <_{11} 4 <_{11} 5 <_{11} 6 <_{11} 11$.

Successive combination of the relative configurations of all joints in the FC leads to the closure condition for Λ_l

$$f_l = \mathbf{I} \tag{7}$$

with

$$f_l := \mathbf{D}_i^{\sigma_l(i)} \cdot \mathbf{D}_j^{\sigma_l(j)} \cdot \ldots \cdot \mathbf{D}_k^{\sigma_l(k)} \cdot \mathbf{D}_l,$$
$$\text{for } i <_l j <_l \ldots k <_l l, \text{ and } J_i, J_j, \ldots, J_l \in \Lambda_l \tag{8}$$

where $\mathbf{I}$ is the 4×4 identity matrix. If all joints are 1-DOF lower pairs, this can be written as

$$f_l := \exp(\sigma_l(i) q_i \mathbf{Y}_i) \cdot \exp(\sigma_l(j) q_j \mathbf{Y}_j) \cdot \ldots \cdot \exp(\sigma_l(k) q_k \mathbf{Y}_k) \cdot \exp(q_l \mathbf{Y}_l). \tag{9}$$

The expression Eqs. (8) and (9) can be interpreted as the configuration of the terminal body of a kinematic chain comprising the joints $i <_l \ldots <_l l$. The closure condition then requires this terminal body to remain fixed. For this reason this approach is also known as the cut-body method (Samin and Fisette, 2003). The body connecting joints J_i and J_l (the cut-body) is virtually cut, and one half serves as terminal body of the chain. Merging the two halves then leads to the above constraints.

Consider the manipulator example in Fig. 1a with oriented topological graph and FCs in Fig. 9. The system of loop constraints for the FC Λ_{11} and Λ_{12} is respectively $f_{11} = \mathbf{I}$ and $f_{12} = \mathbf{I}$ with

$$f_{11} := \mathbf{D}_{10} \cdot \mathbf{D}_3 \cdot \mathbf{D}_2^{-1} \cdot \mathbf{D}_1 \cdot \mathbf{D}_4 \cdot \mathbf{D}_5 \cdot \mathbf{D}_6^{-1} \cdot \mathbf{D}_{11}$$

$$f_{12} := \mathbf{D}_{10} \cdot \mathbf{D}_3 \cdot \mathbf{D}_2^{-1} \cdot \mathbf{D}_1 \cdot \mathbf{D}_7 \cdot \mathbf{D}_8 \cdot \mathbf{D}_9^{-1} \cdot \mathbf{D}_{12}.$$

In summary, the cut-body formulation requires introduction of

- the oriented topological graph $\mathbf{\Gamma}$ representing the joint orientations,
- the spanning tree $\mathcal{G}$ defining the FCs Λ_l, i.e. the kinematic loops for which closure constraints are introduced, and
- an orientation of the FCs.

Remark 6 *The cut-body formulation involves the configurations, thus the joint variables, of all joints in the FC. The expression Eq. (9) has a major importance for the kinematic analysis of linkages with lower pair joints. Firstly because it is determined solely in terms of the joint screw coordinate vectors $\mathbf{Y}_j$, but secondly and most importantly, because its derivatives of any order can be determined by simple algebraic operation, namely the screw products (Lie brackets) of the instantaneous joint screws (Rico et al., 1999; Selig, 2005; Müller, 2014a). This is the basis for any higher-order kinematic analysis of mechanisms.*

Remark 7 *The constraints Eq. (9) define the variety of admissible configurations of the kinematic loop Λ_l as $V_l := \{\mathbf{q} \in \mathbb{V}^n \mid f_l(\mathbf{q}) = \mathbf{I}\}$, and thus the configuration space of the linkage as*

$$V := \bigcap_{l \in \mathcal{H}} V_l. \tag{10}$$

This configuration space variety is the chief subject in the mobility analysis of mechanisms. Clearly, a systematic method for multi-loop mechanisms shall rest on the identification of topologically independent FCs. This has been introduced by Davis (1981, 2015) adopting the principles of Kirchoff's circuit law for electric networks. It is not yet been used widely, however. This frequently leads to the introduction of redundant constraints when topologically redundant loops are considered.

Remark 8 *The kinematic topology is inextricably connected to the (generic) structural mobility, i.e. the mobility that a generic realization of a mechanism with a given number of bodies and joints possesses. Structural mobility criteria hence estimate a lower bound on the mobility of a particular mechanism. They only require structural information but no information about the topology. It is instructive though, to note how topological information enters these criteria. The best-known mobility criterion is the Chebyshev-Kutzbach-Grübler formula $\delta_{\text{gen}} = g(M-1) - \sum_{J_i \in \Gamma} (g - \delta_i)$ where δ_i is the DOF of joint J_i, and g characterizes the "motion type" of the mechanism. For instance, $g = 3$ for planar and spherical, and $g = 6$ for spatial mechanisms (Angeles, 2003). The number g can be specified without investigating the particular geometry if the motion of the loops form a motion subgroup of SE(3). Then the generic mobility, i.e. for generic geometries, is determined with $g = 1, 2, 3, 4, 6$. Now with $\sum_{J_i \in \Gamma} g = Ng$ this reads*

$$\delta_{\text{gen}} = \sum_{i \in J} \delta_i - g\gamma. \tag{11}$$

In other words for each FC a system of g constraints is imposed. Hence, the generic DOF is determined once the number FCs is known. It was shown in Müller (2009) that this is in fact the correct mobility for generic realizations.

5 Conclusions

The kinematic analysis of a mechanism requires evaluation of the motion of its members, and formulation of a system of generically independent loop closure constraints. Any recursive evaluation of the motion of a mechanism rests on an ordering of bodies and joints. The configuration of a body is given in terms of the configurations of its predecessors that form a kinematic chain to the ground (reference body). For a multi-loop mechanism this chain is no unique. The spanning tree of the topological graph gives rise to a unique predecessor relation. This is introduced in this paper making use of a root-directed spanning tree (a tree such that there is an oriented path from any body to the ground). If the mechanism comprises lower pair joints only, the configuration is then recursively expressible by the product of exponentials (POE) in terms of joint screw coordinates.

The recursive formulation of loop closure constraints also requires an ordering, now within the loop. Here it is important that constraints are formulated for fundamental cycles (FC), i.e. for topologically independent kinematic loops. To this end, fundamental cycles are introduced on the topological graph together with an orientation. Two different constraint formulations are considered: cut-joint and cut-body formulation. The cut-joint formulation allows for a higher kinematic pair in a FC, whereas the cut-body formulation is tailored to linkages with lower pairs.

The basic difference of the proposed topology description compared to the various graph representations is that it does not involve matrix representations. Moreover, the presented notation provides the basis for a systematic higher-order analysis of the mechanism kinematics. This will be reported in forthcoming paper.

Appendix A

Table A1. Nomenclature.

N	number of joints in a mechanism
M	number of bodies in a mechanism
$\Gamma, \mathcal{G}, \mathcal{H}$	topological graph, spanning tree, and cotree in Γ
B_α	vertex representing body $\alpha = 0, \ldots, M-1$
$J_i = (B_\beta, B_\alpha)$	edge representing joint $i = 1, \ldots, N$ between bodies B_β and B_α
$\vec{\Gamma}$	oriented topological graph indicating assigned direction of joint transformations
$\vec{\mathcal{G}}_0$	root-oriented spanning tree, so that there is a unique directed path from the root B_0 to any B_α
$\sigma(i)$	function indicating the direction of joint i relative to the root-oriented tree
γ	number fundamental cycles of Γ
Λ_l	fundamental cycle of Γ assigned to co-tree edge $l \in \mathcal{H}$
q_i	joint variable of 1-DOF lower pair joint i
$\mathbf{q} \in \mathbb{V}^N$	vector comprising all joint variables
$\mathbf{Y}_i$	screw coordinate vector of joint i in the zero reference configuration $\mathbf{q} = \mathbf{0}$
SE(3)	matrix representation of the Lie group of rigid body motions
δ_i	DOF of joint i
δ_{gen}	generic DOF of a mechanism

Acknowledgements. The author acknowledges that this work has been partially supported by the Austrian COMET-K2 program of the Linz Center of Mechatronics (LCM).

References

Angeles, J.: Fundamentals of robotic mechanical systems, 2nd Edn., Springer International Publishing, Swizterland, 2003.

Arabyan, A. and Wu, F.: An improved formulation for constrained mechanical systems, Multibody Syst. Dynam., 2, 49–69, 1998.

Arczewski, K.: Graph theoretical approach – I. determination of kinetic energy for a class of particle systems, J. Franklin Inst., 329, 469–481, 1992a.

Arczewski, K.: Graph theoretical approach – II. determination of generalized forces for a class of systems consisting of particles and springs, J. Franklin Inst., 329, 483–491, 1992b.

Arczewski, K.: Graph theoretical approach – III. equations of motion of a class of constrained particle systems, J. Franklin Inst., 329, 493–510, 1992c.

Brockett, R. W.: Robotic manipulators and the product of exponentials formula, Mathematical Theory of Networks and Systems, Lect. Not. Contr. Inf. Sci., 58, 120–129, 1984.

Carricato, M. and Parenti-Castelli, V.: Singularity-free fully-isotropic translational parallel mechanism, Int. J. Robot. Res., 21, 161–174, 2002.

Davies, T. H.: Kirchhoff's circulation law applied to multi-loop kinematic chains, Mech. Mach. Theory, 16, 171–183, 1981.

Davis, T. H.: A network approach to mechanisms and machines: Some lessons learned, Mech. Mach. Theory, 89, 14–27, 2015.

Jain, A.: Graph theoretic foundations of multibody dynamics, Part I: structural properties, Multibody Syst Dyn., 26, 307–333, 2011a.

Jain, A.: Graph theoretic foundations of multibody dynamics, Part II: Analysis and algorithms, Multibody Syst Dyn., 26, 335–365, 2011b.

Kim, S. and Tsai, L.: Evaluation of a cartesian parallel manipulator, in: Advances in robot kinematics, edited by: Lenarčič, J. and Thomas, F., Springer, Dordrecht, the Netherlands, 2002.

Kong, X. and Gosselin, C.: Type synthesis of linear translational parallel manipulators, in: Advances in robot kinematics, edited by: Lenarčič, J. and Thomas, F., Springer, Dordrecht, the Netherlands, 2002.

McCarthy, J. M.: An Introduction to Theoretical Kinematics, MIT Press, Cambridge, 1990.

Meijaard, J. P.: Applications of the Singular Value Decomposition in dynamics, Comput. Meth. Appl. Mech. Eng., 103, 161–173, 1993.

Müller, A.: Generic Mobility of Rigid Body Mechanisms, Mech. Mach. Theory, 44, 1240–1255, 2009.

Müller, A.: Semialgebraic Regularization of Kinematic Loop Constraints in Multibody System Models, ASME Trans., J. Comput. Nonlin. Dyn., 6, 041010, doi:10.1115/1.4002998, 2011.

Müller, A.: Higher Derivatives of the Kinematic Mapping and some Applications, Mech. Mach. Theory, 76, 70–85, 2014a.

Müller, A.: Derivatives of Screw Systems in Body-fixed Representation, in: Advances in Robot Kinematics (ARK), edited by: Lenarcic, J. and Khatib, O., Springer International Publishing, Switzerland, 123–130, 2014b.

Müller, A.: Implementation of a Geometric Constraint Regularization for Multibody System Models, Arch. Mech. Eng., 61, 365–383, doi:10.2478/meceng-2014-0021, 2014c.

Müller, A. and Rico, J. M.: Mobility and Higher Order Local Analysis of the Configuration Space of Single-Loop Mechanisms, in: Advances in Robot Kinematics, edited by: Lenarcic, J. J. and Wenger, P., Springer Netherlands, 215–224, 2008.

Murray, R. M., Li, Z., and Sastry, S. S.: A Mathematical Introduction to Robotic Manipulation, CRC Press, Boca Raton, 1994.

Park, F. C.: Computational Aspects of the Product-of-Exponentials Formula for Robot Kinematics, IEEE Trans. Aut. Contr., 39, 643–647, 1994.

Ploen, S. R. and Park, F. C.: A Lie group formulation of the dynamics of cooperating robot systems, Rob. Auton. Syst., 21, 279-287, 1997.

Rico, J. M., Gallardo, J., and Duffy, J.: Screw theory and higher order kinematic analysis of open serial and closed chains, Mech. Mach. Theory, 34, 559–586, 1999.

Rico, J. M., Cervantes-Sánchez, J. J., and Gallardo, J.: Velocity and Acceleration Analyses of Lower Mobility Platforms via Screw Theory, 32nd ASME Mechanisms and Robotics Conf., 3–6 August 2008, Brooklyn, NY, 2008.

Samin, J.-C. and Fisette, P.: Symbolic Modeling of Multibody Systems, Springer Netherlands, 2003.

Selig, J.: Geometric Fundamentals of Robotics, in: Monographs in Computer Science Series, Springer-Verlag, New York, 2005.

Simoni, R., Melchiades Doria, C., and Martins, D.: Symmetry and invariants of kinematic chains and parallel manipulators, Robotica, 31, 61–70, 2013.

Uicker, J. J., Ravani, B., and Sheth, P. N.: Matrix Methods in the Design Analysis of Mechanisms and Multibody Systems, Cambridge University Press, 2013.

Wittenburg, J.: Dynamics of Systems of Rigid Bodies, B. G. Teubner, Stuttgart, 1977.

Wittenburg, J.: Dynamics of Multibody Systems, 2nd Edn., Springer-Verlag, Berlin, Heidelberg, 2008.

Wohlhart, K.: Screw Spaces and Connectivities in Multiloop Linkages, in: On Advances in Robot Kinematics, Springer Netherlands, 97–104, 2004.

Wojtyra, M. and Fraczek, J.: Solvability of reactions in rigid multibody systems with redundant nonholonomic constraints, Multibody Syst. Dyn., 30, 153–171, 2013.

12

Effect of cutting speed parameters on the surface roughness of Al5083 due to recrystallization

Elias Rezvani[1], Hamid Ghayour[2], and Masoud Kasiri[2]

[1]Department of Mechanical Engineering, Faculty of Engineering, Najafabad Branch, Islamic Azad University, Najafabad, Iran

[2]Advanced Materials Research Center, Faculty of Materials Engineering, Najafabad Branch, Islamic Azad University, Najafabad, Iran

Correspondence to: Hamid Ghayour (ghayour_ham@iust.ac.ir)

Abstract. In the present study, the effect of machining parameters and recrystallization on surface quality of Al5083 has been investigated. In order to achieve minimum surface roughness of aluminum 5083 samples, statistical test design method of "full factorial" was used. In order to achieve the phenomenon of recrystallization, aluminum 5083 samples were set under 50 % cold rolling mechanical operations. Then, the rolled samples were annealed for 2.5 to 240 min at 250 °C. The stress-strain curves were obtained from tensile tests. Then, dry machining was carried out on the original and crystallized samples under the same conditions. Results of surface roughness, tensile, and microstructure tests indicated the reduction of surface roughness in the crystallized sample.

1 Introduction

Aluminum alloys are widely used in aerospace, marine, and automotive industries as material for lightweight structures. In this regard, the hard alloy of aluminum 5083 is preferred due to its acceptable strength, good corrosion resistance and weld ability. Obviously, the development of the capacity of plasticity in this material will increase its potential fo r these applications (Sukumar et al., 2014; CRTD, 1994). Surface roughness is one of the factors affecting the mechanical properties such as corrosion resistance, abrasion resistance, flexibility, fatigue behavior, etc. Therefore, it is one of the effective parameters that cannot be neglected in designing.

Figure 1 shows the fish bone diagram of the parameters affecting surface roughness (Routara et al., 2009).

A number of empirical studies have concentrated on the surface roughness of aluminum alloys. Kiswanto et al. (2014) carried out research on the impact of spindle speed, feed-rate, and time of machining on surface roughness of aluminum 100 alloy by micro-milling operations. They concluded that surface roughness decreases by increasing the spindle speed, while it increases as the time of machining passes and the feed-rate increases (Kiswanto et al., 2014). Results of machining of aluminum alloys using designing test methods indicate that minimum surface roughness can be achieved by maximum cutting speed and minimal feed-rate (Pinar, 2013; Wang and Chang, 2004; Tammineni and Yedula, 2014; Rawangwong et al., 2012, 2014). Clack et al. examined the surface roughness of Aluminum 6061-T8 using genetic algorithm. The practical results of this study were consistent with genetic algorithm (Colak et al., 2007). The optimization process of the aluminum surface roughness was examined in light of machining parameters (Sukumar et al., 2014, Arokiadass et al., 2011, Oktem et al., 2005; Nair and Govindan, 2013; Vakondios et al., 2012; Zhang et al., 2007). Mahesh and Rajesh (2014) carried out 27 experiments to optimize the parameter of CNC machining process on aluminum 7075-T6 alloy using Taguchi fuzzy logic (Mahesh and Rajesh, 2014). The results of high-speed milling on aluminum alloy shows that a minimum surface roughness can be achieved by maximum cutting speed and a minimum feed-rate, and that the cutting speed has the highest impact on the quality of surface (Wang et al., 2013; Xiuli et al., 2010). Premnath et al. (2012) carried out a study on the surface

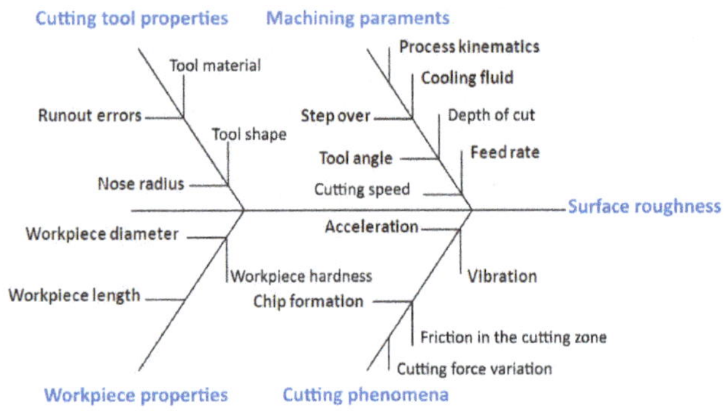

Figure 1. Fishbone diagram of the parameters that affect surface roughness [3].

roughness of combined aluminum surface by using predicted response of surface. The milling insert was tungsten carbide, and the variable parameters were rotational speed, and various compositions of aluminum (Premnath et al., 2012) Amran et al. did a study about the impact of machining parameters on the surface roughness of aluminum alloys during the drilling process. In their study, 20 tests were conducted with parameter variables including speed spindle, and drill diameter by surface response method (Amran et al., 2013). Kuttolamadom et al. (2010) studied the effect of feed-rate on the surface roughness of aluminum 6061 (Kuttolamadom et al., 2010). Lo et al. (2005) investigated the surface roughness of aluminum 6061 by three face-millings of various kinds by the Taguchi method and laser scattering. They finally concluded that high-speed steel face-mill had the best quality in terms of surface roughness (Lo et al., 2005).

Even though a large number of researches have been carried out on the parameters that affect surface roughness of Al5083 (Maeng et al., 2003), however, the simultaneous effect of recrystallization and cutting speed on surface roughness is still a challenge for researchers. Accordingly, this study aims to investigate the effect of machining and recrystallization parameters on Al5083 surface roughness. For this purpose, at first a series of tests were done on the aluminum 5083 sample to achieve optimal machining parameters for minimal surface roughness. Then, the effect of cutting speed on the surface roughness of crystallized aluminum 5083 sample was evaluated. Finally, the results of experiments were analyzed by using roughness measurement test, tensile measuring test, and also the images of microstructures.

2 Materials and methods

2.1 Sample preparation and microstructural study

Commercially available Al5083 plate with chemical composition as shown in Table 2 has been taken as the starting material in this research. For metallography and microstructural investigation, at first, samples with dimensions of 1 cm × 1 cm were prepared by using the abrasive paper No. 100 to 2000. Then the samples were polished with diamond and alumina powders. After polishing, all samples were etched in chemical solution containing 50 mL of Poulton's reagent was mixed with 25 mL of HNO_3 and 40 mL of a composition of 3 g of chromic acid with 10 mL of water (Singh et al., 2013) for 1 min.

In order to study microstructure, optical microscope (Model: Olympus) was employed.

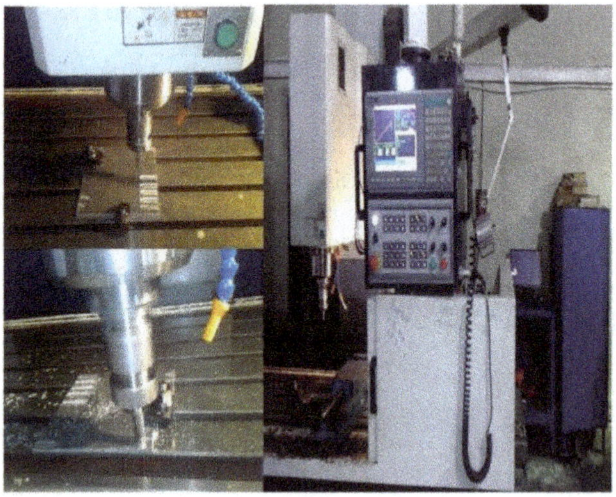

Figure 2. The CNC milling machine.

2.2 Mechanics of the cutting tool

For the tensile test the designing of the samples was done based on the ASTM E8 standard by using Solid Works 2013 software, and the programming for machining was done by using Power Mill 10 software. Then, the three-axis NC milling machine (Syntec 10A model) with the maximum spindle speed of 9000 rpm, feed-rate of 10 mm min^{-1}, 7.5 KW motor drive, and Siemens control was used for machining the samples. Figure 2 shows the machine used in this study.

The tools used for dry machining included carbide tools coated with TiAIN (Seco model) with a diameter of 12 mm, the tool tip radius of 0.8 mm, and a spiral angle of 30°.

In this study, a similar tool was used for both experiments so that the tool wear parameter does not affect the results.

2.3 Roughness and hardness measurements

Many factors affect the machined surface roughness such as workpiece variables, tool variables and machining process variables. One example of workpiece variables is hardness, and some examples of tools variables include quality, tip radius and geometry of cutting. The cutting speed and depth of cutting are also regarded as machining variables (Li and

Table 1. Characteristics of roughness measurement machine.

Hommelwerke T8000 profilometer	
Pick-up type	TK300
Measuring range	80 μm
Assessment length	4.80 mm
Speed	0.15 mm s^{-1}
Filter	M1 DIN 4777

Table 2. Chemical Composition of Al5083 (wt %).

Al	Cu	Si	Fe	Cr	Mn	Mg
Bal.	0.02	0.08	0.05	0.1	0.66	4.1

Table 3. Parameters and levels of test.

Factors	Unit	Level 1	Level 2	Level 3
Cutting speed	m min^{-1}	150	225	300
Feed rate	mm min^{-1}	100	150	200
Depth of cut	mm	0.5	1	1.5

Kishawy, 2006). In this study, in order to evaluate the surface roughness of the machined samples a roughness gauges device was used and its characteristics are given in Table 1. Hardness measuring was done by a universal hardness tester (uv1 model). At least three hardness measurements were done in each area and the average Rockwell B hardness was reported.

2.4 Design of experiment

In order to achieve a more accurate model of mutual impacts of three independent parameters of speed, feed-rate and depth of cutting on the dependent factor of surface roughness, full factorial experimental design was used. All the possible combinations of surface were considered, and the number of required tests equals the number of surfaces to the power of number of parameters ($3^3 = 27$). One surface was also considered for each parameter. Table 3 shows the investigated parameters of the test. In addition, Minitab 16 software was used to examine the impacts of data.

3 Results

The results obtained from the average surface roughness in Table 4 indicate that plastic deformation was facilitated and the friction decreased with the increase of the cutting speed. As a result, the asperities and the filled edge of surface created by machining was reduced to a minimum and the surface roughness was reduced. The improvement in surface quality was obtained by maximum cutting speed of 300 m min^{-1}. By increasing the feed-rate, the surface roughness increased so

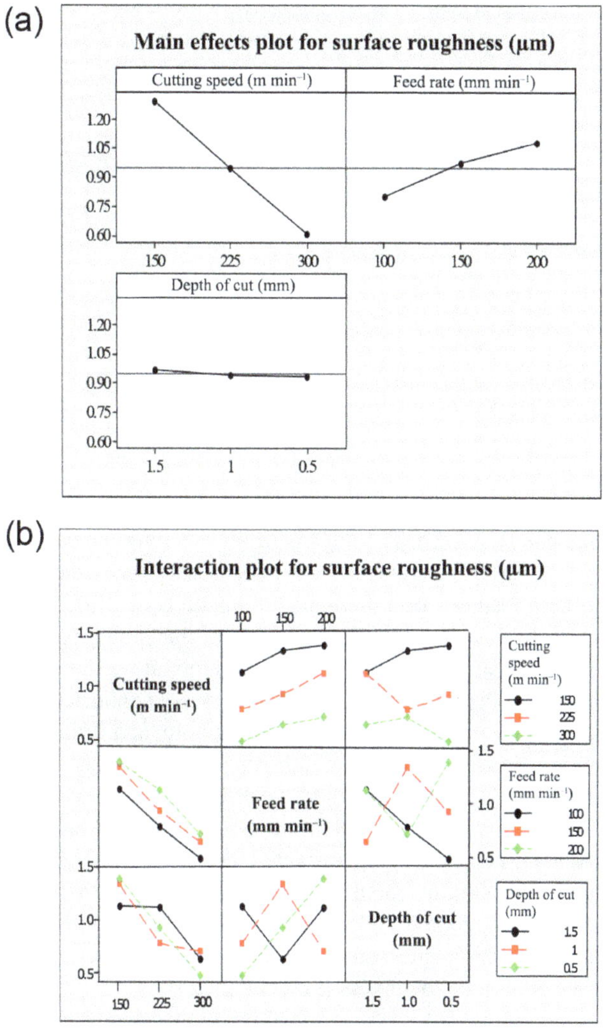

Figure 3. Individual parameters of machining affecting the surface quality: **(a)** main effects and **(b)** interactions.

that with the least feed-rate, i.e. 100 mm min^{-1}, the best quality of the surface was achieved. Also the depth of cut had no regular impact on the surface roughness of aluminum sample but with the least cutting depth, i.e. 0.5 mm, the best quality of surface was obtained. The average roughness for the original, rolled, and crystallized samples with different speeds is shown in Table 5.

4 Discussion

The main and mutual impacts of independent parameters on the surface roughness factor, which was designed by the Minitab 16 software, are shown in Fig. 3. As can be seen in this figure, cutting speed has the highest impact on surface roughness, which is due to the high difference in the height of points. Thus, by increasing of the cutting speed, the quality of surface roughness improved. As expected, with increasing of the feed-rate, the surface roughness increased and it was

Table 4. The results of average surface roughness by full factorial method.

Trial no.	Cutting speed ($mm\,min^{-1}$)	Feed rate ($mm\,min^{-1}$)	Depth of cut (mm)	Ra1 (μm)	Ra2 (μm)	Ra3 (μm)	Rmean (μm)
1	150	100	0.5	0.98	1.07	0.95	1.00
2	150	100	1	1.12	1.18	1.15	1.15
3	150	100	1.5	1.34	1.24	1.26	1.28
4	150	150	0.5	1.31	1.3	1.29	1.30
5	150	150	1	1.34	1.39	1.29	1.34
6	150	150	1.5	1.38	1.4	1.39	1.39
7	150	200	0.5	1.36	1.35	1.40	1.37
8	150	200	1	1.42	1.38	1.40	1.40
9	150	200	1.5	1.41	1.37	1.51	1.43
10	225	100	0.5	0.63	0.67	0.62	0.64
11	225	100	1	0.77	0.81	0.85	0.81
12	225	100	1.5	0.9	0.95	0.82	0.89
13	225	150	0.5	0.83	0.88	0.84	0.85
14	225	150	1	0.95	0.93	0.97	0.95
15	225	150	1.5	0.99	1	0.98	0.99
16	225	200	0.5	1.02	1.05	1.08	1.05
17	225	200	1	1.11	1.15	1.07	1.11
18	225	200	1.5	1.18	1.22	1.29	1.23
19	300	100	0.5	0.42	0.37	0.44	0.41
20	300	100	1	0.49	0.46	0.52	0.49
21	300	100	1.5	0.5	0.54	0.55	0.52
22	300	150	0.5	0.58	0.62	0.6	0.60
23	300	150	1	0.58	0.6	0.56	0.58
24	300	150	1.5	0.71	0.78	0.73	0.74
25	300	200	0.5	0.66	0.65	0.64	0.65
26	300	200	1	0.7	0.73	0.64	0.69
27	300	200	1.5	0.72	0.83	0.79	0.78

Table 5. The average roughness for the original, rolled, and crystallized samples with different speeds.

Trial no.	Cutting speed ($m\,min^{-1}$)	Feed rate ($mm\,min^{-1}$)	Depth of cut (mm)	Ra (μm) without rolled	Ra (μm) 50 % rolled	Ra (μm) annealed at 250 °C for 10 min	Ra (μm) annealed at 250 °C for 120 min	Ra (μm) annealed at 250 °C for 240 min
1	150	100	0.5	1	0.74	0.93	1.41	1.43
2	225	100	0.5	0.64	0.44	0.59	1.11	1.08
3	300	100	0.5	0.41	0.24	0.3	0.84	0.82

observed that the depth of cutting has little effect on surface smoothness.

According to Fig. 4, high hardness (72 RB) is observed in the 50 % cold-rolled sample due to the increased density of the sample. Hardness was reduced from min 2.5 to 5, but an increase can be seen in hardness of the samples from min 5 to 10: it increases from 55.87 RB in the original sample to its maximum of 61.9 RB after min 10. The hardness of samples also decreased from min 10 to 240.

Images of metallography in Fig. 5 show that the new coaxial grains are not formed in min 2.5.

A decrease can be seen in the size of grains from min 5 to 10, so that the size of grains reached to its minimum of 27 microns in min 10. The size of grains also increases from min 10 to 240 and finally reaches to 85 microns where it remains almost constant. Based on the results of hardness and metallographic images, recrystallization phenomenon occurred in min 10.

The reason for hardness variations can be explained in light of recrystallization phenomenon. After 50 % cold working on the sample, due to dramatic increase in density of dislocations and also residual energy in the sample, the hardness

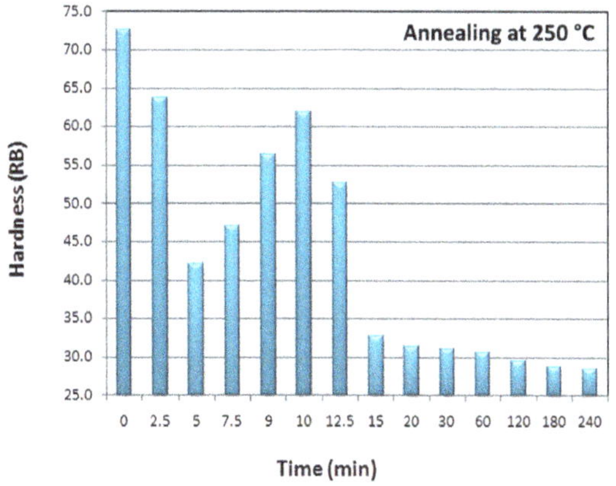

Figure 4. Hardness testing vs. annealing at 250 °C for the 50 % cold rolled sample at different times.

value reaches 72 RB. After annealing at 250 °C for 5 min, the internal energy decreases. At the same time, the density of dislocations decreases as well, which leads to the reduction of hardness to about 42 RB. As can be seen from the figure displaying microstructure (Fig. 5a), at 5 min the new recrystallized grains have not been formed yet and the decrease of hardness is due to the decrease in density of dislocations. By increasing the annealing time to 10 min, the sample hardness remarkably increases to 61.9 RB, which is due to the formation of recrystallized grains and decrease of grain size. According to Petch–Hall relation ($\sigma_Y = \sigma_i + K_Y/D^{1/2}$), the strength and hardness increase as the grain size becomes smaller because the grain boundaries act as barriers against the movement of dislocations, hence leading to the rise of strength as well as hardness.

It is worth considering that in order for recrystallization to be initiated, a minimum cold work is required. The more the percentage of cold working, the higher the energy and the greater the number of formed grains. Also, the structure will be more fine-grained. In our study, the amount of cold work was 50 %. As can be seen in Fig. 5b, the fine recrystallized grains are formed. These grains are smaller compared to those in Fig. 5a.

With increase of the annealing time from 10 to 240 min, the sample hardness dramatically declines to less than 30 RB, which is due to the extreme decrease of dislocations density and growth of recrystallized grains (Fig. 5c–g).

Dry machining was done under constant conditions for the samples annealed at 250 °C similar to the original sample and with optimized parameters. By investigation of Fig. 6, it becomes clear that the minimal surface roughness can be seen in min 10 which is due to the recrystallization phenomenon. Surface roughness increases from min 10 to 240 and remains almost constant.

As can be seen in Fig. 7, in the 50 % rolled sample, high strength and little strain can be observed. Due to the high strain and low strength, higher surface roughness is seen

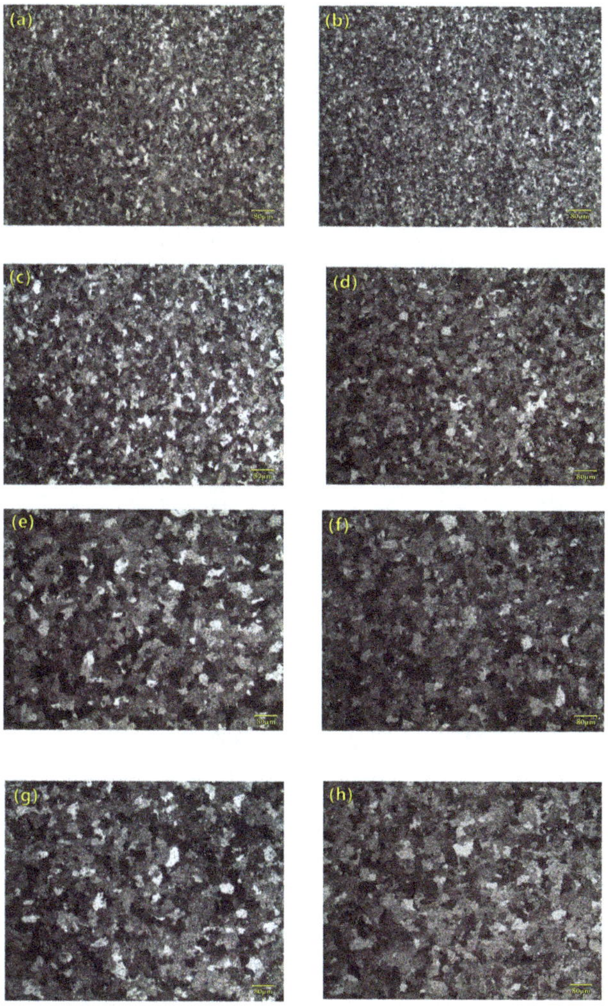

Figure 5. Metallographic samples annealed at 250 °C at different times: **(a)** 5 min, **(b)** 10 min, **(c)** 15 min, **(d)** 30 min, **(e)** 60 min, **(f)** 120 min, **(g)** 180 min and **(h)** 240 min.

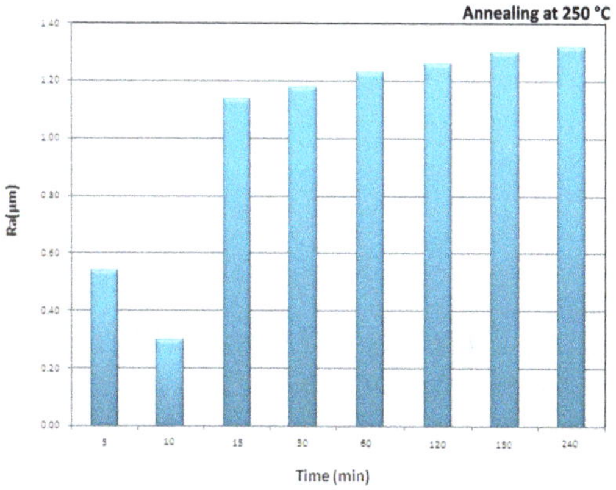

Figure 6. Surface roughness of the samples annealed at 250 °C at different times.

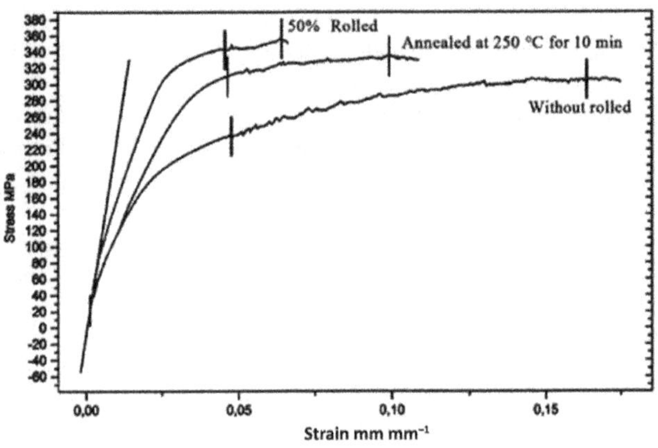

Figure 7. The strain-stress test for the original, crystallized, and 50 % cold rolled samples.

in the original sample compared to the crystallized sample; however, better surface roughness is seen in the crystallized sample.

The machining was done on the crystallized samples with different cutting speeds and optimized parameters of feed-rate and constant cutting depth. By comparison of roughness measurements in Table 5 it becomes clear that surface roughness decreases as the speed increases. In addition, by comparing roughness of the original sample with the crystallized one, it can be concluded that the surface roughness in the crystallized sample is less than the original sample due to the increase of hardness and the decrease in the grain size.

5 Conclusions

The effect of the parameters of the cutting speed, feed-rate and depth of cutting on the surface quality of aluminum 5083 was achieved by "full factorial" design method. Also, the effect of the cutting speed on the surface roughness of crystallized aluminum 5083 was investigated. The results of the analysis are as follows:

1. The minimum roughness (0.41 micron) was observed in experiment 19. From the results obtained from machining, it can be said that in order to reach an ideal surface roughness, maximum cutting speed ($300\,\mathrm{mm\,min^{-1}}$), minimum feed-rate ($100\,\mathrm{mm\,min^{-1}}$), and minimum cutting depth (0.5 mm) should be taken into consideration among the current machining parameters.

2. With 50 % of cold working and annealing temperature of 250 °C, recrystallization starts at 10 min.

3. In the recrystallized sample, because of minimum grain size and maximum hardness (61.9 RB), the best surface quality (i.e. minimum roughness: 0.41 μm) was achieved.

Acknowledgements. The authors are thankful to Islamic Azad University of Najafabad for help and support.

References

Amran, M. A., Salmah, S., Hussein, N. I. S., Izamshah, R., Hadzley, M., Sivaraos, Kasim, M. S., and Sulaiman, M. A.: Effects of machine parameters on surface roughness using response surface method in drilling process, MITC, 18–20 November 2013, Malaysia, 2013.

Arokiadass, R., Palaniradja, K., and Alagumoorthi, N.: Surface roughness prediction model in end milling of Al/SiCP MMC by carbide tools, Int. J. Eng. Sci. Technol., 3, 78–87, 2011.

Colak, O., Kurbanoglu, C., and Kayacan, M. C.: Milling surface roughness prediction using evolutionary programming methods, Materials Design, 28, 657–666, 2007.

CRTD: Aluminium product applications in transportation and industry, ASME Centre for Research and Technology Development, 29 pp., 1994.

Kiswanto, G., Zariatin, D. L., and Ko, T. J.: The effect of spindle speed, feed-rate and machining time to the surface roughness and burr formation of Aluminium Alloy 1100 in micro-milling operation, J. Manufact. Process., 31, 231–242, 2014.

Kuttolamadom, M. A., Hamzehlouia, S., and Laine Mears, M.: Effect of Machining Feed on Surface Roughness in Cutting 6061 Aluminium, International Centre for Automotive Research, 2010.

Li, L. and Kishawy, H. Y.: A model for cutting forces generated during machining with self-propelled rotary tools, Int. J. Mach. Tools Manufact., 46, 1388–1394, 2006.

Lo, S. P., Chiu, J. T., and Lin, H. Y.: Rapid measurement of surface roughness for face-milling aluminium using laser scattering and the Taguchi method, Int. J. Adv. Manuf. Technol., 26, 1071–1077, 2005.

Maeng, D. Y., Lee, J. H., and Hong, S. I.: The effect of transition elements on the superplastic behavior of Al-Mg alloys, Mat. Sci. Eng., A357, 188–195, 2003.

Mahesh, T. P. and Rajesh, R.: Optimal Selection of process parameters in CNC end milling of Al 7075-T6 aluminium alloy using a Taguchi-Fuzzy approach, AMME, 27–29 May 2014, Egypt, 2014.

Nair, A. and Govindan, P.: Multiple Surface Roughness Characteristics Optimization in CNC End Milling of Aluminium using PCA, Int. J. Res. Mech. Eng. Technol., 3, 17–21, 2013.

Oktem, H., Erzurumlu, T., and Kurtaran, H.: Application of response surface methodology in the optimization of cutting conditions for surface roughness, J. Mater. Process. Technol., 170, 11–16, 2005.

Pinar, A. M.: Optimization of Process Parameters with Minimum Surface Roughness in the Pocket Machining of AA5083 Aluminium Alloy via Taguchi Method, Arab. J. Sci. Eng., 38, 705–714, 2013.

Premnath, A. A., Alwarsamy, T., Abhinnav, T., and Krishnakant, C. A.: Surface Roughness Prediction by Response Surface Methodology in Milling of Hybrid Aluminium composites, Interna-

tional Conference on Modelling Optimisation and Computing, 10–11 April 2012, Kumarakoil, 2012.
Rawangwong, S., Chatthong, J., Boonchouytan, W., and Burapa, R.: An Investigation of Optimum Cutting Conditions in Face Milling Aluminium Semi Solid 2024 Using Carbide Tool, 10th EMSES2012, 5–8 December 2012, Muang, Ubon-Ratchathani, Thailand, 2012.
Rawangwong, S., Chatthong, J., Boonchouytan, W., and Burapa, R.: Influence of Cutting Parameters in Face Milling Semi-Solid AA 7075 Using Carbide Tool Affected the Surface Roughness and Tool Wear, 11th EMSES2014, 21 December 2014, Thailand 2014.
Routara, B. C., Bandyopadhyay, A., and Sahoo, P.: Roughness modelling and optimization in CNC end milling using response surface method: effect of workpiece material variation, Int. J. Adv. Manuf. Technol., 40, 1166–1180, 2009.
Singh, D., Rao, P. N., and Jayaganthan, R.: Microstructures and impact toughness behavior of Al 5083 alloy processed by cry rolling and afterwards annealing, Int. J. Miner. Metal. Mater., 20, 34–42, 2013.
Sukumar, M. S., Ramaiah, P. V., and Nagarjuna, A.: Optimization and Prediction of Parameters in Face Milling of Al-6061 Using Taguchi and ANN Approach, 12th GCMM, 8–10 December 2014, Vellore, India, 2014.
Tammineni, L. and Yedula, H. P. R.: Investigation of influence of milling parameters on surface roughness and flatness, Int. J. Adv. Eng. Technol., 6, 2416, 2014.
Vakondios, D., Kyratsis, P., Yaldiz, S., and Antoniadis, A.: Influence of milling strategy on the surface roughness in ball end milling of the aluminium alloy Al7075-T6, Measurement, 45, 1480–1488, 2012.
Wang, M. Y. and Chang, H. Y.: Experimental study of surface roughness in slot end milling AL2014-T6, Int. J. Mach. Tools Manufact., 44, 51–57, 2004.
Wang, T., Xie, L. J., Wang, X. B., Jiao, L., Shen, J. W., Xu, H., and Nie, F. M.: Surface integrity of high speed milling of Al/SiC/65p aluminium matrix composites, 14th CIRP CMMO2013, 13–14 June 2013, Turin, Italy, 2013.
Xiuli, F., Yongzhi, P., Yi, W., and Xing, A.: Research on Predictive Model Surface Roughness in High Speed Milling for Aluminium Alloy 7050-T7451, IEEE, 5–6 June 2010, Wuhan, 186–189, 2010.
Zhang, J. Z., Chen, J. C., and Kirby, E. D.: Surface roughness optimization in an end-milling operation using the Taguchi design method, J. Mater. Process. Technol., 184, 233–239, 2007.

13

A two-stage calibration method for industrial robots with joint and drive flexibilities

M. Neubauer[1], H. Gattringer[1], A. Müller[1], A. Steinhauser[1], and W. Höbarth[2]

[1]Institute of Robotics, Johannes Kepler University Linz, Altenbergestr. 69, 4040 Linz, Austria
[2]Bernecker + Rainer Industrie Elektronik Ges.m.b.H., B & R Str. 1, 5142 Eggelsberg, Austria

Correspondence to: M. Neubauer (matthias.neubauer_1@jku.at)

Abstract. Dealing with robot calibration the neglection of joint and drive flexibilities limit the achievable positioning accuracy significantly. This problem is addressed in this paper. A two stage procedure is presented where elastic deflections are considered for the calculation of the geometric parameters. In the first stage, the unknown stiffness and damping parameters are identified. To this end the model based transfer functions of the linearized system are fitted to captured frequency responses of the real robot. The real frequency responses are determined by exciting the system with periodic multisine signals in the motor torques. In the second stage, the identified elasticity parameters in combination with the measurements of the motor positions are used to compute the real robot pose. On the basis of the estimated pose the geometric calibration is performed and the error between the estimated end-effector position and the real position measured with an external sensor (laser-tracker) is minimized. In the geometric model, joint offsets, axes misalignment, length errors and gear backlash are considered and identified. Experimental results are presented, where a maximum end-effector error (accuracy) of 0.32 mm and for 90 % of the poses a maximum error of 0.23 mm was determined (Stäubli TX90L).

1 Introduction

One of the main characteristics of industrial robots is their positioning accuracy, strongly depending on the sensor resolution and the geometric parameters. The calculation of the real geometric parameters is called geometric robot calibration and is crucial for accurate robot movements. In literature different calibration methods exist (see Khalil and Dombre, 2004 or Siciliano and Khatib, 2008). However, they are dealing with kinematic models only, neglecting the effects of flexibilities in the joints and drives and thus cause a systematic error in the calculation of the real geometric parameters. This systematic error limits the achievable positioning accuracy of the robot essentially. Only a few publications where flexibilities of the robot in the calibration are considered exist, see Whitney et al. (1986), Gong et al. (2000) or Khalil and Besnard (2002). In the paper Whitney et al. (1986) similar to our contribution a two stage method is proposed. They first identified a compliance model by performing multiple experiments with different external forces and then secondly the geometric calibration is performed by including displacements according to the identified model from stage one. The papers of Gong et al. (2000) and Khalil and Besnard (2002) include the elastic displacements in the calculation of the geometric parameters but assumed that the stiffness is known. The goal of this contribution is to present a procedure, which in the first stage identifies the main flexibilities of our robot without external hardware. To this end, the methods presented in Hardeman (2008) and Wernholt (2007) are implemented which both performed frequency domain identifications of industrial robots. In the second stage, the elastic deflections are considered in the geometric calibration leading to reliable geometric parameters.

In Sect. 2 the dynamic model of our 6-axis articulated robot with joint and drive elasticities is derived. Section 3 deals with the modeling of the geometric error parameters, like joint offsets, axes misalignment, length errors and gear backlash. Subsequently, in Sect. 4 the elasticity parameters are identified, using a frequency domain approach as in the work of Hardeman (2008) and Wernholt (2007). The identification of the robots frequency response matrix is presented,

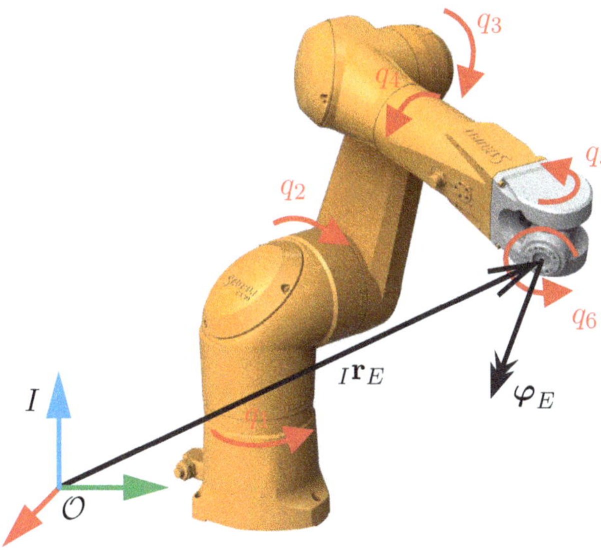

Figure 1. Arm coordinates of the Stäubli TX90L.

where the computation of suitable excitation signals are of prime importance. Then the matching between the transfer matrix of the dynamic model and the identified robot frequency response matrix is discussed. Finally in Sect. 5 the computation of the geometric parameters is addressed and the results for an industrial robot Stäubli TX90L are presented.

2 Dynamic modeling

This section deals with derivation of a dynamic model for the 6-axis industrial robot depicted in Fig. 1. Looking at the mechanical setup, the flexibilities of the first three joints and drives influence the positioning accuracy of the robots end-effector essentially. Hence, a finite bearing and drive stiffness is modeled leading to 4 degrees of freedom (DOF) for each axis $i = \{1, 2, 3\}$. The 4 DOF are composed of the motor position $q_{i\mathrm{M}}$, the small bearing distortions q_{ix} and q_{iy} and the arm rotation q_i (see Fig. 2). The bearing distortions and the arm rotation are combined in the vector $\boldsymbol{q}_i^T = [q_{ix}, q_{iy}, q_i]$ representing the arm orientation. The modeling of flexible joints is motivated by the work of Hardeman (2008) which showed that the bearing and drive stiffness is at the same order of magnitude. Elastic effects of the last three axes (the spherical wrist) are of minor importance w.r.t. the end-effector positioning accuracy and are thus modeled without flexibilities. For the derivation of the dynamic model, the transformation from the body fixed frame of the previous arm (p) into the body fixed frame of the current arm (c) is necessary. Hence, the rotation matrix

$$\mathbf{R}_{\mathrm{cp}} = \mathbf{R}_\alpha\big|_{\alpha=q_{ix}} \mathbf{R}_\beta\big|_{\beta=q_{iy}} \mathbf{R}_\gamma\big|_{\gamma=q_i} \tag{1}$$

is introduced, where successive rotations about the main axes are performed. The rotation matrix for a rotation about the x, y or z axis is given by

$$\mathbf{R}_\alpha = \begin{bmatrix} 1 & 0 & 0 \\ 0 & \cos(\alpha) & \sin(\alpha) \\ 0 & -\sin(\alpha) & \cos(\alpha) \end{bmatrix},$$

$$\mathbf{R}_\beta = \begin{bmatrix} \cos(\beta) & 0 & -\sin(\beta) \\ 0 & 1 & 0 \\ \sin(\beta) & 0 & \cos(\beta) \end{bmatrix} \quad \text{or} \tag{2}$$

$$\mathbf{R}_\gamma = \begin{bmatrix} \cos(\gamma) & \sin(\gamma) & 0 \\ -\sin(\gamma) & \cos(\gamma) & 0 \\ 0 & 0 & 1 \end{bmatrix}, \tag{3}$$

respectively. The modeling is based on the paper of Öhr et al. (2006) where consecutive rotations are suggested to model joint and drive compliance. In total, the robot is represented by the vector $\boldsymbol{q}^T = [\boldsymbol{q}_{\mathrm{M}}^T, \boldsymbol{q}_{\mathrm{A}}^T, \boldsymbol{q}_{\mathrm{SW}}^T]$ consisting of the motor coordinates $\boldsymbol{q}_{\mathrm{M}}^T = [q_{1M}, q_{2M}, q_{3M}]$ the arm coordinates $\boldsymbol{q}_{\mathrm{A}}^T = [\boldsymbol{q}_1^T, \boldsymbol{q}_2^T, \boldsymbol{q}_3^T]$ and the coordinates of the spherical wrist $\boldsymbol{q}_{\mathrm{SW}}^T = [q_4, q_5, q_6]$. For the dynamic model, the bearing stiffness is modeled with linear springs and dampers c_{ix}, c_{iy} and d_{ix}, d_{iy} respectively and the drive stiffness and damping with the parameters c_i and d_i for axis $i = \{1, 2, 3\}$. For further details about the dynamical modeling see Öhr et al. (2006). Finally, the dynamic model is derived with the Projection Equation, see Bremer (2008) leading to the equation of motion

$$\mathbf{M}(\boldsymbol{q})\ddot{\boldsymbol{q}} + \boldsymbol{g}\left(\boldsymbol{q}, \dot{\boldsymbol{q}}, \boldsymbol{p}_{\mathrm{elast}}\right) = \mathbf{B}_{\mathrm{M}}\boldsymbol{\tau}_{\mathrm{M}} + \mathbf{B}_{\mathrm{SW}}\boldsymbol{\tau}_{\mathrm{SW}} = \begin{bmatrix} \mathbf{I} \\ \mathbf{0} \\ \mathbf{0} \end{bmatrix} \boldsymbol{\tau}_{\mathrm{M}} + \begin{bmatrix} \mathbf{0} \\ \mathbf{0} \\ \mathbf{I} \end{bmatrix} \boldsymbol{\tau}_{\mathrm{SW}} \tag{4}$$

with the mass matrix $\mathbf{M}(\boldsymbol{q}) \in \mathbb{R}^{15\times15}$, the motor torques of the first three axis $\boldsymbol{\tau}_{\mathrm{M}} \in \mathbb{R}^{3\times1}$ and the motor torques of the spherical wrist $\boldsymbol{\tau}_{\mathrm{SW}} \in \mathbb{R}^{3\times1}$. Throughout the whole paper, the identity matrix is represented by $\mathbf{I}$ and the zero matrix/vector by $\mathbf{0}$. The vector $\boldsymbol{g}(\boldsymbol{q}, \dot{\boldsymbol{q}}, \boldsymbol{p}_{\mathrm{elast}})$ contains the Coriolis, centrifugal, gravitation and friction forces as well as the stiffness and damping parameters which are going to be identified and are combined in the vector

$$\boldsymbol{p}_{\mathrm{elast}}^T = \big[c_{1x}, c_{1y}, c_1, c_{2x}, c_{2y}, c_2, c_{3x}, c_{3y}, c_3, d_{1x}, d_{1y}, d_1, d_{2x}, d_{2y}, d_2, d_{3x}, d_{3y}, d_3\big].$$

It is apparent from Eq. (4), that we are dealing with an underactuated system.

2.1 Linearized dynamic model

Basis for the calculation of the transfer matrix is the linearization of the dynamic model. A static equilibrium $\boldsymbol{q}_0$ ($\dot{\boldsymbol{q}}_0 = \mathbf{0}, \ddot{\boldsymbol{q}}_0 = \mathbf{0}$) is used as linearization point. The computation of the static equilibrium is presented in Sect. 2.3. For the

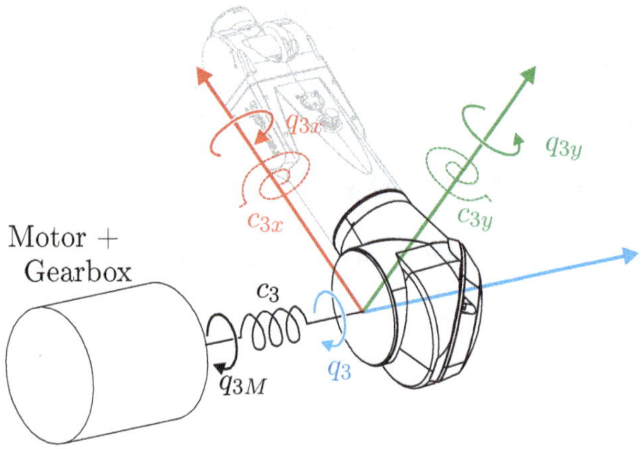

Figure 2. Model of the third axis.

linearization, a Taylor series expansion of Eq. (4) at the equilibrium configuration $\boldsymbol{q}_0$ up to order 1 is carried out, leading to

$$\mathbf{M}_{\text{lin}}\Delta\ddot{\boldsymbol{q}} + \mathbf{D}_{\text{lin}}\Delta\dot{\boldsymbol{q}} + \mathbf{K}_{\text{lin}}\Delta\boldsymbol{q} = \mathbf{B}_{\text{M}}\Delta\boldsymbol{\tau}_{\text{M}} + \mathbf{B}_{\text{SW}}\Delta\boldsymbol{\tau}_{\text{SW}} - \left(\boldsymbol{g}\left(\boldsymbol{q}_0, \mathbf{0}, \boldsymbol{p}_{\text{elast}}\right) - \mathbf{B}_{\text{M}}\boldsymbol{\tau}_{\text{M0}} - \mathbf{B}_{\text{SW}}\boldsymbol{\tau}_{\text{SW0}}\right) \quad (5)$$

with the mass matrix

$\mathbf{M}_{\text{lin}} = \mathbf{M}\left(\boldsymbol{q}_0\right)$, the damping matrix $\mathbf{D}_{\text{lin}} = \left.\frac{\partial \boldsymbol{g}\left(\boldsymbol{q}, \dot{\boldsymbol{q}}, \boldsymbol{p}_{\text{elast}}\right)}{\partial \dot{\boldsymbol{q}}}\right|_{\boldsymbol{q}=\boldsymbol{q}_0, \dot{\boldsymbol{q}}=0}$,

the stiffness matrix $\mathbf{K}_{\text{lin}} = \left.\frac{\partial \mathbf{M}(\boldsymbol{q})\ddot{\boldsymbol{q}} + \boldsymbol{g}\left(\boldsymbol{q}, \dot{\boldsymbol{q}}, \boldsymbol{p}_{\text{elast}}\right)}{\partial \boldsymbol{q}}\right|_{\boldsymbol{q}=\boldsymbol{q}_0, \dot{\boldsymbol{q}}=0, \ddot{\boldsymbol{q}}=0}$

and $\boldsymbol{\tau}_{\text{M}} = \boldsymbol{\tau}_{\text{M0}} + \Delta\boldsymbol{\tau}_{\text{M}}$ and $\boldsymbol{\tau}_{\text{SW}} = \boldsymbol{\tau}_{\text{SW0}} + \Delta\boldsymbol{\tau}_{\text{SW}}$. The vectors $\boldsymbol{\tau}_{\text{M0}}$ and $\boldsymbol{\tau}_{\text{SW0}}$ represent the constant motor torques in the static configuration $\boldsymbol{q}_0$. Thus, if $\boldsymbol{q}_0$ is a valid static configuration the term $\boldsymbol{g}(\boldsymbol{q}_0, \mathbf{0}, \boldsymbol{p}_{\text{elast}}) - \mathbf{B}_{\text{M}}\boldsymbol{\tau}_{\text{M0}} - \mathbf{B}_{\text{SW}}\boldsymbol{\tau}_{\text{SW0}}$ vanishes, and we get the linearized equations of motion

$$\mathbf{M}_{\text{lin}}\Delta\ddot{\boldsymbol{q}} + \mathbf{D}_{\text{lin}}\Delta\dot{\boldsymbol{q}} + \mathbf{K}_{\text{lin}}\Delta\boldsymbol{q} = \mathbf{B}_{\text{M}}\Delta\boldsymbol{\tau}_{\text{M}} + \mathbf{B}_{\text{SW}}\Delta\boldsymbol{\tau}_{\text{SW}}. \quad (6)$$

2.2 Motor transfer matrix

For the identification of the modeled elasticity parameters, see Sect. 4, the transfer matrix from the motor torques $\boldsymbol{\tau}_{\text{M}}$ to the motor accelerations $\ddot{\boldsymbol{q}}_{\text{M}}$ is necessary (Only the first three axes are included). By using the motor accelerations the double integrator behavior in the transfer matrix is avoided. Starting with the linearized robot dynamics of Eq. (6), the system is reduced to the coordinates $\boldsymbol{q}_{\text{M}}$ by using the selection matrix $\mathbf{F}_{\text{M}}^T = [\mathbf{I}, \mathbf{0}]$ and the relation $\Delta\boldsymbol{q}_{\text{M}} = \mathbf{F}_{\text{M}}^T\Delta\boldsymbol{q}$. Using the principle of virtual work, we get the linearized motor equations

$$\mathbf{F}_{\text{M}}^T\mathbf{M}_{\text{lin}}\mathbf{F}_{\text{M}}\Delta\ddot{\boldsymbol{q}}_{\text{M}} + \mathbf{F}_{\text{M}}^T\mathbf{D}_{\text{lin}}\mathbf{F}_{\text{M}}\Delta\dot{\boldsymbol{q}}_{\text{M}} + \mathbf{F}_{\text{M}}^T\mathbf{K}_{\text{lin}}\mathbf{F}_{\text{M}}\Delta\boldsymbol{q}_{\text{M}} = \underbrace{\mathbf{F}_{\text{M}}^T\mathbf{B}_{\text{M}}}_{I}\Delta\boldsymbol{\tau}_{\text{M}}. \quad (7)$$

The motor torques of the spherical wrist $\boldsymbol{\tau}_{\text{SW}}$ vanish because $\mathbf{F}_{\text{M}}^T\mathbf{B}_{\text{SW}} = \mathbf{0}$. Applying the Laplace transformation to Eq. (7) we get

$$\mathbf{F}_{\text{M}}^T\mathbf{M}_{\text{lin}}\mathbf{F}_{\text{M}}\Delta\overline{\boldsymbol{a}}_{\text{M}} + \mathbf{F}_{\text{M}}^T\mathbf{D}_{\text{lin}}\mathbf{F}_{\text{M}}\Delta\overline{\boldsymbol{a}}_{\text{M}}\frac{1}{s} + \mathbf{F}_{\text{M}}^T\mathbf{K}_{\text{lin}}\mathbf{F}_{\text{M}}\Delta\overline{\boldsymbol{a}}_{\text{M}}\frac{1}{s^2} = \Delta\overline{\boldsymbol{\tau}}_{\text{M}} \quad (8)$$

with $\Delta\boldsymbol{a}_{\text{M}} = \Delta\ddot{\boldsymbol{q}}_{\text{M}}$. The vectors $\Delta\overline{\boldsymbol{a}}_{\text{M}}$ and $\Delta\overline{\boldsymbol{\tau}}_{\text{M}}$ are the Laplace transformed values of $\Delta\boldsymbol{a}_{\text{M}}$ and $\Delta\boldsymbol{\tau}_{\text{M}}$, respectively. Thus the transfer matrix leads to

$$\mathbf{G}^{[\boldsymbol{q}_0]} = \Delta\overline{\boldsymbol{a}}_{\text{M}}\Delta\overline{\boldsymbol{\tau}}_{\text{M}}^{-1} = \left(\mathbf{F}_{\text{M}}^T\mathbf{M}_{\text{lin}}\mathbf{F}_{\text{M}}s^2 + \mathbf{F}_{\text{M}}^T\mathbf{D}_{\text{lin}}\mathbf{F}_{\text{M}}s + \mathbf{F}_{\text{M}}^T\mathbf{K}_{\text{lin}}\mathbf{F}_{\text{M}}\right)^{-1}\mathbf{I}s^2 \quad (9)$$

where the superscript $\boldsymbol{q}_0$ denotes the linearization point.

2.3 Determination of a static equilibrium

The calculation of a static equilibrium $\boldsymbol{q}_0$ for our underactuated system, see Eq. (4) is neccessary for two different reasons. First, for the correct evaluation of the transfer matrix by using the correct linearization point, and second for the evaluation of the real robot pose and thus for the real end-effector position and orientation in a static configuration.

The motor position $\boldsymbol{q}_{\text{M}}$ and the coordinates of the spherical wrist $\boldsymbol{q}_{\text{SW}}$ are measured. Hence, only the static values of the arm coordinates $\boldsymbol{q}_{\text{A}}$ have to be computed. Therefore the linear system Eq. (5) is reduced to the coordinates of interest i.e. the arm coordinates $\boldsymbol{q}_{\text{A}}$, by again applying the principal of virtual work and the relation $\Delta\boldsymbol{q}_{\text{A}} = \mathbf{F}_{\text{A}}^T\Delta\boldsymbol{q}$ with $\mathbf{F}_{\text{A}}^T = [\mathbf{0}_{9\times3}, \mathbf{I}, \mathbf{0}_{9\times3}]$. Assuming a static solution – time derivatives of $\Delta\boldsymbol{q}$ are zero – we get

$$\mathbf{F}_{\text{A}}^T\mathbf{K}_{\text{lin}}\mathbf{F}_{\text{A}}\Delta\boldsymbol{q}_{\text{A}} = \mathbf{F}_{\text{A}}^T\mathbf{B}_{\text{M}}\Delta\boldsymbol{\tau}_{\text{M}} + \mathbf{F}_{\text{A}}^T\mathbf{B}_{\text{SW}}\Delta\boldsymbol{\tau}_{\text{SW}} - \mathbf{F}_{\text{A}}^T\left(\boldsymbol{g}\left(\boldsymbol{q}_0, \mathbf{0}, \boldsymbol{p}_{\text{elast}}\right) - \mathbf{B}_{\text{M}}\boldsymbol{\tau}_{\text{M0}} - \mathbf{B}_{\text{SW}}\boldsymbol{\tau}_{\text{SW0}}\right). \quad (10)$$

Since the arm coordinates are not actuated $\mathbf{F}_{\text{A}}^T\mathbf{B}_{\text{M}} = \mathbf{F}_{\text{A}}^T\mathbf{B}_{\text{SW}} = \mathbf{0}$, Eq. (10) leads to

$$\mathbf{F}_{\text{A}}^T\mathbf{K}_{\text{lin}}\mathbf{F}_{\text{A}}\Delta\boldsymbol{q}_{\text{A}} = -\mathbf{F}_{\text{A}}^T\boldsymbol{g}\left(\boldsymbol{q}_0, \mathbf{0}, \boldsymbol{p}_{\text{elast}}\right). \quad (11)$$

If $\boldsymbol{q}_0$ is a static pose of our robot, then $\boldsymbol{g}(\boldsymbol{q}_0, \mathbf{0}, \boldsymbol{p}_{\text{elast}}) = \mathbf{0}$ otherwise the displacement $\Delta\boldsymbol{q}_{\text{A}}$ is calculated by

$$\Delta\boldsymbol{q}_{\text{A}} = -\left(\mathbf{F}_{\text{A}}^T\mathbf{K}_{\text{lin}}\mathbf{F}_{\text{A}}\right)^{-1}\mathbf{F}_{\text{A}}^T\boldsymbol{g}\left(\boldsymbol{q}_0, \mathbf{0}, \boldsymbol{p}_{\text{elast}}\right) \quad (12)$$

according to Eq. (11). Thus the static solution $\boldsymbol{q}_0$ can be calculated iteratively by

$$\boldsymbol{q}_0^{(i+1)} = \boldsymbol{q}_0^{(i)} + \mathbf{F}_{\text{A}}\Delta\boldsymbol{q}_{\text{A}} \quad (13)$$

with the iteration index i. Because the stiffness of our robot is quite high, the initial solution

$$\boldsymbol{q}_0^{(0)} = [q_{1\mathrm{M}}, q_{2\mathrm{M}}, q_{3\mathrm{M}}, 0, 0, q_{1\mathrm{M}}, 0, 0, q_{2\mathrm{M}}, 0, 0, q_{3\mathrm{M}}, q_4, q_5, q_6]^T \quad (14)$$

is obvious and often one iteration is sufficient to obtain a valid static solution. Having a valid static solution calculated, a geometric model is needed to evaluate the corresponding end-effector position and orientation.

3 Geometric model

The forward kinematics describes the end-effectors position ${}_I\boldsymbol{r}_\mathrm{E}$ and orientation $\boldsymbol{\varphi}_\mathrm{E}$ (see Fig. 1). Furthermore, it depends on the arm and wrist coordinates, $\boldsymbol{q}_\mathrm{A}$ and $\boldsymbol{q}_\mathrm{SW}$ respectively, and also on the geometric parameters. In this paper, the geometric parameters are separated into the known nominal values $\boldsymbol{p}_\mathrm{nom}$ which describe the nominal kinematics of the robot and the unknown geometric error parameters $\boldsymbol{p}_\mathrm{ge}$ comprising joint offsets, axes misalignment, length errors, and gear backlash. To model a joint offset p_0 for a rotation around the x axis with the DOF q the rotation matrix $\mathbf{R}$ leads to

$$\mathbf{R} = \mathbf{R}_\alpha|_{\alpha=q+p_0}. \quad (15)$$

Axes misalignment, also for the example of a rotation around the x axis are introduced by adapting the rotation matrix $\mathbf{R}$ to

$$\mathbf{R} = \mathbf{R}_\alpha|_{\alpha=q}\,\mathbf{R}_\beta\big|_{\beta=p_\beta}\,\mathbf{R}_\gamma\big|_{\gamma=p_\gamma} \quad (16)$$

with p_β and p_γ as misalignment angles. To include length errors the connection vector $\boldsymbol{r}$ which describes the links of our robot, is extended by the length error parameters $p_{\mathrm{l}x}$, $p_{\mathrm{l}y}$ and $p_{\mathrm{l}z}$ leading to

$$\boldsymbol{r} = \begin{bmatrix} l_x + p_{\mathrm{l}x} \\ l_y + p_{\mathrm{l}y} \\ l_z + p_{\mathrm{l}z} \end{bmatrix}. \quad (17)$$

Note, the parameters l_x, l_y and l_z are the nominal values of $\boldsymbol{r}$. In contrast to the previous errors, gear backlash is not a geometric error and depend on the pose of the robot. In order to identify the direction in which the clearance is present, the center of mass of each body must be considered. To avoid the evaluation of the body dynamics the decision is based on the sign of the motor torque τ_M in each axis leading to the rotation matrix

$$\mathbf{R} = \mathbf{R}_\alpha|_{\alpha=q+\mathrm{sign}(\tau_\mathrm{M})p_\mathrm{BL}} \text{ with } \mathrm{sign}(\tau_\mathrm{M}) = \begin{cases} -1 & \tau_\mathrm{M} < 0 \\ 0 & \tau_\mathrm{M} = 0 \\ 1 & \tau_\mathrm{M} > 0 \end{cases}. \quad (18)$$

In the final geometric model of our robot, joint offsets, axes misalignments and length errors combined in the vector $\boldsymbol{p}_i = [p_{0i}, p_{\beta i}, p_{\gamma i}, p_{\mathrm{l}xi}, p_{\mathrm{l}yi}, p_{\mathrm{l}zi}]^T$ for all axes $i = \{1, \ldots, 6\}$ as well as for the end-effector tool $i = E$ are included. Gear backlash $p_{\mathrm{BL}i}$ is only modeled for the second and third axes $i = \{2, 3\}$. The offset of the inertial frame of the robot and the external sensor (laser-tracker) is modeled by the length errors $p_{\mathrm{l}xI}$, $p_{\mathrm{l}yI}$ and $p_{\mathrm{l}zI}$. Thus finally, 47 error parameters combined in the vector $\boldsymbol{p}_\mathrm{ge} = [p_{\mathrm{l}xI}, p_{\mathrm{l}yI}, p_{\mathrm{l}zI}, \boldsymbol{p}_1^T, \boldsymbol{p}_2^T, p_\mathrm{BL2}, \boldsymbol{p}_3^T, p_\mathrm{BL3}, \boldsymbol{p}_4^T, \boldsymbol{p}_5^T, \boldsymbol{p}_6^T, \boldsymbol{p}_\mathrm{E}^T]^T$ describe the end-effector position ${}_I\boldsymbol{r}_\mathrm{E}(\boldsymbol{q}_\mathrm{A}, \boldsymbol{q}_\mathrm{SW}, \boldsymbol{p}_\mathrm{nom}, \boldsymbol{p}_\mathrm{ge})$ and orientation $\boldsymbol{\varphi}_\mathrm{E}(\boldsymbol{q}_\mathrm{A}, \boldsymbol{q}_\mathrm{SW}, \boldsymbol{p}_\mathrm{nom}, \boldsymbol{p}_\mathrm{ge})$ of the robot. Since the end-effector position and orientation depend on the arm coordinates, and the arm coordinates highly depend on the elasticity parameters their identification is treated in the next section.

4 Identification of elasticity parameters

For the identification of elasticity parameters, first the non-parametric frequency response matrix (FRM) see Pintelon and Schoukens (2012) of the real robot is determined, and second the transfer matrix calculated in Eq. (9) is adapted to the real one by adapting the elasticity parameters.

4.1 Identification of the non-parametric frequency response matrix

In standard operation mode, the robot is controlled by a feedback PD controller. To improve the identification accuracy, the influence of the PD controller is reduced by using small feedback gains. To ensure that the robot maintains the position, feed forward torque calculated from the inverse dynamics is added. The measurement setup is depicted in Fig. 3. The system is excited with periodic motor torques while simultaneously the actual motor torque and motor position is measured. The obtained measurement data is then transformed into the frequency domain and used for the FRM calculation. To reduce the effects of stick-slip transitions a sine wave with an amplitude of 3° serves as reference signal $\boldsymbol{q}_\mathrm{Md}$ for the position controller. For the excitation of the system, normalized random multisine signals are used which leads to a separation of the FRM in three parts

$$\mathbf{G}(j\omega_k) = \mathbf{G}_\mathrm{BLA}(j\omega_k) + \mathbf{G}_\mathrm{S}(j\omega_k) + \mathbf{N}_G(j\omega_k) \quad (19)$$

with $\mathbf{G}_\mathrm{BLA}(j\omega_k)$ the best linear approximation (BLA), $\mathbf{G}_\mathrm{S}(j\omega_k)$ the stochastic nonlinear contributions and $\mathbf{N}_G(j\omega_k)$ the errors due to the output noise (see Pintelon and Schoukens, 2012). We are interested in $\mathbf{G}_\mathrm{BLA}(j\omega_k)$ which is used for the identification of the elasticity parameters in Sect. 4.2.

4.1.1 Calculation of the FRM

Assuming an appropriate excitation of the system, the measured motor torques and motor positions can be transformed to the frequency domain using the fast Fourier transformation (FFT). Note a signal x is represented in the frequency

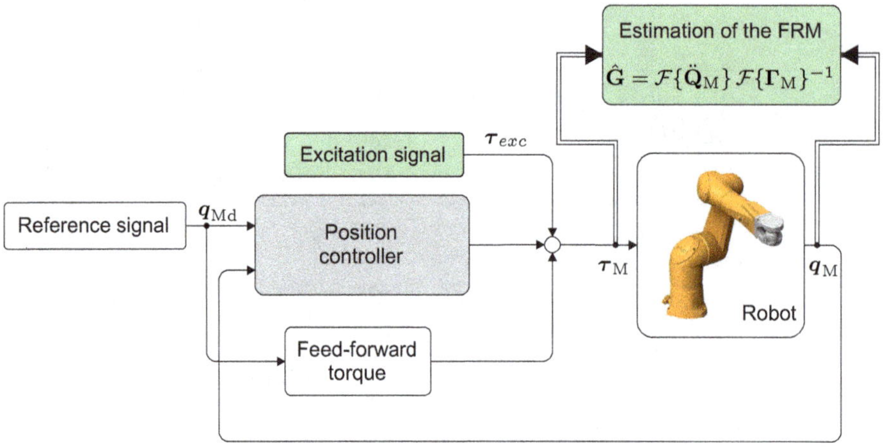

Figure 3. Block diagram of the control concept.

domain as $\overline{x}$ or $\mathcal{F}\{x\}$. In order to reduce the nonlinear contributions $\mathbf{G}_\mathrm{S}$, N_r different excitation signals and to reduce the noise contribution $\mathbf{N}_G$, N_p periods are recorded. The FRM for an excitation signal r in pose $\boldsymbol{q}$ is calculated by

$$\hat{\mathbf{G}}^{[r,q]} = \mathcal{F}\left\{\ddot{\mathbf{Q}}_\mathrm{M}\right\}\mathcal{F}\left\{\mathbf{\Gamma}_\mathrm{M}\right\}^{-1} \tag{20}$$

with $\ddot{\mathbf{Q}}_\mathrm{M} = [\ddot{\boldsymbol{q}}_{\mathrm{M},1}\,\ddot{\boldsymbol{q}}_{\mathrm{M},2}\,\ddot{\boldsymbol{q}}_{\mathrm{M},3}]$ and $\mathbf{\Gamma}_\mathrm{M} = [\boldsymbol{\tau}_{\mathrm{M},1}\,\boldsymbol{\tau}_{\mathrm{M},2}\,\boldsymbol{\tau}_{\mathrm{M},3}]$ resulting from measurement data for 3 orthogonal excitations, see Sect. 4.1.2. For different excitation signals, also slightly different FRM results are computed which have to be combined using averaging techniques. The choice of the averaging method depend on the signal to noise ratio (SNR) of the measurement data and the measurement setup. A comparison of different strategies is discussed in the paper of Wernholt and Moberg (2008). For our robot, the simple arithmetic averaging is useful, which is given by

$$\hat{\mathbf{G}}^{[q]} = \frac{1}{N_\mathrm{r}}\sum_{r=1}^{N_\mathrm{r}}\hat{\mathbf{G}}^{[r,q]}. \tag{21}$$

4.1.2 Synthesis of the excitation signal

The quality of the FRM highly depends on the excitation signal. For frequency domain identification, periodic signals are very useful because they do not suffer from leakage effects. Furthermore the frequency resolution and power spectrum can be customized for the robot resulting in a minimum measurement time by a maximum quality of the signals (see Pintelon and Schoukens, 2012). Odd random phase multisine signals feature all these properties and are given by

$$\tau_{i,\mathrm{exc}}(t) = \sum_{k=0}^{N_\mathrm{f}-1} A_k\cos(\omega_k t + \phi_k),\, \omega_k = (2k+1)\omega_0,$$
$$\phi_k \in [0, 2\pi), \tag{22}$$

consisting of N_f different frequencies of odd multiplicity of the basis frequency ω_0. We selected uniform distributed random phases ϕ_k and a constant amplitude spectrum A_k. Because the FRM of the first three axes should be identified, a excitation signal for each motor has to be generated. Furthermore, for the FRM calculation with Eq. (20) the motor torque matrix $\mathbf{\Gamma}_\mathrm{M}$ must have full rank. To get this property, three sets of orthogonal excitation signals are calculated. Therefore, three different odd random phase multisine signals according to Eq. (22) are generated and combined in a diagonal matrix leading to

$$\overline{\mathbf{D}}_R^{[r]}(k) = \begin{bmatrix} \overline{\tau}_{1,\mathrm{exc}}(k) & 0 & 0 \\ 0 & \overline{\tau}_{2,\mathrm{exc}}(k) & 0 \\ 0 & 0 & \overline{\tau}_{3,\mathrm{exc}}(k) \end{bmatrix}. \tag{23}$$

Matrix $\overline{\mathbf{D}}_R^{[r]}$ is generated in the frequency domain and k represents the discrete frequency index. The columns of $\overline{\mathbf{D}}_R^{[r]}(k)$ are orthogonalized using the matrix

$$T_{pq} = n_u^{-1/2} e^{j2\pi(p-1)(q-1)/n_u} \tag{24}$$

with n_u the number of input signals ($n_u = 3$). This yields the excitation matrix

$$\mathbf{\Gamma}_\mathrm{exc}^{[r]}(k) = \overline{\mathbf{D}}_R^{[r]}(k)\mathbf{T}. \tag{25}$$

Each column of $\mathbf{\Gamma}_\mathrm{exc}^{[r]}(k)$ represents one set of excitation signals and is orthogonal to the other ones. Using the matrix $\mathbf{T}$ in combination with the multisine signals $\overline{\mathbf{D}}_R^{[r]}(k)$ an arbitrary number of different but orthogonal excitation signals can be calculated.

4.1.3 Measurement procedure for the FRM identification

To calculate the FRM of our robot, N_r different excitation signals are calculated, each for N_p periods. Note, at least the first period of the measurement data can not be used for the FRM calculation because the system has to be in steady state. To reduce nonlinear contributions each excitation signal is

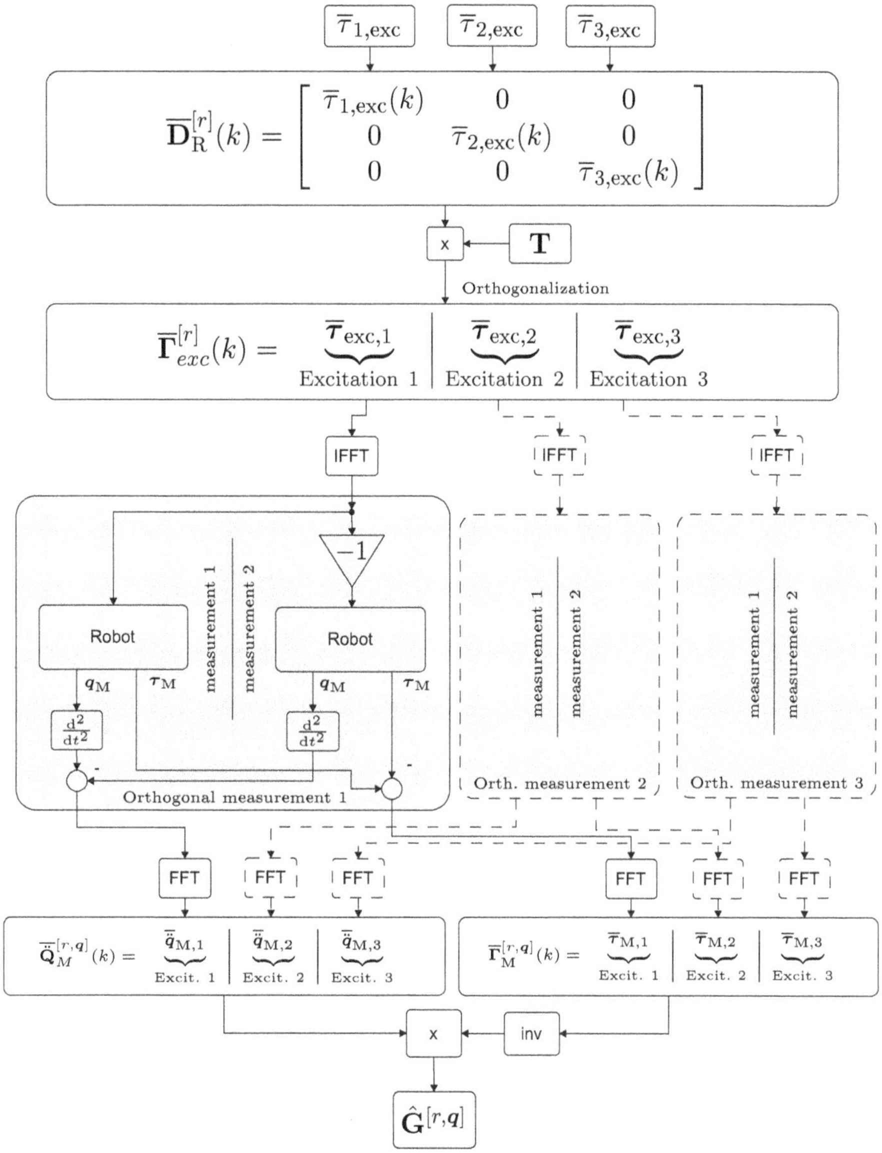

Figure 4. Measurement schedule for the FRM calculation of one realization.

applied once with positive and negative sign but with the same reference trajectory $\boldsymbol{q}_{\text{Md}}$. Then the measurement data of these two excitations are subtracted from each other and are used for FRM calculation, as suggested in Hardeman (2008). Figure 4 shows a graphical representation of the procedure. It starts with the calculation of three independent excitation signals $\overline{\tau}_{i,\text{exc}}$ using Eq. (22), and it ends with an accurate estimate of the FRM $\hat{\mathbf{G}}^{[r,q]}$ for one excitation r in pose $\boldsymbol{q}$. The procedure is performed N_r times, leading to N_r slightly different FRM results which are in a next step averaged using Eq. (21). Our experiments have shown that $N_\text{r} = 4$ different realizations each with 4 periods, two periods for the calculation of the transfer matrix and two periods to get in steady state are convenient. The basis frequency of the excitation signal (see Eq. 22) is defined to $\omega_0 = 2\pi\ 0.25\ \text{rad s}^{-1}$. Hence the signal has a period of 4 s. The excitation of the system for obtaining the transfer matrix for a single realization requires 96 s. This time is composed of the duration of the 3 orthogonal excitations (see Fig. 4) where each consists of a measurement with positive and negative sign with 4 periods. Thus all measurements (4 different realizations) are completed after 384 s. This yields $\hat{\mathbf{G}}^{[q]}$ which is shown in Fig. 5 for the pose $\boldsymbol{q}P1$ of Fig. 6. To get a global representation of the robot, the FRM is identified in 3 different joint configurations. The selected poses are depicted in Fig. 6. These poses represent extrema regarding the moment of inertia in the considered axes. By using poses with extrema in the inertia, the obtained transfer matrices are assumed to be sufficiently different to identify global valid parameters. For all poses the FRM is

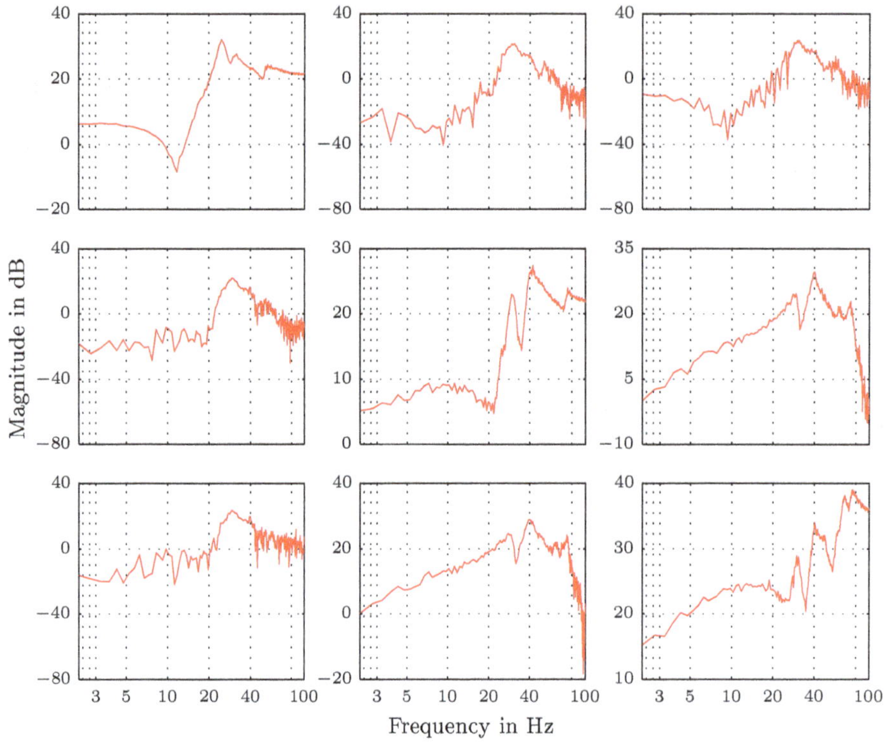

Figure 5. Calculated FRM $\hat{\mathbf{G}}^{[qP1]}$ with $N_r = 4$.

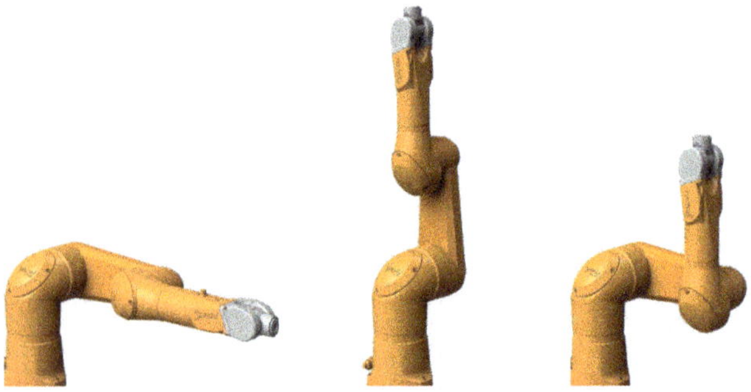

Figure 6. Poses for the FRM calculation represented by $\boldsymbol{q}P1$, $\boldsymbol{q}P2$ and $\boldsymbol{q}P3$ from left to right respectively.

identified and used for the elasticity parameter identification of our robot.

4.2 Identification of the elasticity parameters

To identify the elasticity parameters, the parametric transfer matrix of our robot – derived in Sect. 2.1 and given by Eq. (9) – is fitted to the measured one, by changing the parameters $\boldsymbol{p}_{\text{elast}}$. The fitting procedure performs a minimization of the cost functional

$$J(\boldsymbol{p}_{\text{elast}}) = \sum_{q \in \{\boldsymbol{q}P1, \boldsymbol{q}P2, \boldsymbol{q}P3\}} \sum_{k=1\ldots N_\omega} \left[\boldsymbol{\Delta}^{[q]}(\omega_k, \boldsymbol{p}_{\text{elast}})\right]^* \mathbf{W}^{[q]}(\omega_k) \left[\boldsymbol{\Delta}^{[q]}(\omega_k, \boldsymbol{p}_{\text{elast}})\right] \quad (26)$$

with the complex vector

$$\boldsymbol{\Delta}^{[q]}(\omega_k, \boldsymbol{p}_{\text{elast}}) = \log\left(\text{vec}\left(\hat{\mathbf{G}}^{[q]}(\omega_k)\right)\right) - \log\left(\text{vec}\left(\mathbf{G}^{[q]}(\omega_k, \boldsymbol{p}_{\text{elast}})\right)\right). \quad (27)$$

The minimization is carried out by applying a genetic algorithm followed by a gradient based minimization algorithm. The vector $\boldsymbol{\Delta}^{[q]}(\omega_k, \boldsymbol{p}_{\text{elast}})$ represents the complex error between the measured FRM and the transfer matrix of the linearized system see Eq. (9). The superscript $*$ denotes the conjugate transpose matrix. To get a good fit in the region of interest, a weighting matrix $\mathbf{W}^{[q]}(\omega_k)$ is introduced. Especially the error in the vicinity of poles and zeros in the diagonal elements is amplified, leading to reasonable system parameters. Figure 7 shows both, the non-parametric FRM $\hat{\mathbf{G}}^{[qP1]}$ and the optimized parametric transfer function $\mathbf{G}^{[qP1]}$ in the first pose of the robot. It can be seen, that the coupling between the first and the other two axes is quite low and thus the measured and identified transfer functions agree not very well. Furthermore, the weighting matrix was tuned to get a good fit in the diagonal elements of the transfer matrices of poses $\boldsymbol{q}P1$, $\boldsymbol{q}P2$ and $\boldsymbol{q}P3$. Figures 8 and 9 show the diagonal ele-

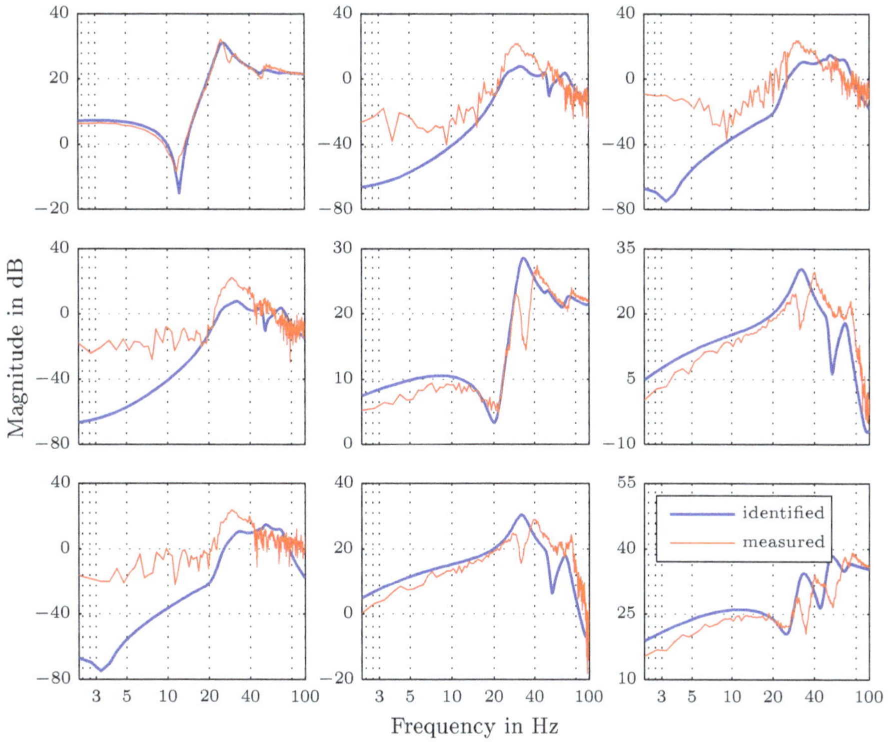

Figure 7. Comparison of the non-parametric FRM $\hat{\mathbf{G}}^{[qP1]}$ and the identified parametric one $\mathbf{G}^{[qP1]}$.

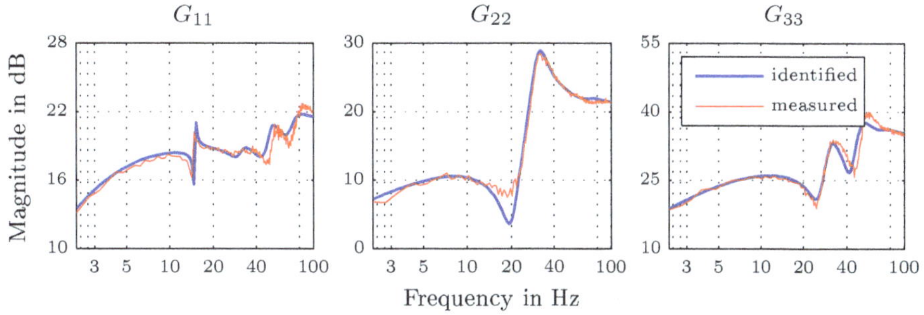

Figure 8. Comparison of the measured non-parametric FRM $\hat{\mathbf{G}}^{[qP2]}$ and the identified parametric one $\mathbf{G}^{[qP2]}$.

ments of the non-parametric and optimized parametric transfer matrices of the poses $\boldsymbol{q}P2$ and $\boldsymbol{q}P3$. Unfortunately, not all poles and zeros agree exactly, which is mainly attributed to nonlinearities in the stiffness parameters. Overall, a good fit between the non-parametric and parametric transfer functions is achieved and the identified parameters are listed in Table 1. Note, if only one pose is matched the accordance of the identified and measured FRM is very high but the parameters are not reliable anymore.

5 Geometric calibration

With the identified dynamic model of the last section the real robot pose due to the motor positions can be calculated. This estimated poses in combination with the real end-effector position, measured by an external sensor (laser-tracker), is used to calibrate the geometric model of Sect. 3. Our external sensor can not measure the end-effector orientation and thus the geometric calibration is discussed by using end-effector positions only.

5.1 Calculation of the estimated geometric parameters

The estimated end-effector position ${}_I\boldsymbol{r}_{\mathrm{E}}^{[q]}(\boldsymbol{q}_{\mathrm{A}}, \boldsymbol{q}_{\mathrm{SW}}, \boldsymbol{p}_{\mathrm{nom}}, \boldsymbol{p}_{\mathrm{ge}})$ can be evaluated by inserting the calculated arm coordinates $\boldsymbol{q}_{\mathrm{A}}$ in Eq. (13) and the measured spherical wrist coordinates $\boldsymbol{q}_{\mathrm{SW}}$. Assuming small geometric errors we start with the initial vector of $\boldsymbol{p}_{\mathrm{ge}}^{(0)} = \mathbf{0}$. Thus the end-effector error for a set of N poses follows

Table 1. Identified elasticity parameters.

kNm rad^{-1}	kNm rad^{-1}	kNm rad^{-1}	kNms rad^{-1}	kNms rad^{-1}	kNms rad^{-1}
c_{1x} 349.9	c_{1y} 753.5	c_1 143.7	d_{1x} 45.9	d_{1y} 192.1	d_1 115.1
c_{2x} 797.7	c_{2y} 697.1	c_2 250.7	d_{2x} 52.1	d_{2y} 211.2	d_2 70.6
c_{3x} 181.4	c_{3y} 149.2	c_3 57.1	d_{3x} 47.6	d_{3y} 39.1	d_3 32.9

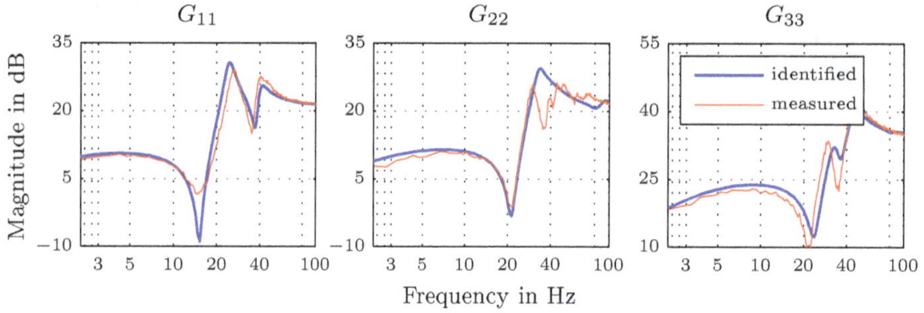

Figure 9. Comparison of the measured non-parametric FRM $\hat{\mathbf{G}}^{[qP3]}$ and the identified parametric one $\mathbf{G}^{[qP3]}$.

$$e_{\text{geo}} = \begin{bmatrix} {}_I r_{\text{E,m}}^{[q1]} - {}_I r_{\text{E}}^{[q1]} \\ \vdots \\ {}_I r_{E,m}^{[qN]} - {}_I r_{E}^{[qN]} \end{bmatrix} \tag{28}$$

with ${}_I r_{\text{E,m}}^{[q]}$ the end-effector position measured by the laser-tracker in pose q of the robot. Performing a Taylor series expansion on e_{geo} at $p_{\text{ge}} = p_{\text{ge}}^{(i)}$ of order 1 we get

$$e_{\text{geo,lin}} = \Theta \Delta p_{\text{ge}} + e_{\text{geo}}|_{p_{\text{ge}}^{(i)}} \tag{29}$$

with $\Theta = \frac{\mathrm{d}e_{\text{geo}}}{\mathrm{d}p_{\text{ge}}}|_{p_{\text{ge}}^{(i)}}$. The superscript (i) indicates the current set of parameters. Unfortunately, not all geometric error parameters are independent, i.e. rank of Θ is less than the number of parameters p_{ge}. These linear dependencies must be eliminated. Therefore a numerical regularization algorithm which is based on a QR-decomposition is applied (see Gautier, 1991). 30 independent parameters out of the 47 modeled error parameters are found. The parameters p_{ge} are calculated by minimizing the error e_{geo}, using e.g. the Levenberg–Marquardt algorithm. The quality of the calibration result highly depends on the poses.

5.2 Calculation of optimal poses for the calibration

In order to get a good excitation of the geometric parameters the calibration poses are very important. In the paper of Sun and Hollerbach (2008a) the choice of the observability index for the pose calculation and in Sun and Hollerbach (2008b) and Zhuang (1994) selection algorithm for the computation of optimal poses are presented. We used the algorithm presented in Sun and Hollerbach (2008b) where the minimum singular value of the covariance matrix $\Lambda = \Theta^T \Theta$ is maximized. Finally, 40 optimal poses are calculated with a minimum singular value of $\sigma_{\text{min}} = 0.46$. Exemplarily, 3 out of the 40 poses are depicted in Fig. 10. For all optimal poses the end-effector is measured with a laser tracker and the actual motor positions are measured by the motor encoders. Then, the estimated arm angles q_{A} are evaluated using Eq. (13) and the error of Eq. (28) is minimized using the Levenberg-Marquardt algorithm. The absolute positioning accuracy for the uncalibrated and calibrated case is depicted in Fig. 11. The positioning accuracy for the calibrated case is shown in Fig. 12. For the calibration poses, a positioning accuracy of about 0.1 mm is obtained which is in the range of the repeatability of our robot, and thus a very good result.

5.3 Evaluation of the absolute positioning accuracy

With the geometric parameters p_{ge} identified in the last section, the positioning accuracy in the whole workspace should be evaluated. Therefore 150 random poses in the whole workspace of the robot are generated and measurements with the laser-tracker. The calculated end-effector position due to the estimated robot configuration is compared to the measured position and the obtained absolute positioning accuracy is shown in Fig. 13. For 150 arbitrary poses an absolute error less than 0.32 mm and in 90 % of the poses an error less than 0.23 mm is obtained. If the elastic deflections are not considered in the calibration, the maximum error for the

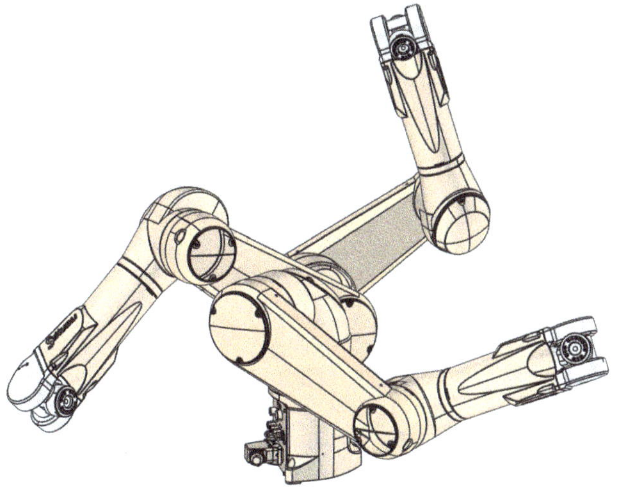

Figure 10. Three of the 40 optimal poses.

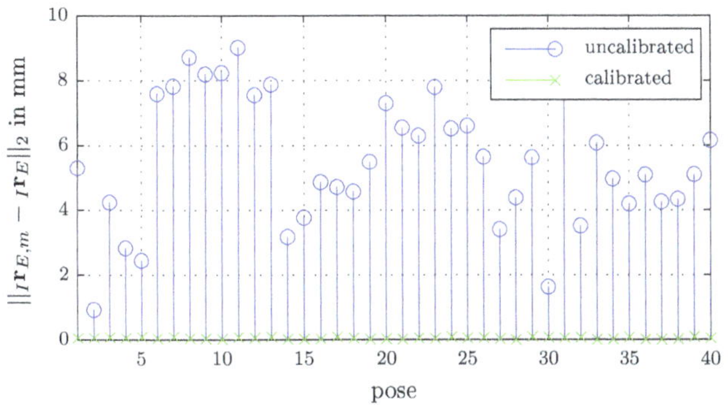

Figure 11. Absolute error of the end-effector for the calibration poses.

same poses is 0.7 mm which thus justifies the effort of the presented procedure.

6 Conclusions

This paper presents the identification of geometric parameters by including stiffness and damping parameters in the calibration process. Not only the inclusion of the elasticity parameters also their identification was discussed. The identification was carried out in the frequency domain, where the transfer matrix of the linearized system was adapted to the real robots transfer matrix by changing the elasticity parameters. Regarding the calculation of the real robots transfer matrix, the excitation signal is crucial and thus it is discussed in detail. The identification of globally valid parameters requires different poses. In this paper, the poses were selected manually. The calculation of optimal poses for the elasticity parameter identification is part of future work. With the identified elasticity parameters, the real robots configuration is estimated using the system dynamics and the geometric calibration is performed. Experimental results for a Stäubli TX90L are reported. Thereupon geometric model parameters were identified which lead to a positioning accuracy of 0.32 mm in the whole workspace.

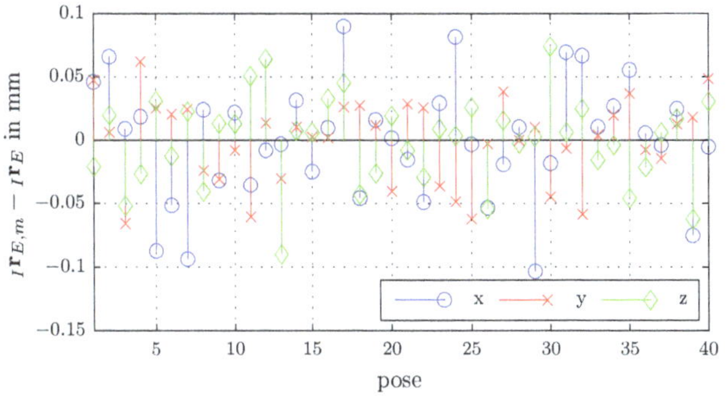

Figure 12. Error of the end-effector for the calibration poses in the inertial frame.

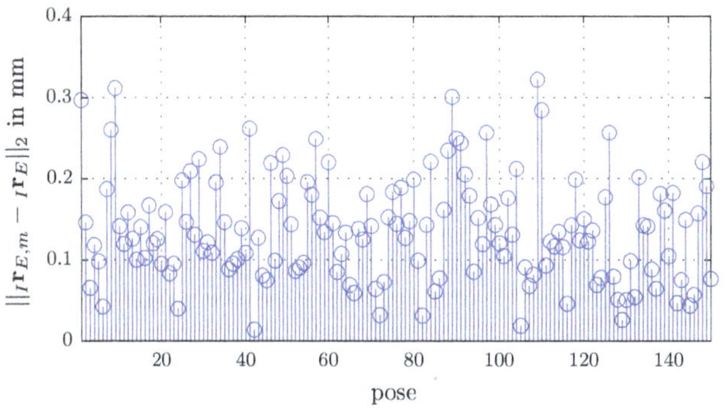

Figure 13. Absolute error of the end-effector for the 150 verification poses.

Acknowledgements. This work has been supported by the Austrian COMET-K2 program of the Linz Center of Mechatronics (LCM), and was funded by the Austrian federal government and the federal state of Upper Austria.

References

Bremer, H.: Elastic Multibody Dynamics: A Direct Ritz Approach, Springer Netherlands, 2008.

Gautier, M.: Numerical Calculation of Base Inertial Parameters of Robots, J. Robot. Syst., 8, 485–506, 1991.

Gong, C., Yuan, J., and Ni, J.: Nongeometric error identification and compensation for robotic system by inverse calibration, Int. J. Mach. Tools Manufact., 40, 2119–2137, 2000.

Hardeman, T.: Modeling and Identification of Industrial Robots Including Drive and Joint Flexibilities, PhD thesis, University of Twente, Twente, the Netherlands, 156 pp., 2008.

Khalil, W. and Besnard S.: Geometric Calibration of Robots with Flexible Joints and Links, J. Intell. Robot. Syst., 34, 357–379, 2002.

Khalil, W. and Dombre, E.: Modeling, Identification and Control of Robots, Kogan Page Science, London, UK, 2004.

Öhr, J., Moberg, S., Wernholt, E., Hanssen, S., Pettersson, J., Persson, S., and Sander-Tavallaey, S.: Identification of Flexibility Parameters of 6-axis Industrial Manipulator Models, in: Proceedings of ISMA2006, 18–20 September 2006, Leuven, Belgium, 3305–3314, 2006.

Pintelon, R. and Schoukens, J.: System Identification: A Frequency Domain Approach, John Wiley & Sons, Inc., Hoboken, New Jersey, 2012.

Siciliano, B. and Khatib, O.: Handbook of Robotics, Springer-Verlag, Berlin, Heidelberg, Germany, 2008.

Sun, Y. and Hollerbach, J. M.: Observability index selection for robot calibration, in: Proceedings on the IEEE International Conference on Robotics and Automation, 19–23 May 2008, Pasadena, USA, 831–836, 2008a.

Sun, Y. and Hollerbach, J. M.: Active robot calibration algorithm, in: Proceedings on the IEEE International Conference on Robotics and Automation, 19–23 May 2008, Pasadena, USA, 1276–1281, 2008b.

Wernholt, E.: Multivariable Frequency-Domain Identification of Industrial Robots, PhD thesis, Institute of Technology, University Linköping, Linköping, 204 pp., 2007.

Wernholt, E. and Moberg, S.: Experimental Comparison of Methods for Multivariable Frequency Response Function Estimation, in: Proceedings of the 17th IFAC World Congress, 6–11 July 2008, Coex, Korea, 15359–15366, 2008.

Whitney, D. E., Lozinski, C. A. and Rourke, J. M.: Industrial Robot Forward Calibration Method and Results, J. Dynam. Syst. Meas. Contr., 108, 1–8, 1986.

Zhuang, H.: Optimal selection of measurements configurations for robot calibration using simulated annealing, in: Proceedings on the IEEE International Conference on Robotics and Automation, 8–13 May 1994, San Diego, USA, 393–398, 1994.

14

Synthesis of PR-/RP-chain-based compliant mechanisms – design of applications exploiting fibre reinforced material characteristics

U. Hanke[1], E.-C. Lovasz[2], M. Zichner[1], N. Modler[1], A. Comsa[1], and K.-H. Modler[1]

[1]Faculty of Mechanical Engineering and Machine Science, TU Dresden, Dresden, Germany
[2]Department of Mechatronics, Politehnica University of Timisoara, Timisoara, Romania

Correspondence to: U. Hanke (uwe.hanke@tu-dresden.de)

Abstract. Compliant mechanisms have several advantages, especially their smaller number of elements and therefore less movable joints. The flexural members furthermore allow an integration of special functions like balancing or locking. Synthesis methods based on the rigid body model (Howell, 2001; Sönmezv, 2008) or topology optimisation (Zhou and Mandala, 2012) provide practical applications from the advantages of compliant elements. Beside these methods, a much simpler approach is the geometric-based synthesis (Ehlig et al., 2013) which is focused on solving guidance tasks by using RR-chain[1]-based compliant linkages. More compact compliant linkages can be build up by using only PR[2] or RP[3] chains. Therefore a tool is needed to extend the RR-chain-based approach. The necessary analysis of the compliant beam element can be done by numerical analysis and through experiments. Due to the validity of the Bernoulli beam model the elastic similitude can be specialised and a more general synthesis of compliant beam elements can be created. Altogether a generalised synthesis method can be created for handling different linkage structures as well integrating beam elements derived numerically or by measurement. The advances in this method are applied in the synthesis for a cupholder mechanism made of fiber reinforced material.

1 Introduction

Classical linkage structures need additional elements (e.g. springs) for balancing or locking, which gives rise to the structure complexity. These features can be directly implemented by using compliant linkages. This function integration used in a compliant cupholder mechanism (Ehlig et al., 2013) lead to a simple two part assembly (RR chain coupled to a beam element (B): RRB mechanism; Fig. 1). The integrated compliant section allows for the implementation of locking so that additional springs are not required. The applied geometric-based synthesis method allows for its direct use in the design process, providing maximum design freedom regarding position and size of the mechanism.

Much simpler mechanisms can be derived by introducing P joints. The intermediate link can be completely eliminated. Hence for assembling such near monolithic mechanism only one bolt (R joint) is needed. To make use of this advantage, the synthesis of compliant mechanisms has to be extended to P joints. The advantage of the geometric method should form the basis when implementing RP/PR chains.

Specialising the similitude theory provides additional benefits in deriving a set of compliant beam elements and in the direct use of experimental data for the mechanism synthesis.

2 Synthesis

The compliant beam section B_{0B} of the mechanisms shown in Fig. 1c is integrated in the guidance link AB and acts like a spring element. For the synthesis model, the large deflections

[1]one link with two rotational joints (R)

[2]one link with one frame fixed prismatic joint (P) and one moving rotational joint (R)

[3]one link with one frame fixed rotational joint (R) and one moving prismatic joint (P)

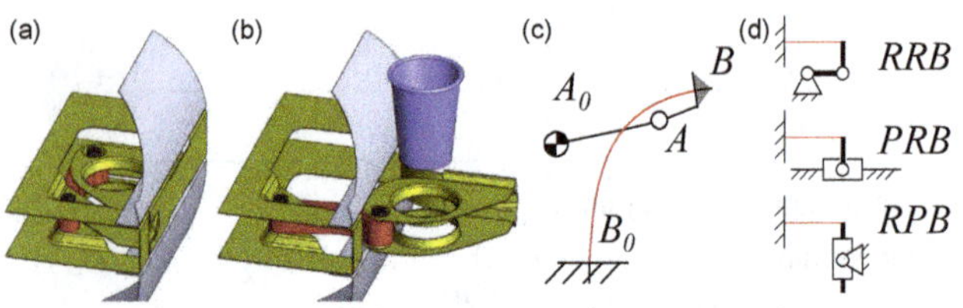

Figure 1. Compliant cupholder mechanism (Ehlig et al., 2013): **(a)** closed position **(b)** open position, **(c)** kinematic scheme of the RRB mechanism, **(d)** RRB/PRB and RPB mechanism.

of this beam section are introduced by a force/moment load, defined through the coupled joint pair. This can be an RR, PR or RP joint pair. The direct implementation into a graphical synthesis method is problematic. By using the elastic similitude modified for the case of the Bernoulli beam model, a graphical implementation is possible. The result is a synthesis method with two serial coupled synthesis steps (Ehlig et al., 2013):

- synthesis of the RR, PR and RP chains and
- analysis or experiment of the compliant beam element for different load conditions coupled by a similarity transformation for one solution of this set.

The synthesis of the RR, PR and RP chains defines the load configuration of the beam element. For the kinematic solution of the beam element, two parameters can be used to define the load case. These parameters are direct coupled to the scaling and orientation of the beam element.

2.1 Graphical two-pose synthesis of PR and RP chains

RR, PR and RP chains are link systems with three links: a, b, f. The following example will include f as frame link, a as guidance link and b as intermediate link. The graphical algorithm used to solve the two-pose task is the construction of the relative pole triangle for a given vertex of this triangle (Luck and Modler, 1990; McCarthy, 2000). In the case of an RR chain, three parameters can be arbitrarily chosen. The triangle of the PR and RP chains degenerates into a triangle with two parallel sides and one vertex at infinity. Therefore the synthesis of these special chains reduces to two parameters. By introducing the rules (see Appendix A), the two-pose synthesis of the PR and RP chains can be completed through the synthesis steps listed in Table 1 referenced in Fig. 2.

The scheme in Fig. 2 is statically balanced in position 1, when the force action line between link a and f is equal to link line b_1. This fact is of fundamental importance for the implementation of the compliant beam element. The shortest distance to b_1 measured from P_{12}^{af} is given by vector $\boldsymbol{r}$.

2.2 Synthesis of the compliant beam element using the elastic similitude

The deflection of the compliant beam depends on the material parameters (Young's modulus E), geometric parameters (the cross section with the area moment of inertia I and beam length L) and the load condition (force F, force direction Φ and distance d).

The mathematical description of large deflection behaviour of cantilevered beams is the subject of several scientific publications (Bisshopp and Drucker, 1943; Venanzi et al., 2005; Kimball, 2002; Campanile and Hasse, 2008; Zhang and Chen, 2013; De Bona and Zelenika, 1997). The following statements and equations are based upon the reasoning of Howell (Bisshopp and Drucker, 1943; Howell, 2001; Zhang and Chen, 2013; Campanile and Hasse, 2008).

The Bernoulli–Euler beam theory is the vital element when dealing with large deflection analysis. It says that for constant material parameter E, I the curvature κ is proportional to the internal beam moment:

$$\kappa = \frac{\mathrm{d}\vartheta}{\mathrm{d}s} = \frac{M}{EI}. \tag{1}$$

A constant internal moment means a constant curvature with radius $r = EI/M$ (Howell, 2001; Modler et al., 2013). This hypothesis is, according to Fig. 3, equal to $d \to \infty$. For all other cases the internal moment reads $M = -F \cdot d + F(x \sin\Phi + y \cos\Phi)$. Hence it makes sense to take the second-order form of Eq. (1) (Holst et al., 2011) and use the energy method (De Bona and Zelenika, 1997) to establish the elliptic integral

$$\frac{1}{\sqrt{2}}\sqrt{\frac{F}{EI}}L = \int_{\vartheta_1=0}^{\vartheta_2} \frac{\mathrm{d}\vartheta}{\sqrt{\cos(\Phi+\vartheta)+C}}, \tag{2}$$

where L is the beam length and C an integration constant. The constant C is directly coupled to the initial conditions on the clamped side of the beam: $C = Fd^2/(2EI) - \cos\Phi$. For a set of known parameters F, d, E, I the integral (2) can be resolved numerically (Howell, 2001; Campanile and Hasse, 2008; Kimball, 2002; Venanzi et al., 2005).

Using the similitude theory (Kerle, 2006; Weber, 1930) leads to a more general way of handling compliant beam elements. The idea behind this synthesis is the direct implementation of a precalculated or measured beam element. The elastic similitude is defined by the Hooke number:

$$Ho = \frac{F}{EL^2}. \tag{3}$$

In this case, a scaling of dimensions affects all dimensions and therefore the section modulus I as well. For the compliant beam element, bending is the major load case. So it is possible to specialise the similitude to beams and use a modified Hooke number:

Table 1. Synthesis of the RP and PR chains.

Step	PR chain	RP chain
1	Set moving joint A_1^{ab}	Set rotatory ground joint A_0^{bf}
2	Draw line a_1 defined by joint A_1^{ab} and pole P_{12}^{af}	Draw line f_1 defined by joint A_0^{bf} and pole P_{12}^{af}
3	Draw line f_1 defined by $P_{12}^{af}, \alpha_{12}^{af}$ and line a_1	Draw line a_1 defined by $P_{12}^{af}, \alpha_{12}^{af}$ and line f_1
4	Draw line b_1 defined by $A_1^{ab}, \alpha_{12}^{ab} = \alpha_{12}^{af}$ and line a_1	Draw line b_1 defined by $A_0^{bf}, \alpha_{12}^{bf} = \alpha_{12}^{af}$ and line f_1
5	The direction of translation of the prismatic joint $A_0^{bf\infty}$ is perpendicular to line b_1	The direction of translation of the prismatic joint $A_1^{ab\infty}$ is perpendicular to line b_1

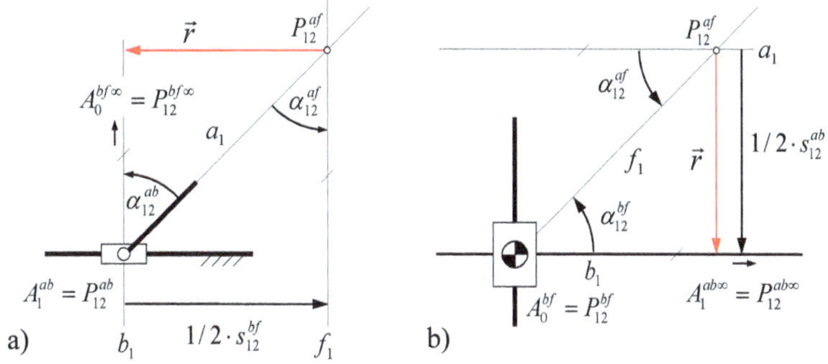

Figure 2. Two-pose synthesis of the **(a)** PR chain, **(b)** RP chain.

$$mHo = \frac{FL^2}{EI}, \tag{4}$$

where the section modulus I can be independently scaled. The square root of this number is the left hand side of Eq. (2) (see $\sqrt{\frac{F}{EI}}L = \sqrt{\text{mHo}}$). Hence Eq. (2) reads

$$\sqrt{\text{mHo}(C)} = \int_{\vartheta_1=0}^{\vartheta_2} \frac{d\vartheta}{\sqrt{\cos(\Phi+\vartheta)+C}}. \tag{5}$$

Each modified Hooke number mHo is a unique value for a compliant beam dependent on the angles ϑ_1, ϑ_2 and the load condition C. Therefore deriving a set of deflected beams is described by a set of different $mHo(C)$ as the result of equation (5). The compliant beam is fully defined by naming three of the four variables F, E, I, L.

The two positions of the non-deflected and the deflected beam end define the pole of rotation P_{12}^{af}. This pole is coupled to the load condition of the beam. This load condition is defined by two parameters: the load direction Φ and the distance d, defining the moment introduced at the beam end. These parameters define the line f_F of the action of the force F^{af}. The shortest distance between the pole P_{12}^{af} and f_F is given by the vector $\boldsymbol{r}$ (Figs. 3a, 4a, b). This vector is used to find the dimensions of the compliant beam element by using a similarity transformation to scale and orient the beam according to the derived PR and RP chain.

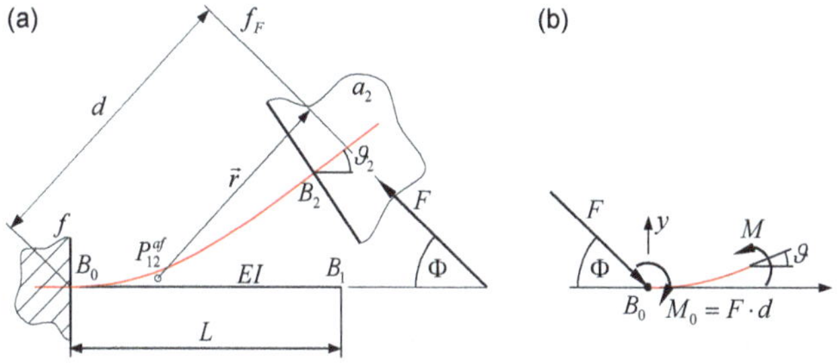

Figure 3. Cantilevered compliant beam under external load: **(a)** beam model, **(b)** internal moment M.

The elastic similitude makes no difference if the deflections are calculated (Fig. 4a) or measured (Fig. 4b). The rough measurement with the simple test bench still give results useful for further synthesis. Here, a more advanced tool kit (Limaye et al., 2012) could bring more exact results. Errors due to poor beam clamping and friction can be omitted. Doing measurements using a cantilevered beam is a fast and easy way to get to reliable solutions. In this special case the implementation is done by the use of the MATLAB programming environment. The geometric algorithms are based on object-oriented programming with three objects: points, lines and circles with their related methods, and the beam is implemented by solving Eq. (1) numerically as an initial value problem for a normalised beam model.

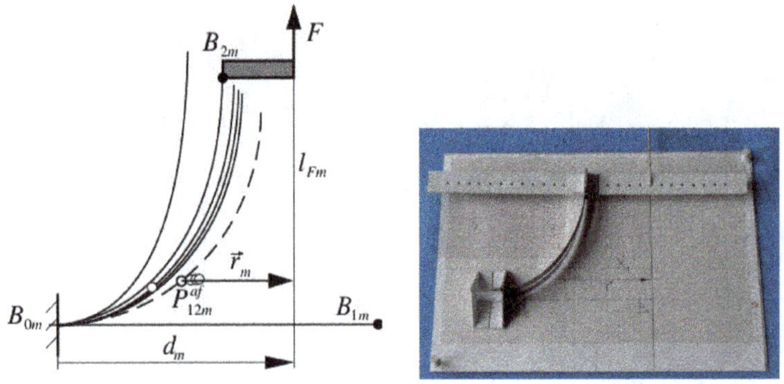

a) b)

Figure 4. Deflection analysis of a cantilevered beam: **(a)** calculation for different load sets, **(b)** measurement using a simple test bench.

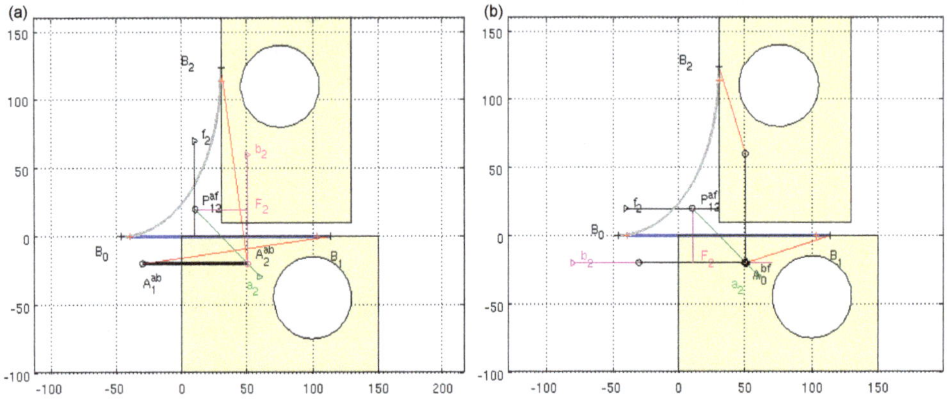

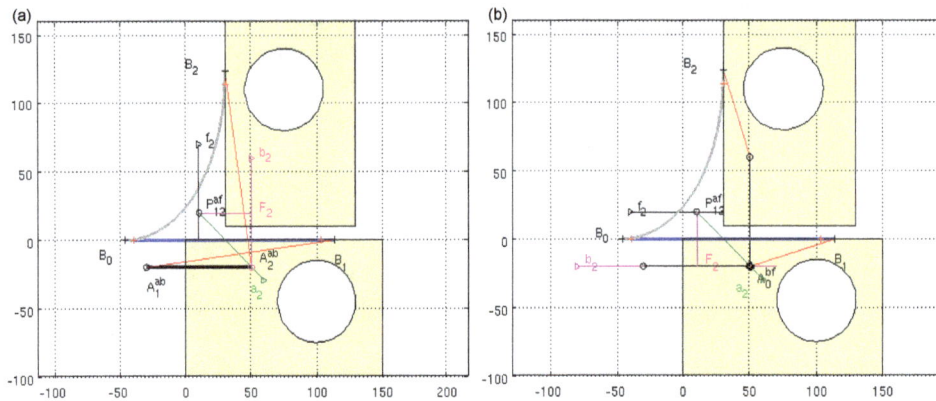

Figure 5. Task solution for compliant: **(a)** PRB linkage, **(b)** RPB linkage.

3 Application

The described synthesis method is a tool used in designing a cupholder mechanism as one can find in vehicles (Barej et al., 2012). This special application is a feature in luxury cars for holding a cup while driving. It has to be in reaching distance of the driver. When the cupholder is not in use it can be stored/hidden behind the board. We aim to reduce the number of parts via the integration of locking functions within the compliance of the mechanism structure, and we discuss practical differences to the RR-chain-based mechanism shown in Ehlig et al. (2013).

The synthesis task is a two-pose task referring to Ehlig et al. (2013) (Table 2). In the initial position a_1 the cupholder should be hidden behind the front board. The second position a_2 offers a panel with a circular hole to place the cup. When the holder is not in use, the front end should be hidden behind the surface of the board. All joints should be located behind the board.

The derived PRB linkage (Prismatic joint, Rotatory joint and compliant Beam element) given in Fig. 5a is characterised by the small distance of the moving joint A^{ab} to the front end of the board. The beam element is relatively large, compared to the RR chain solution given in Ehlig et al. (2013). Extending this distance will enlarge the beam and therefore lead to difficulties in the linkage design and its stability. Thermoplastic fibre reinforced material (FRM) enables the adjustment of its mechanical properties. Material with high strength, coupled with a low Young's modulus E is extremely appropriate for use in compliant structures. The very compact design shown in Fig. 6 is constructible.

The use of thermoplastic matrix material enables easy post-processing by thermal bending. Using preprocessed FRM preforms (Fig. 7a), the final processing in manufacturing this linkage is a three-step folding process (Fig. 7b, c, d) with the final assembly of one bolt.

Table 2. Synthesis parameters and results for the RPB and PRB mechanisms.

	a_1	a_2	P_{12}^{af}	PR:A_1^{ab}	PR:A_2^{ab} RP:A_0^{bf}	B_0	B_1	B_2
x	0.0	20.0	10.0	−30.0	50.0	−46.2	113.5	30.0
y	10.0	10.0	20.0	−20.0	−20.0	−0.0	−0.0	123.5
φ	0.0	90.0						

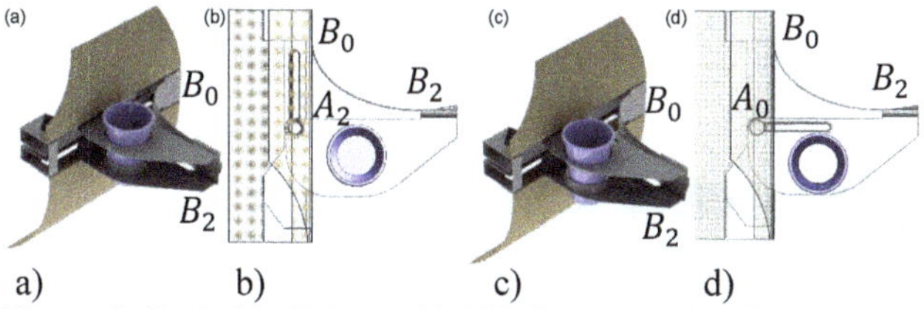

Figure 6. Cupholder linkage: **(a)** RPB linkage model, **(b)** RPB linkage transparent top view, **(c)** PRB linkage model, **(d)** PRB linkage transparent top view.

As seen in Fig. 5b and Fig. 6c, d the RPB linkage is similar to the PRB linkage discussed. The differences between these solutions are in design and kinematic behaviour. Both linkages are less complex than the RR-chain-based solution given in Ehlig et al. (2013). Instead of two bolts and one link arm, the PRB and RPB linkages only use one bolt to complete the linkage.

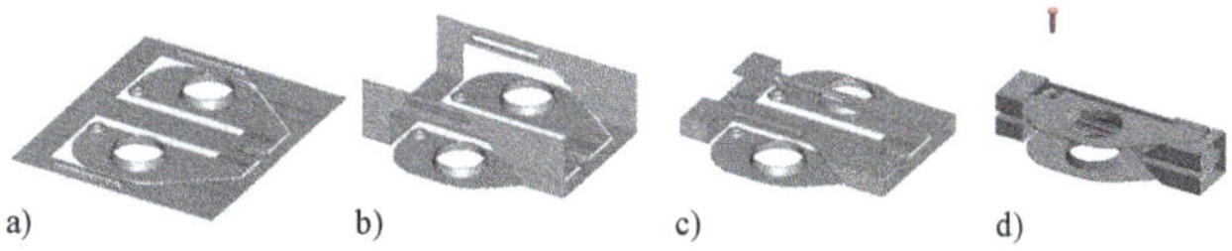

Figure 7. Processing and assembly of a PRB linkage: **(a)** preform, **(b)** first folding step, **(c)** second folding step, **(d)** bolt assembly.

4 Conclusions

Extending the graphical-based synthesis approach using elastic similitude to PR-/RP-chain-based compliant linkages widens the range of compliant linkages to include more compact mechanisms. The drawback is one less synthesis parameter compared to the RRB linkage: two parameters defining the PR/RP chain and one the size of the beam element are three free parameters that allow geometric synthesis for PR-/RP-chain-based mechanisms with a compliant beam element. The extension of the synthesis scheme given in Ehlig et al. (2013) to PR-/RP-chain-based compliant linkages widens the range of compliant linkages to include near monolithic structures. These structures are less complex and characterised by a simplified design. Future research work should focus on coupled beam structures as the BRB linkage as a further step to in designing monolithic mechanisms.

Appendix A: Index rules

Using graphical synthesis methods, it is useful to name links with small letters and joints/poles as big letters. Motions are defined by link positions and the reference link from where this movement is measured. In handling definite positions we have to distinguish between link pairs and position pairs. Therefore all kinematic symbols are marked with indices P_{12}^{af} (af: link index $\rightarrow a$ moves in reference to f; 12: pose index $\rightarrow$ position 1 moves to position 2). This nomenclature leads to the rules given in Fig. A1.

rotations $\rightarrow$ pole, angle

a in reference to $f, \rightarrow P_{12}^{af}, \alpha_{12}^{af}$

b in reference to $f, \rightarrow P_{12}^{bf}, \alpha_{12}^{bf}$

a in reference to $b, \rightarrow P_{12}^{ab}, \alpha_{12}^{ab}$

$a_1\left(P_{12}^{af}, P_{12}^{ab}\right)$

$b_1\left(P_{12}^{bf}, P_{12}^{ab}\right)$

$f_1\left(P_{12}^{af}, P_{12}^{bf}\right)$

angles

$\alpha_{12}^{af} = \angle P_{12}^{ab} P_{12}^{af} P_{12}^{bf}$

$\alpha_{12}^{bf} = \angle P_{12}^{ab} P_{12}^{bf} P_{12}^{af}$

$\alpha_{12}^{ab} = \angle P_{12}^{af} P_{12}^{ab} P_{12}^{bf}$

sign rule

$\alpha_{12}^{af} = -\alpha_{12}^{fa}$

sum

$\alpha_{12}^{af} + \alpha_{12}^{fb} + \alpha_{12}^{ba} = 0$

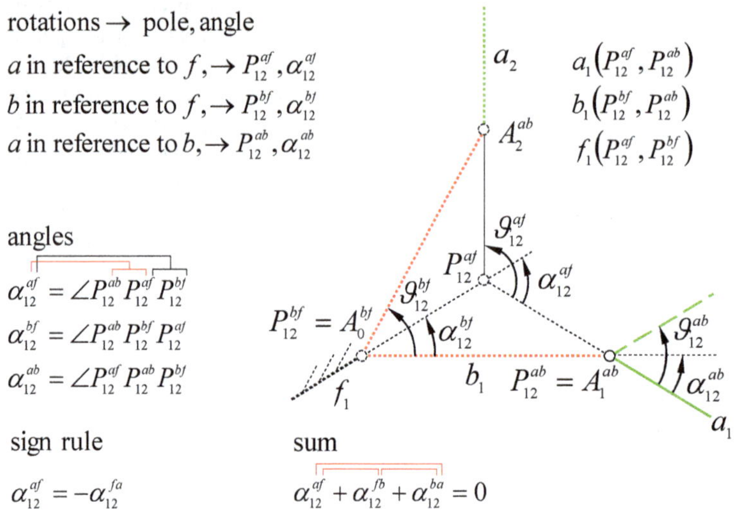

Figure A1. Rules and nomenclature for the graphical synthesis demonstrated at a two-pose RR chain example.

Acknowledgement. The authors would like to express their gratitude to the Deutsche Forschungsgemeinschaft (DFG), which supports the research leading to this publication as part of the scope of the subproject D2 of the Collaborative Research Centre SFB 639 "Textile-Reinforced Composite Components in Function-Integrating Multi-Material Design for Complex Lightweight Applications".

References

Barej, M. and Hüsing, M., and Corves, B.: Teaching Mecanism Theory - From Hands-on Analysis to Virtual Modeling, in: New Trends in Mechanism and Machine Science, edited by: Viadero, F. and Ceccarelli, M., Mechanism and Machine Science, 7, 703–710, Springer Netherlands, 2001.

Bisshopp, K. E. and Drucker, D. C.: Large Deflection of Cantilever Beams, Q. Appl. Math., 3, 272–275, 1945.

Campanile, L. F. and Hasse, A.: A simple and effective solution of the elastica problem, Proceedings of the Institution of Mechanical Engineers, Part C: J. Mech. Eng. Sci., 222, 2513–2516, 2008.

De Bona, F.; Zelenika, S.: A generalized elastica-type approach to the analysis of large displacements of spring-strips, **211**, 509-517, Journal of Mechanical Engineering Science, (1997).

Ehlig, J., Hanke, U., Lovasz, E.-C. Zichner, M., and Modler, K.-H.: Geometrical synthesis approach for compliant mechanisms – Design of applications exploiting fibre reinforced material characteristics, in: New Advances in Mechanisms, Transmissions and Applications, edited by: Petuya, V., Pinto, C., and Lovasz, E.-C., Mechanisms and Machine Science, 17, 215–224, Springer Netherlands, 2013.

Holst, G. L., Teichert, G. H., and Jensen, B. D.: Modeling and Experiments of Buckling Modes and Deflection of Fixed-Guided Beams in Compliant Mechanisms, J. Mech. Design, 133, 051002, doi:10.1115/1.4003922, 2011.

Howell, L. l.: Compliant Mechanisms, Wiley-Interscience, New York, 480 pp., 2001.

Kerle, H.: Zur Entwicklung von Baureihen für Getriebe und von belastbaren Getriebemodellen auf der Grundlage der Ähnlichkeitsmechanik, Bewegungstechnik VDI-Getriebetagung 2006, VDI-Berichte 1966, VDI-Verlag, Düsseldorf, 2006.

Kimball, C. and Tsai, L.-W.: Modeling of Flexural Beams Subjected to Arbitrary End Loads, J. Mech. Design, 124, 223–235, 2002.

Limaye, P., Ramu, G., Pamulapati, S., and Ananthasuresh, G. K.: A compliant mechanism kit with flexible beamsand connectors along with analysis and optimal synthesis procedures, Mech. Mach. Theory, 49, 21–39, 2012.

Luck, K. and Modler, K.-H.: Getriebetechnik – Analyse Synthese Optimierung, Springer-Verlag, Wien New York, 1990.

McCarthy, J. M.: Geometric Design of Linkages, Springer-Verlag, New York, 2000.

Modler, N., Modler, K.-H., Hufenbach, W. A., Jaschinski, J., Zichner, M., Hanke, U., and Ehlig, J.: Optimization of a Test Bench for Testing Compliant Elements Under Shear-Force-Free Bending Load, Procedia Materials Science, 2, 130–136, 2013

Sönmez, Ü. and Tutum, C. C.: A Compliant Bistable Mechanism Design Incorporating Elastica Buckling Beam Theory and Pseudo-Rigid-Body Model, 130, 042304, doi:10.1115/1.2839009, J. Mech. Design, 2008.

Venanzi, S., Giesen, P., and Parenti-Castelli, V.: A novel technique for position analysis of planar compliant mechanisms, Mech. Mach. Theory, 40, 1224–1239, 2005.

Weber, M.: Das Allgemeine Ähnlichkeitsprinzip der Physik und sein Zusammenhang mit der Dimensionslehre und der Modellwissenschaft, Jahrbuch der Schiffbautechnischen Gesellschaft, 31. Bd., Kap. XIV, 274–354, Springer-Verlag, Berlin, 1930.

Zhang, A. and Chen, G.: A Comprehensive Elliptic Integral Solution to the Large Deflection Problems of Thin Beams in Compliant Mechanisms, Journal of Mechanisms and Robotics, 5, 021006, doi:10.1115/1.4023558, 2013.

Zhou, H. and Mandala, A. R.: Topology Optimization of Compliant Mechanisms Using the Improved Quadrilateral Discretization Model, Journal of Mechanisms and Robotics, 4, 021007, doi:10.1115/1.4006194, 2012.

15

Coupled dynamic model and vibration responses characteristic of a motor-driven flexible manipulator system

Y. F. Liu[1], W. Li[1], X. F. Yang[1], Y. Q. Wang[1], M. B. Fan[1], and G. Ye[2]

[1]School of Mechatronic Engineering, China University of Mining and Technology, Xuzhou, China
[2]School of Mechanical & Electrical Engineering, Jiangsu Normal University, Xuzhou, China

Correspondence to: W. Li (liweicumt@163.com)

Abstract. Motor-driven flexible manipulator systems (MDFMSs) are widely used in industry robot fields. During the dynamic modeling, the separate investigation method which neglects the dynamic behavior of the driving motor will cause an error in the dynamic analysis of the flexible manipulator, especially with high-speed operations. This paper proposes a coupled dynamic model of the MDFMS in which the driving motor and flexible manipulator are considered as an integrated system, which can clearly reflect the influence of the dynamic effect of the driving motor. Based on the proposed dynamic model, the vibration responses of the flexible manipulator under different velocities, accelerations and structure parameters, as well as the effect mechanism of the driving motor on the vibration responses, are investigated. The results obtained in this paper contribute to the structure design, motion optimization and dynamic analysis of flexible manipulators.

1 Introduction

Flexible manipulators have been actively developed and widely used, particularly in aerospace and robot fields because of their lightweight feature (Ge et al., 1998; Dwivedy and Eberhard, 2006), which can meet the purposes of high-speed operation and high energy utilization efficiency. Due to their constructional features, however, the damp and stiffness of the flexible manipulators are generally lower; as a result, undesirable elastic deformations and residual vibrations will be easily excited during the operation tasks of the flexible manipulators, such as starting, transforming gestures and stopping, which are inevitable and have a significant influence on the operation precision and structural life of precision operating systems, especially in high-speed operations (Dwivedy and Eberhard, 2006; Maria et al., 2013). In order to control the flexible manipulators more accurately, numerous researchers have studied the dynamic modeling of the flexible manipulators in order to investigate the dynamic behaviors (Mohsen et al., 2004; Qiu, 2012; Kerem et al., 2009; Qiu et al., 2012).

As a typical coupled system, the flexible manipulator can be modeled as a flexible arm and a driving base (Siciliano and Khatib, 2008). Generally, the flexible arm can be described as a flexible beam to reflect its lower damp and stiffness as well as its elasticity properties, while the driving base is defined by translational motions or rotational motions for driving the flexible arm and executing the tasks. In these existing studies, however, the dynamic investigations have mainly focused on the flexible arm and the dynamic behavior of the driving base has been neglected. Due to the fact that the flexible arm is coupled with the driving base, this separate investigation method will cause an error in the dynamic analysis (Zhang et al., 2010; Li et al., 2008). On the other hand, as a typical electromechanical integrated system, the motor-driven flexible manipulator system (MDFMS) conveys the driving motor, transmission mechanism and flexible arm and will exhibit complex dynamic behaviors because of the coupled effect between each component (Siciliano and Khatib, 2008), which will influence the performance of the flexible manipulator and lead to certain motion fluctuations in the driving force and torque or the motion velocities and accelerations of the driving base, especially for systems using harmonic

drives (Zhao et al., 2014). The main reason for this is the effect of the motor parameters and mechanism inertias (Li et al., 2008; Liou et al., 1991; Andreaus and Casini, 2001; Liu et al., 2015), such as the unbalanced mass of motor in different rotational speeds and the backlash between the splines (Tuttle and Seering, 1996). However, it is worth noting that the driving force and torque or the motion velocities and accelerations of the driving base in the existing studies are assumed to be constant without any fluctuations (Li et al., 2008; Wei et al., 2006). From the above analysis we can see that this ideal assumption deviates from the actual cases and will cause an error in the dynamic analysis for a precision system. Thus, in order to conduct an accurate dynamic analysis for the flexible manipulator, the dynamic modeling should also consider the dynamic behavior of the driving base. The observation by Smaili et al. (1996) indicated that their devised dynamic model, which treats the linkage and its drive train as a system, can provide a more accurate estimation for the dynamic response of a mechanism during start-up. Considering the driving motor and the linkage mechanism as an integrated system, Li et al. (2008) investigated the nonlinear vibration of a three-phase AC motor-linkage mechanism system.

From reviewing these current studies, it can be seen that most of the related investigations considering the dynamic effect of the driving motor mainly focus on the linkage mechanism systems and that few surveys have studied the flexible manipulator. Moreover, as the flexible manipulator has actual motions and structure parameters, the effect mechanism of the dynamic effect on the vibration responses characteristic of the flexible manipulator under different motions and structure parameters, as well as the strategies to suppress the influence of the driving motor, has not yet been reported. The objectives of this paper are to establish the coupled dynamic model of the MDFMS considering the dynamic effect of the driving motor and investigate the dynamic property and vibration responses characteristic of the flexible manipulator under different motion velocities, accelerations and structure parameters. Also, based on the proposed dynamic model and analysis, the effect mechanism of the dynamic effect of the driving motor on the vibration characteristic and the relevant suppressing strategies are presented, which is meaningful for the motion optimization and vibration control for the flexible manipulator. The remainder of this paper is organized as follows. Section 2 presents the coupled dynamic model of the MDFMS. The analysis results based on the dynamic model are presented and discussed in Sect. 3. Finally, Sect. 4 concludes the paper.

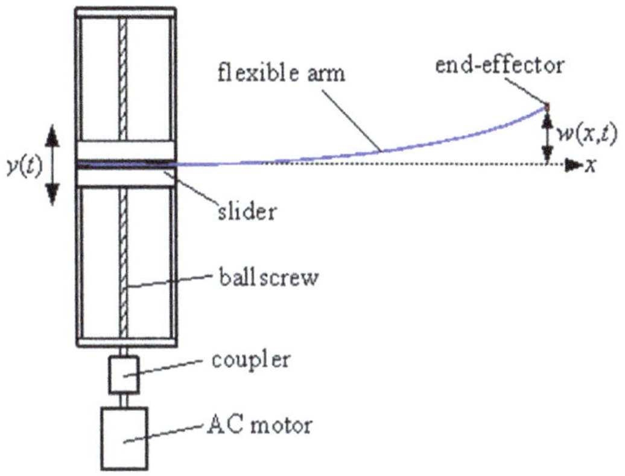

Figure 1. Schematic diagram of the MDFMS.

2 Coupled dynamic model of the MDFMS

A schematic diagram of the MDFMS is shown in Fig. 1, which gives a description of the comprising components: the driving motor connects the ball screw system through a coupler, and the flexible arm together with the end effector on its tip is clamped on the moving slider of the ball screw system. Due to the damp and stiffness being lower, the flexible arm and the end effector will be easily excited by elastic deformations and residual vibrations during the operation tasks. These undesirable elastic deformations and residual vibrations conflict with the precision demand of the end effector. In this section, to analyze the dynamic performance of the flexible manipulator, the coupled dynamic model taking the driving motor and the flexible arm as an integrated system is firstly established, and the dynamic equations obtained will be used for the following investigation of the characteristic vibration responses.

2.1 Dynamic equation of the motor and flexible manipulator system

As an integrated system, the dynamic equation of the MDFMS should consider the behaviors of the driving motor and the flexible arm. In the dynamic modeling, the flexible arm is modeled as a flexible beam to describe its lower damp and stiffness properties, and it is assumed that the flexible beam satisfies the Bernoulli–Euler beam assumption, and the transverse vibrations are the main motions considered. Thus, the transverse vibration displacement of the flexible manipulator can be expressed as

$$Y(x,t) = y(t) + w(x,t), \tag{1}$$

where $y(t)$ denotes the displacement of the moving slider and $w(x,t)$ denotes the transverse vibration displacement of the flexible arm; according to the assumed modes method (Singiresu, 2004), this yields

$$w(x,t)=\sum_{i=1}^{n\to\infty} W_i(x)q_i(t), \tag{2}$$

where $q_i(t)$ denotes the ith generalized coordinate and $W_i(x)$ denotes the ith orthogonal mode shapes, which can be written as (Singiresu, 2004)

$$W_i(x)=\sin k_i x-\sinh k_i x+\zeta\,(\cos k_i x-\cosh k_i x). \tag{3}$$

Here $\zeta=-\frac{\sin k_i L+\sinh k_i L}{\cos k_i L+\cosh k_i L}$, $k_i^4=\rho A\omega_i^2/EI$, in which ω_i is the ith natural frequency of the flexible arm; L, ρ, A and E are the length, mass density, cross-sectional area and Young's modulus of the flexible arm, respectively; and I denotes the cross-sectional moment of inertia about the neural axis and can be expressed as $I=bh^3/12$, where b and h are the width and thickness of the flexible arm, respectively.

The kinetic energy of the MDFMS is

$$E_k=\frac{1}{2}m_s\left(\frac{dy}{dt}\right)^2+\frac{1}{2}m_t\left[\frac{dy}{dt}+\frac{\partial w(L,t)}{\partial t}\right]^2+\frac{1}{2}\int_0^L\rho A\left(\frac{dy}{dt}+\frac{\partial w}{\partial t}\right)^2 dx. \tag{4}$$

Here, the first part of the equation denotes the kinetic energy of the moving slider, the second part denotes the kinetic energy of the end effector, and the third part denotes the kinetic energy of the flexible arm; m_s and m_t are the mass of the moving slider and the end effector, respectively.

The potential energy of the MDFMS mainly considers the elastic potential energy of the flexible arm and can be expressed as

$$E_p=\frac{1}{2}\int_0^L EI\left[\frac{\partial^2 w(x,t)}{\partial x^2}\right]^2 dx. \tag{5}$$

According to Eq. (2), Eqs. (4) and (5) can be further written as

$$E_k=\frac{1}{2}\sum_i\left[m_t W_i^2(L)+M_i\right]q_i^{\bullet 2}+\sum_i\left[m_t W_i(L)+\int_0^L\rho A W_i(x)dx\right]y^{\bullet}q_i^{\bullet}+\frac{1}{2}(m_s+\rho AL+m_t)y^{\bullet 2}, \tag{6}$$

$$E_p=\frac{1}{2}\sum_i K_i q_i^2, \tag{7}$$

where $(\bullet)$ denotes the time derivative and M_i and K_i denote the ith generalized mass and generalized stiffness, respectively, and can be defined as

$$M_i=\int_0^L\rho A\left(\frac{dW_i}{dx}\right)^2 dx, \tag{8}$$

$$K_i=\int_0^L EI\left(\frac{d^2W_i}{dx^2}\right)^2 dx. \tag{9}$$

Equations. (6) and (7) can be substituted into following Lagrange equation (Gross et al., 2011):

$$\frac{d}{dt}\left(\frac{\partial T}{\partial q_i^{\bullet}}\right)-\frac{\partial T}{\partial q_i}=Q_i, \tag{10}$$

where $T=E_k-E_p$ and Q_i represents the generalized force undergone in the flexible manipulator and satisfies the following relationship:

$$Q_i=\int_0^L f(x,t)W_i(x)dx. \tag{11}$$

Here $f(x,t)$ denotes the external force generated at the clamped end ($x=0$) of the flexible arm and can be expressed as

$$f(0,t)=m_s y^{\bullet\bullet}(t). \tag{12}$$

By combining Eqs. (11) and (12), the generalized force can be obtained as

$$Q_i=m_s y^{\bullet\bullet}(t)\int_0^L W_i(x)dx. \tag{13}$$

Through substitution of Eq. (13) into Eq. (10), the dynamic equation of the motor and flexible manipulator system can be obtained as

$$\left[m_t W_i^2(L)+M_i\right]q_i^{\bullet\bullet}(t)+K_i q_i(t)=\left[(m_s-\rho A)\int_0^L W_i(x)dx-m_t W_i(L)\right]y^{\bullet\bullet}(t). \tag{14}$$

According to the Duhamel integral (Singiresu, 2004), the vibration response equation of the flexible arm is obtained as

$$q_i(t)=B_{1i}\cos\beta_i t+B_{2i}\sin\beta_i t+\frac{\xi_i}{K_i}\int_0^t y^{\bullet\bullet}(\tau)\sin\beta_i(t-\tau)d\tau, \tag{15}$$

where

$$\beta_i^2 = \frac{K_i}{m_t W_i^2(L) + M_i}, \tag{16}$$

$$\xi_i = (m_s - \rho A)\int_0^L W_i(x)\mathrm{d}x - m_t W_i(L). \tag{17}$$

B_{1i} and B_{2i} are coefficients determined by the initial conditions and can be expressed as

$$B_{1i} = \frac{\rho A}{M_i}\int_0^L w(x,0)W_i(x)\mathrm{d}x,$$

$$B_{2i} = \frac{\rho A}{M_i \beta_i}\int_0^L w^{\bullet}(x,0)W_i(x)\mathrm{d}x. \tag{18}$$

On the other hand, for the ball screw transmission system, the displacement of the moving slider $y(t)$ in Eq. (15) satisfies

$$y^{\bullet}(t) = \frac{D}{2}\varphi^{\bullet}(t), \tag{19}$$

where D is the diameter of the ball screw and $\phi(t)$ is the angle displacement of the electromotor shaft.

In this case, Eq. (15) further becomes

$$q_i(t) = B_{1i}\cos\beta_i t + B_{2i}\sin\beta_i t + \frac{D\xi_i}{2K_i}\int_0^t \varphi^{\bullet\bullet}(\tau)\sin\beta_i(t-\tau)\mathrm{d}\tau. \tag{20}$$

From Eq. (20), we can obtain the result that the vibration responses of the flexible arm are related to the angle displacement of the driving motor, namely the dynamic behavior of the driving motor or the dynamic equation of the electromotor shaft should be also considered in the investigation of the dynamic analysis of the flexible manipulator.

2.2 Vibration displacement equation of the flexible manipulator

Section 2.1 shows that the dynamic equation of the electromotor shaft should be derived to determine the coupled dynamic equation of the system. According to the electromechanical dynamic (Wach, 2011), the dynamic equation of the electromotor shaft can be given as

$$J_M\varphi^{\bullet\bullet}(t) = M_{dc} - M_{fz} - M_0. \tag{21}$$

Here M_{dc} denotes the electromagnetic torque of the electromotor relating to the types and control strategies of the motor; M_{fz} denotes the load torque of the electromotor, which is mainly caused by the moving slider in the slideway pair and the ball screw pair; M_0 denotes the torque loss of the electromotor caused by the power loss in the transmission system, which is much smaller than M_{dc}; and J_M denotes the rotational inertia of the electromotor shaft, which can be expressed as

$$J_M = J_R + J_b + J_s. \tag{22}$$

Here J_R is the rotational inertia of the rotor determined by the motor type and structure, and J_b is the rotational inertia of the ball screw, which can be written as

$$J_b = \frac{\pi\rho_b D^4 L_b}{32}, \tag{23}$$

where ρ_b and L_b are the mass density and length of the ball screw, respectively.

J_s is the rotational inertia of the load which can be determined using the law of conservation of energy. The conversion rotational inertia of the load satisfies

$$\frac{1}{2}J_s\varphi^{\bullet 2}(t) = \frac{1}{2}m_s y^{\bullet 2}(t). \tag{24}$$

Through combination of Eqs. (19) and (24), J_s can be obtained as

$$J_s = \frac{1}{4}m_s D^2. \tag{25}$$

To determine the dynamic equation of the electromotor shaft as shown in Eq. (21), the electromagnetic torque and the load torque of electromotor should also be determined. As indicated above, M_{dc} is related to the types and control strategies of the motor. In this paper, a three-phase AC motor is considered and M_{dc} can be expressed as

$$M_{dc} = \frac{P_M}{2\pi f_1}p, \tag{26}$$

where f_1 is the power frequency, p is the number of pole pairs of the motor; P_M denotes the electromagnetic power $P_M = 3I_2'^2 r_2'/s$, where s is the slip ratio; r_2' is the conversion resistance of the rotor; and I_2' is the conversion current of the rotor, which can be further written as

$$I_2' = \frac{U_v}{\sqrt{\left(r_1 + \frac{r_2'}{s}\right)^2 + \left(x_1 + x_2'\right)^2}}, \tag{27}$$

where U_v is the voltage of the motor, r_1 and x_1 are the resistance and reactance of the stator, respectively, and x_2' is the conversion reactance of the rotor.

Substituting Eq. (27) into Eq. (26) yields

$$M_{dc} = \frac{3p}{2\pi f_1}\cdot\frac{U_v^2\frac{r_2'}{s}}{\left(r_1 + \frac{r_2'}{s}\right)^2 + \left(x_1 + x_2'\right)^2}. \tag{28}$$

Equation (28) shows that M_{dc} relates to the voltage, power frequency and structure parameters of the motor. By deriving

M_{dc} with respect to s and assigning the result as zero, we obtain

$$s_m = \frac{r_2'}{\sqrt{r_1^2 + (x_1 + x_2')^2}}. \tag{29}$$

Here s_m denotes the critical slip ratio.

Through combining Eqs. (28) and (29), the maximum electromagnetic torque can be given as

$$M_{max} = \frac{3p}{4\pi f_1} \cdot \frac{U_v^2}{\left[r_1 + \sqrt{r_1^2 + (x_1 + x_2')^2}\right]}. \tag{30}$$

By combining Eqs. (28) and (30), the result can be obtained that

$$\frac{M_{dc}}{M_{max}} = \frac{2r_2'\left[r_1 + \sqrt{r_1^2 + (x_1 + x_2')^2}\right]}{s\left[\left(r_1 + \frac{r_2'}{s}\right)^2 + (x_1 + x_2')^2\right]}. \tag{31}$$

According to Eq. (29), Eq. (31) can be simplified as

$$\frac{M_{dc}}{M_{max}} = \frac{2 + 2\frac{r_1}{r_2'}s_m}{\frac{s}{s_m} + \frac{s_m}{s} + 2\frac{r_1}{r_2'}s_m}. \tag{32}$$

Generally $r_1 \approx r_2'$; substituting this relation into Eq.(32) yields

$$M_{dc} = \frac{2\lambda_m M_N}{s_m}s. \tag{33}$$

Here λ_m denotes the overload factor of the motor and can be expressed as $\lambda_m = M_{max}/M_N$, $M_N = 9550\, P_N/n_N$, where P_N and n_N denote the rated power and the rated speed of the motor, respectively.

For the constant load torque system, λ_m and s_m satisfy the following relationship:

$$s_m = s_N\left(\lambda_m + \sqrt{\lambda_m^2 - 1}\right). \tag{34}$$

Generally $\lambda_m = 1.6 \ll 2.2$, and by combining Eqs. (33) and (34), M_{dc} can be subsequently determined.

For the MDFMS, the load torque of the electromotor is mainly generated by the moving slider, and during the motion, the friction force that the moving slider suffered in the slideway pair can be expressed as

$$F_s = \mu m_s g, \tag{35}$$

where μ is the friction coefficient of the interface and $g = 9.8\,\mathrm{m\,s^2}$.

The other part of the load torque is the friction torque generated in the ball screw pair, which can be written as

$$M_b = \frac{\mu_0 N D}{2}, \tag{36}$$

where μ_0 is the friction coefficient of the ball screw pair, which is relatively small for its transmitting style, and N is the load on the ball screw generated by the moving slider and workpiece – this paper mainly considers that of the moving slider; thus $N = m_s g$.

According to the law of energy conservation, we obtain

$$M_{fz}\varphi(t)^{\bullet}\eta_t = F_s y^{\bullet}(t) + M_b. \tag{37}$$

Here η_t is the energy transmission efficiency.

Through combination of Eqs. (19), (35) and (37), M_{fz} can be obtained as

$$M_{fz} = \frac{(\mu + \mu_0)m_s g D}{2\eta_t}. \tag{38}$$

Because the torque loss of the electromotor is relatively small, M_0 can be neglected for the purpose of simplifying the analysis.

With substitution of Eqs. (33) and (38) into Eq. (21), the dynamic equation of the electromotor shaft becomes

$$J_M\varphi^{\bullet\bullet} = \frac{2\lambda_m M_N}{s_m}s - \frac{(\mu + \mu_0)m_s g D}{2\eta_t}. \tag{39}$$

Here, $s = 1 - n/n_1$, n denotes the rotation speed of the motor, $n = 60\varphi^{\bullet}/(2\pi)$, n_1 is the synchronous rotation speed of the motor, and $n_1 = 60 f_1/p$. Thus,

$$s = 1 - \frac{p}{2\pi f_1}\varphi^{\bullet}. \tag{40}$$

Substituting Eq. (40) into Eq. (39) yields

$$J_M\varphi^{\bullet\bullet} + \frac{\lambda_m M_N}{s_m} \cdot \frac{p}{\pi f_1}\varphi^{\bullet} = \frac{2\lambda_m M_N}{s_m} - \frac{(\mu + \mu_0)m_s g D}{2\eta_t}. \tag{41}$$

According to Eq. (41), the rotation acceleration of the motor shaft can be obtained as

$$\varphi^{\bullet\bullet} = \left(\frac{2\pi f_1}{p}\alpha^2 - \alpha\gamma\right)e^{-\alpha t}, \tag{42}$$

where $\alpha = \frac{\lambda_m M_N}{J_M s_m} \cdot \frac{p}{\pi f_1}$, $\gamma = \frac{(\mu+\mu_0) m_s g D}{2 J_M \eta_t}$.

With Eq. (42) substituted into Eq. (20), vibration responses of the flexible arm can be written as

$$q_i(t) = B_{1i}\cos\beta_i t + B_{2i}\sin\beta_i t + \frac{D\xi_i}{2K_i}\int_0^t \left(\frac{2\pi f_1}{p}\alpha^2 - \alpha\gamma\right)e^{-\alpha t}\sin\beta_i(t-\tau)\mathrm{d}\tau. \tag{43}$$

With Eq. (43) substituted into Eq. (2), vibration response displacement of the flexible arm can be subsequently obtained as

$$w(x,t) = \sum_{i=1}^{n\to\infty} \left[\begin{array}{l} B_{1i}\cos\beta_i t + B_{2i}\sin\beta_i t \\ + \frac{D\xi_i}{2K_i}\left(\frac{2\pi f_1}{p}\alpha^2 - \alpha\gamma\right)\left(\begin{array}{l} \frac{\alpha}{\alpha^2+\beta_i^2}\sin\beta_i t - \frac{\beta_i}{\alpha^2+\beta_i^2}\cos\beta_i t \\ + \frac{\beta_i}{\alpha^2+\beta_i^2}e^{-\alpha t} \end{array} \right) \end{array} \right] W_i(x), \tag{44}$$

where

$$f_1 = \frac{p}{\pi D(1-s)} y^{\bullet}(t). \tag{45}$$

Equation (44) shows the vibration responses characteristic of the flexible manipulator considering the dynamic behavior of the driving motor. The situation where the flexible manipulator has velocities and accelerations is considered in this paper. For the ideal situation that the slider has constant velocities which ignore the dynamic effect of the driving motor, the velocity of the moving slider can be expressed as

$$v = y^{\bullet}(t). \tag{46}$$

By substituting Eqs. (19), (20) and (46) into Eq. (2), the vibration response displacement of the flexible arm can be obtained as

$$w(x,t) = \sum_{i=1}^{n\to\infty} (B_{1i}\cos\beta_i t + B_{2i}\sin\beta_i t)\, W_i(x). \tag{47}$$

By combining Eqs. (19), (41) and (45), s in Eq. (44) can be determined as

$$s_{\mathrm{v}} = \frac{(\mu+\mu_0)m_{\mathrm{s}}gDs_{\mathrm{m}}}{4\lambda_{\mathrm{m}}M_N\eta_{\mathrm{t}}}. \tag{48}$$

Similarly, for the ideal situation that the slider has constant accelerations, the acceleration of the moving slider can be expressed as

$$a = y^{\bullet\bullet}(t). \tag{49}$$

With Eqs. (19), (20) and (49) substituted into Eq. (2), the vibration response displacement of the flexible arm can be obtained as

$$w(x,t) = \sum_{i=1}^{n\to\infty} \left[B_{1i}\cos\beta_i t + B_{2i}\sin\beta_i t + \frac{a\xi_i}{\beta_i K_i}(1-\cos\beta_i t)\right] W_i(x). \tag{50}$$

By combining Eqs. (19), (41) and (49), s in Eq. (44) can be determined as

$$s_{\mathrm{a}} = \frac{(\mu+\mu_0)m_{\mathrm{s}}gDs_{\mathrm{m}}}{4\lambda_{\mathrm{m}}M_N\eta_{\mathrm{t}}} + \frac{J_{\mathrm{M}}s_{\mathrm{m}}}{\lambda_{\mathrm{m}}M_N D}a. \tag{51}$$

Equations (47) and (50) show the vibration responses of the flexible manipulator ignoring the dynamic effect of the driving motor. With substitution of Eqs. (48) and (51) into Eq. (44), the vibration responses of the flexible manipulator considering the dynamic effect of the driving motor can be subsequently investigated. Through comparison investigation of Eqs. (44), (47) and (50), the dynamic effect of the driving motor on the vibration response characteristic and the effect mechanism under different motion velocities and accelerations can be investigated.

3 Results and discussion

In this section, numerical simulations are conducted to investigate the influence of the dynamic effect of the driving motor on the vibration responses characteristic of the flexible manipulator under different motion velocities, accelerations and structure parameters.

The parameters of the flexible arm are length $L = 0.800$ m, width $b = 0.080$ m, thickness $h = 0.002$ m, Young's modulus $E = 197$ GPa and mass density $\rho = 7850\,\mathrm{kg\,m^{-3}}$. The driving motor is a three-phase AC motor with the parameters as follows: pole pairs $p = 3$, rated power $P_N = 2.2$ kW, rated rotation speed $n_N = 1430\,\mathrm{r\,min^{-1}}$, overload factor $\lambda_{\mathrm{m}} = 1.8$, and rotational inertia of the rotor $J_{\mathrm{R}} = 0.0054\,\mathrm{kg\,m^{-2}}$. The diameter and length of the ball screw are $D = 10$ mm and $L_{\mathrm{b}} = 800$ mm. The mass of the slider $m_{\mathrm{s}} = 500$ g, and the mass of the end effector $m_{\mathrm{t}} = 50$ g. The energy transmission efficiency $\eta_{\mathrm{t}} = 0.9$, the friction coefficient of the slideway $\mu = 0.004$, and the friction coefficient of the ball screw $\mu_0 = 0.001$.

Figure 2 shows the vibration responses of the MDFMS under velocity motions and acceleration motions. During the simulations, the motion velocity and acceleration are assigned as $v = 0.05\,\mathrm{m\,s^{-1}}$ and $a = 1.0\,\mathrm{m\,s^{-2}}$. It can be seen that the vibration responses considering the dynamic effect of the driving motor are different to the ideal situation ignoring the dynamic effect of the driving motor. The response amplitudes are larger than those of the ideal situation, especially in the initial time, for the acceleration motions. This indicates that the dynamic effect of the driving motor increases the vibration responses and the ideal situation will cause an error in the dynamic analysis of the flexible manipulator.

Figure 3 shows the frequency responses of the MDFMS under velocity motions. During the simulations, the motion velocity is assigned as $v = 0.05\,\mathrm{m\,s^{-1}}$. The result indicates that the frequency responses considering the dynamic effect of the driving motor are similar to the ideal situation. It should be noted that a consistent result can also be obtained for the MDFMS that has acceleration motions. Figures 2 and 3 provide a conclusion that the dynamic effect of the driving motor has a significant influence on the dynamic characteristic of the flexible manipulator, primarily in that it increases the vibration response amplitude.

To further investigate the effect mechanism of the driving motor on the vibration responses, vibration responses of the MDFMS under different motion velocities and accelerations

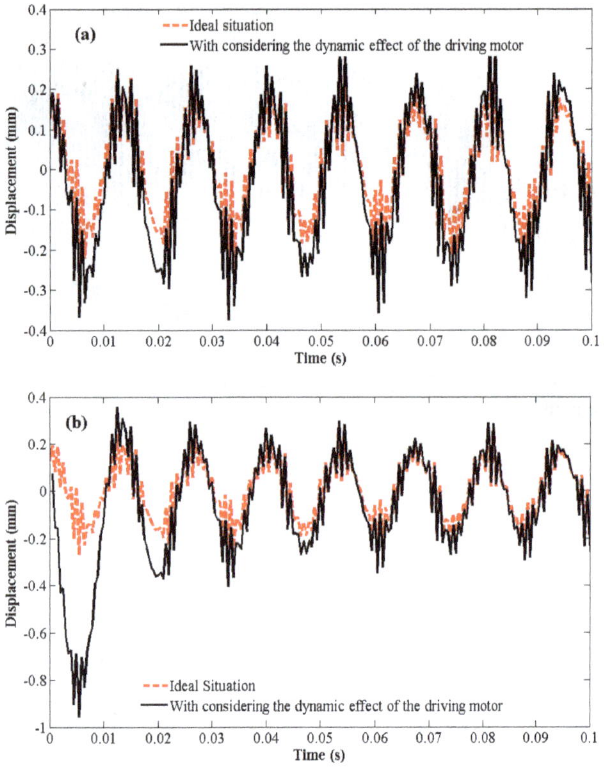

Figure 2. Vibration responses of the MDFMS under velocity **(a)** and acceleration **(b)** motions.

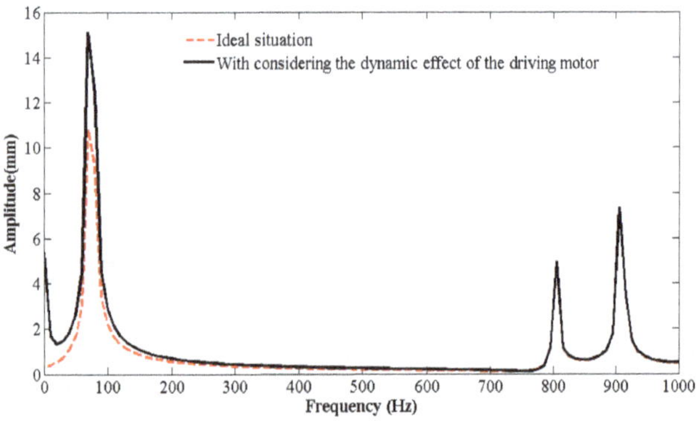

Figure 3. Frequency responses of the MDFMS under velocity motions.

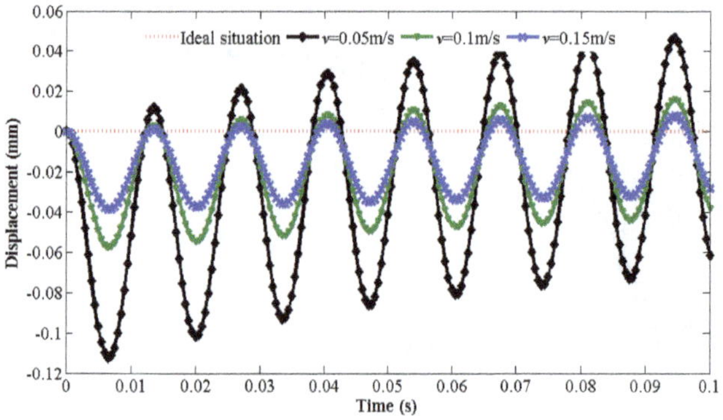

Figure 4. Vibration responses of the MDFMS under different motion velocities.

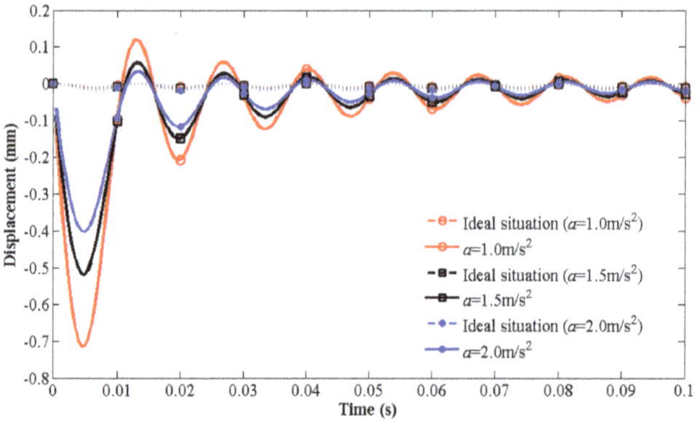

Figure 5. Vibration responses of the MDFMS under different motion accelerations.

are presented in Figs. 4 and 5. With the purpose of simplifying the analysis process, only the results of steady-state responses are considered in the following investigations.

Figure 4 shows the vibration responses of the MDFMS under different motion velocities. During the simulations, the motion velocities are assigned as $v = 0.05$, 0.1 and $0.15\,\mathrm{m\,s^{-1}}$. It can be seen that, for the ideal situation ignoring the dynamic effect of the driving motor, the steady-state responses of different motion velocities are consistent and show minor changes, as described by the dotted red line, which can also be obtained from Eq. (47). However, considering the dynamic effect of the driving motor, the velocities exhibit noticeable influence on the vibration responses and the response amplitudes are larger than the ideal situations; moreover, as the velocities increase, the response amplitudes become smaller and the influence decreases.

The vibration responses of the MDFMS under different motion accelerations are presented in Fig. 5. During the simulations, the motion accelerations are assigned as $a = 1.0$, 1.5 and $2.0\,\mathrm{m\,s^{-2}}$. It can be found that the vibration responses considering the dynamic effect of the driving motor are obviously larger than the ideal situation, and the results show an obvious difference, especially in the initial time; as time progresses, the difference decreases and the results gradually approach the ideal situation. Moreover, if we do not consider the dynamic effect of the driving motor, the influence of accelerations on the vibration responses is not obvious. However, when the dynamic effect of the driving motor is considered, the accelerations show a considerable influence; as the accelerations increase, the difference of the results and the influence of the driving motor decrease, which is similar to the velocity motion situation.

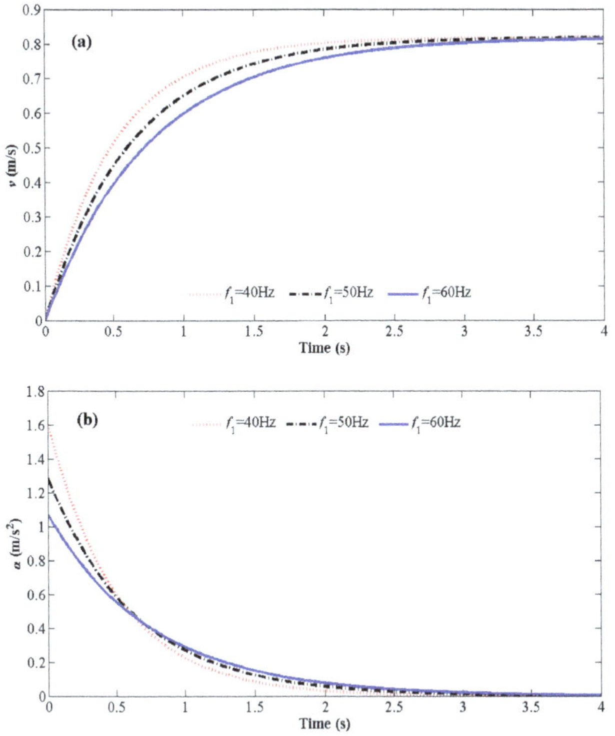

Figure 6. Velocity (**a**) and acceleration (**b**) characteristics of the MDFMS under varied power frequencies.

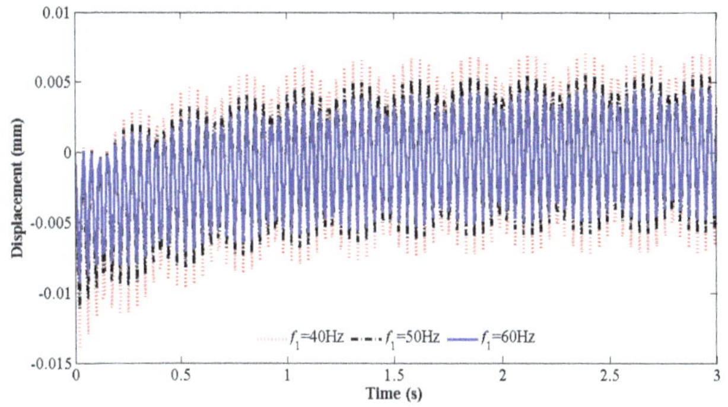

Figure 7. Vibration responses of the MDFMS under varied power frequencies.

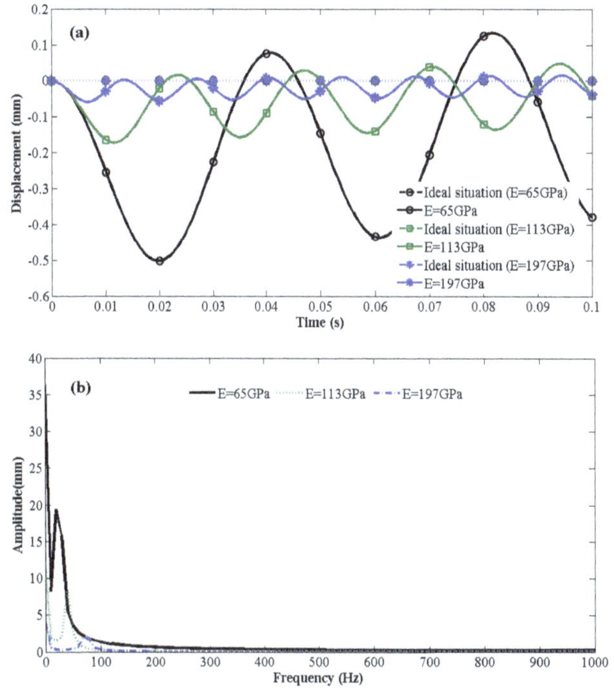

Figure 8. Effect of Young's modulus on the vibration responses of the MDFMS: (**a**) response amplitudes and (**b**) response frequencies.

Figures 4 and 5 demonstrate that the influence of the dynamic effect of the driving motor is more considerable, especially in the initial time, and the influence decreases as the velocities and accelerations increase, which indicates that larger velocities and accelerations have a suppressing effect on the influence of the driving motor. As we know, for a constant moment load, the velocity and acceleration characteristics and the response time of the system are determined by the power frequency. Therefore, to illustrate this phenomenon, Fig. 6 shows the velocity and acceleration characteristics of the MDFMS under different power frequencies assigned as $f_1 = 40$, 50 and 60 Hz. We can obtain the result that, for lower power frequencies, the velocity response and the acceleration response are more rapid, and the vibration responses are more intense, as shown in Fig. 7, which can also be obtained from Eq. (28,) which gives a description that lower power frequencies can obtain larger electromagnetic torques; as a result, if the load torques are constant, the response processes will become more rapid.

The flexible arm and the end effector are the main components of the operation system; both have a significant influence on the structural service and dynamic performance. Generally, the flexible arm can be structured using lightweight materials of different Young's modulus, and the end effector can be established with different masses. Figure 8 shows the vibration responses of the flexible arm with a different Young's modulus, assigning $E = 65$, 113 and 197 GPa. The result reveals that Young's modulus will change the dynamic performance and have considerable influence on the vibration responses. If the flexible arm is structured using a smaller Young's modulus, the vibration response amplitudes increase and the response frequencies become lower, which indicates that the vibration responses are enhanced and lower-frequency vibrations are easily excited. Moreover, in this case the influence of the driving motor will become more noticeable and should be given more consideration.

Figure 9 shows the influence of the end effector on the vibration responses. During the simulations, the mass of the end effector is assigned as $m_t = 10$, 20 and 50 g. We can see that the vibration responses vary with the mass of the end

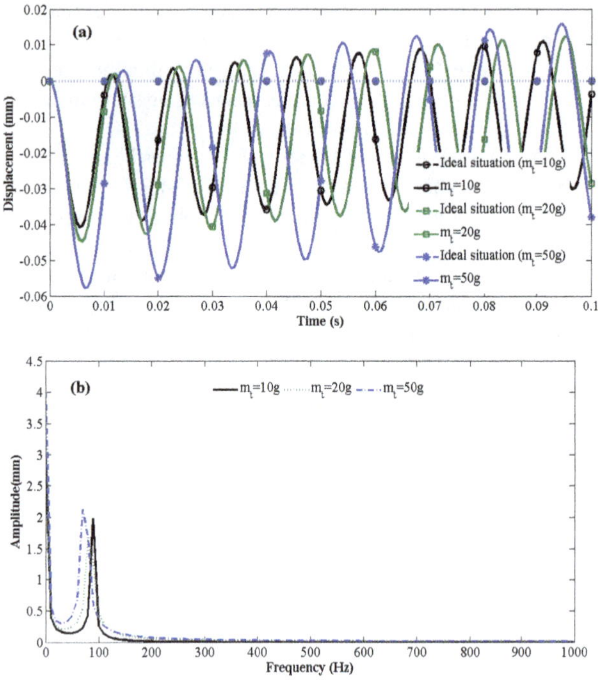

Figure 9. Effect of the end effector on the vibration responses of the MDFMS: **(a)** response amplitudes and **(b)** response frequencies.

effector; for lighter end effectors, the response amplitudes are smaller, and as mass increases, the response amplitudes increase while the response frequencies decrease. This result can also be obtained from Eq. (16). This indicates that a lightweight end effector contributes to suppressing the vibration responses of the arm, which is meaningful for the structure design and vibration control of the flexible manipulator.

4 Conclusions

A coupled dynamic model of the MDFMS is established which can clearly reflect the dynamic effect of the driving motor. Based on the proposed coupled dynamic model, the vibration responses of the flexible manipulator under different velocities, accelerations and structure parameters, as well as the effect mechanism of the driving motor on the vibration responses, are investigated. The results demonstrate that the dynamic effect of the driving motor has a significant influence on the dynamic characteristic of the flexible manipulator, primarily in that it increases the vibration responses amplitude; furthermore, the ideal assumption ignoring the dynamic effect of the driving motor, which is generally regarded in recent research, will cause an error in the dynamic analysis, especially in the initial time. The response amplitudes are larger than the ideal situations; as the velocities and accelerations increase, the response amplitudes become smaller and the influence decreases. It can also be observed that larger velocities and accelerations have a suppressing effect on the influence of the driving motor. Moreover, the Young's modulus of the flexible arm and the mass of the end effector have a considerable effect on the vibration responses, which indicates that a smaller Young's modulus increases the vibration response amplitudes and the influence of the driving motor is more noticeable; further, a lightweight end effector contributes to suppressing the influence of the driving motor. The results obtained in this paper are meaningful for the structure design, motion optimization and dynamic analysis of the flexible manipulator.

Acknowledgements. This research work is supported by the National Natural Science Foundation of China (no. 51305444), the Doctoral Fund of Ministry of Education (no. 20120095120013), the Scientific and Technological Project of Jiangsu Province (BY2014028-06), and the project funded by the Priority Academic Program Development of Jiangsu Higher Education Institutions (PAPD). The authors sincerely thank the reviewers for their significant and constructive comments and suggestions, which substantially helped in improving this paper.

References

Andreaus, U. and Casini, P.: Dynamics of friction oscillators excited by a moving base and/or driving force, J. Sound Vibrat., 245, 685–699, 2001.

Dwivedy, S. K. and Eberhard, P.: Dynamic analysis of flexible manipulators, a literature review, Mech. Mach. Theory, 41, 749–777, 2006.

Ge, S. S., Lee, T. H., and Zhu, G.: Asymptotically stable end-point regulation of a flexible SCARA/Cartesian robot, IEEE/ASME T. Mechatron., 3, 138–144, 1998.

Gross, D., Hauger, W., Schröder, J., Wall, W. A., and Govindjee, S.: Engineering Mechanics 3: Dynamics, Springer, London, 2011.

Kerem, G., Bradley, J. B., and Edward, J. P.: Vibration control of a single-link flexible manipulator using an array of fiber optic curvature sensors and PZT actuators, Mechatronics, 19, 167–177, 2009.

Li, Z. J., Cai, G. W., Huang, Q. B., and Liu, S. Q.: Analysis of nonlinear vibration of a motor-linkage mechanism system with composite links, J. Sound Vibrat., 311, 924–940, 2008.

Liou, F. W., Erdman, A. G., and Lin, C. S.: Dynamic analysis of a motor-gear-mechanism system, Mech. Mach. Theory, 26, 239–252, 1991.

Liu, Y. F., Li, W., Yang, X. F., Fan, M. B., Wang, Y. Q., and Lu, E.: Vibration response and power flow characteristics of a flexible manipulator with a moving base, Shock and Vibration, Hindawi Publishing Corporation, New York, 1–8, 2015.

Maria, A. N., Jorge, A. C., Ambrósio, L. M., Roseiro, A. A., and Vasques, C. M. A.: Active vibration control of spatial flexible multibody systems, Multibody Sys. Dynam., 30, 13–35, 2013.

Mohsen, D., Nader, J., Zeyu, L., and Darren, M D.: An observer-based piezoelectric control of flexible cartesian robot arms: theory and experiment, Control Eng. Pract., 12, 1041–1053, 2004.

Qiu, Z. C.: Adaptive nonlinear vibration control of a Cartesian flexible manipulator driven by a ball screw mechanism, Mech. Syst. Signal Proc., 30, 248–266, 2012.

Qiu, Z. C., Shi, M. L., Wang, B., and Xie, Z. W.: Genetic algorithm based active vibration control for a moving flexible smart beam driven by a pneumatic rod cylinder, J. Sound Vibrat., 331, 2233–2256, 2012.

Siciliano, B. and Khatib, O.: Springer Handbook of robotics, Springer, London, 2008.

Singiresu, S. R.: Mechanical Vibration, 4th Edn., Pearson Education Inc., New York, 2004.

Smaili, A., Kopparapu, M. and Sannah, M.: Elastodynamic response of a d.c motor driven flexible mechanism system with compliant drive train components during start-up, Mech. Mach. Theory, 31, 659–672, 1996.

Tuttle, T. D. and Seering, W. P.: A nonlinear model of a harmonic drive gear transmission, IEEE T. Robot. Automat., 12, 368–374, 1996.

Wach, P.: Dynamics and Control of Electrical Drives, Springer, London, 2011.

Wei, K. X., Meng, G., Zhou, S., and Liu, J. W.: Vibration control of variable speed/acceleration rotating beams using smart materials, J. Sound Vibrat., 298, 1150–1158, 2006.

Zhang, X., Mills, J. K., and Cleghorn, W. L.: Investigation of axial forces on dynamic properties of a flexible 3-PRR planar parallel manipulator moving with high speed, Robotica, 28, 607–619, 2010.

Zhao, J. L., Yan, S. Z., and Wu, J. N.: Analysis of parameter sensitivity of space manipulator with harmonic drive based on the revised response surface method, Acta Astronaut., 98, 86–96, 2014.

16

B-spline parameterized optimal motion trajectories for robotic systems with guaranteed constraint satisfaction

W. Van Loock, G. Pipeleers, and J. Swevers

Department of Mechanical Engineering, Division PMA, KU Leuven, 3001 Leuven, Belgium

Correspondence to: W. Van Loock (wannes.vanloock@kuleuven.be)

Abstract. When optimizing the performance of constrained robotic system, the motion trajectory plays a crucial role. In this research the motion planning problem for systems that admit a polynomial description of the system dynamics through differential flatness is tackled by parameterizing the system's so-called flat output as a polynomial spline. Using basic properties of B-splines, sufficient conditions on the spline coefficients are derived ensuring satisfaction of the operating constraints over the entire time horizon. Furthermore, an intuitive relaxation is proposed to tackle conservatism and a supporting software package is released. Finally, to illustrate the overall approach and potential, a numerical benchmark of a flexible link manipulator is discussed.

1 Introduction

The computation of a constrained optimal motion trajectory is a challenging problem in control and has attracted researchers already for several decades. In the 1990s the concept of differential flatness (Fliess et al., 1995) arose, which allows characterizing all the state space trajectories and the corresponding input history by means of a particular set of outputs. Differentially flat systems encompass all linear, controllable systems and many nonlinear systems as well. It quickly gained popularity for solving optimal control problems since in this way, the integration of the system dynamics is avoided. Hence, the problem reduces to finding the best flat output that obeys the boundary conditions and the state and input constraints. To deal with the infinite dimensionality of this problem, a polynomial or spline parameterization for the flat output is often used. To impose state and input constraints classical approaches in the literature (Louembet et al., 2009; Milam et al., 2000) apply a sampling strategy. As a result the constraints are not guaranteed to be satisfied in between the samples such that post-analysis is required for critical constraints. The aim in this paper is to provide constraints that can guarantee constraint satisfaction.

For linear systems, several methods have been proposed in the literature to guarantee constraint satisfaction at all times. Henrion and Lasserre (2006) propose a polynomial parameterization for the flat output, hereby transforming the constrained motion planning problem into a polynomial nonnegativity problem. Subsequently, a sum-of-squares decomposition is sought for using semidefinite programming. Piecewise polynomials can allow for more freedom in the parameterization and the former approach can be straightforwardly extended by searching for sum-of-squares decompositions on the individual polynomial pieces. A similar strategy is followed by Louembet et al. (2010), where a sum-of-squares decomposition is searched for directly in the B-spline basis, but by doing so a conservative solution is determined. Suryawan et al. (2012) also adopt a piecewise polynomial parameterization, but in contrast to the sum-of-squares procedure of Louembet et al. (2010), the authors express the semi-infinite constraints by applying basis function segmentation and using the convex hull property of B-splines, leading to linear constraints. Such an approach yields only sufficient conditions and hence introduces conservatism, which can be quite severe (de Boor and Daniel, 1974).

For nonlinear systems, existing approaches resort to convex approximations of the feasible set. Louembet et al. (2010) require a polytopic inner approximation of the feasible set. Inevitably, this method introduces conservatism in the problem. Moreover, some feasible sets do not admit such a polytopic approximation, e.g. obstacle avoidance constraints. For nonlinear systems that admit a polynomial representation by differential flatness, Suryawan et al. (2012) propose

a strategy to impose the semi-infinite constraints by relying on the convex hull property of splines and only keeping the linear and cubic monomial terms in the polynomial expansion. This way the optimization problem amounts to a simple QP. It should be noted, however, that this approach results in overly conservative constraints. To the best of our knowledge, no adequate method exists for guaranteed constraint satisfaction in nonlinear systems.

This paper aims to develop an optimization approach with guaranteed constraint satisfaction over the entire time horizon for systems that admit a polynomial representation by differential flatness. For systems that are flat but do not have such a polynomial representation, the equations are transformed into polynomial form, either by simple manipulation, a change of variables or approximation. Similar to Suryawan et al. (2012), our method is based on the convex hull property of B-splines. However, we do not require basis function segmentation and additionally we propose an intuitive method to control the conservatism that is introduced.

Section 2 introduces the motion planning problem as well as the concept of differential flatness. The following section proposes a spline parameterization of the flat output and discusses various relevant properties of splines. In Sect. 4, two relaxation strategies are discussed that effectively reduce conservatism. The free end-time problem is discussed as well. Section 5 validates our approach on a numerical benchmark problem. Furthermore, as a complement to the paper, a supporting software tool is released to aid the user in formulating motion planning problems involving splines.

2 Problem formulation

Consider a system governed by the differential equation

$$\dot{\boldsymbol{x}} = \boldsymbol{f}(\boldsymbol{x}, \boldsymbol{u}), \boldsymbol{x}(0) = \boldsymbol{x}_0, \tag{1}$$

with states $\boldsymbol{x}(t) \in \mathbb{R}^{n_x}$ and inputs $\boldsymbol{u}(t) \in \mathbb{R}^{n_u}$. We are interested in finding the control law $\boldsymbol{u}(t)$, $t \in [0, t_\mathrm{f}]$ that steers the system from an initial state $\boldsymbol{x}_0$, at $t = 0$, to a terminal state $\boldsymbol{x}_{t_\mathrm{f}}$, at $t = t_\mathrm{f}$, and that minimizes a performance criterion $g(\boldsymbol{x}, \boldsymbol{u}, t_\mathrm{f})$. At the same time the control law must obey state and input constraints:

$$\boldsymbol{h}(\boldsymbol{x}(t), \boldsymbol{u}(t)) \geq 0, \ \forall t \in [0, t_\mathrm{f}], \tag{2}$$

where the constraint function $\boldsymbol{h} : \mathbb{R}^{n_x} \times \mathbb{R}^{n_u} \to \mathbb{R}^{n_h}$ are assumed to be polynomial in $\boldsymbol{x}$ and $\boldsymbol{u}$.

In this work, we assume the system Eq. (1) is differentially flat (Fliess et al., 1995). This means that there exists a set of variables, called the flat outputs, $\boldsymbol{y} \in \mathbb{R}^{n_u}$ of the form

$$\boldsymbol{y} = \boldsymbol{\phi}\left(\boldsymbol{x}, \boldsymbol{u}, \boldsymbol{u}^{(1)}, \ldots, \boldsymbol{u}^{(q)}\right)$$

such that

$$\boldsymbol{x} = \boldsymbol{\psi}_x\left(\boldsymbol{y}, \boldsymbol{y}^{(1)}, \ldots, \boldsymbol{y}^{(r-1)}\right)$$

and

$$\boldsymbol{u} = \boldsymbol{\psi}_u\left(\boldsymbol{y}, \boldsymbol{y}^{(1)}, \ldots, \boldsymbol{y}^{(r)}\right)$$

for some positive integers q, r. So for a flat system, there exists an algebraic relationship between the states and inputs, and the flat output and its derivatives. Aside from all linear, controllable systems also many nonlinear systems are differentially flat. For more details and a catalog of flat systems the interested reader is referred to Martin et al. (2003) and Lévine (2010).

Differential flatness is particularly interesting when solving optimal control problems since it avoids integration of the system dynamics Eq. (1), an often costly and numerically challenging step. Indeed, by formulating the problem from the first paragraph in terms of the flat output, we arrive at the following optimization problem:

$$\begin{aligned} \underset{\boldsymbol{y}(\cdot)}{\text{minimize}} \quad & g\left(\boldsymbol{\psi}_x\left(\boldsymbol{y}, \ldots, \boldsymbol{y}^{(r-1)}\right), \boldsymbol{\psi}_u\left(\boldsymbol{y}, \ldots, \boldsymbol{y}^{(r)}\right), t_\mathrm{f}\right) \\ \text{subject to} \quad & \boldsymbol{y}^{(j)}(0) = \boldsymbol{y}_0^{(j)}, \ j = 0, \ldots, r-1 \\ & \boldsymbol{y}^{(j)}(t_\mathrm{f}) = \boldsymbol{y}_{t_\mathrm{f}}^{(j)}, \ j = 0, \ldots, r-1 \\ & \boldsymbol{h}\left(\boldsymbol{\psi}_x\left(\boldsymbol{y}, \ldots, \boldsymbol{y}^{(r-1)}\right), \boldsymbol{\psi}_u\left(\boldsymbol{y}, \ldots, \boldsymbol{y}^{(r)}\right)\right) \\ & \geq 0, \forall t \in [0, t_\mathrm{f}], \end{aligned} \tag{3}$$

where the boundary conditions for the flat output, $\boldsymbol{y}_0^{(j)}$ and $\boldsymbol{y}_{t_\mathrm{f}}^{(j)}$, are readily determined from $\boldsymbol{x}_0$ and $\boldsymbol{x}_{t_\mathrm{f}}$.

In solving the above optimization problem, we still face two challenges: (i) instead of a finite set of variables, the optimization variable is a function $\boldsymbol{y}(\cdot)$ and (ii) the constraints must be enforced at all time instances. Therefore, the problem is infinite dimensional with infinitely many constraints. To cope with the infinite dimensionality a fixed parameterization is usually chosen for $\boldsymbol{y}(\cdot)$. As splines provide a good approximation for smooth functions (de Boor, 2001), we will use a polynomial spline parameterization for $\boldsymbol{y}$ in this paper resulting in an optimization problem with few optimization variables that can be solved efficiently. In addition, as shown in the following section, such a parameterization allows us to impose the semi-infinite constraints by a finite number of sufficient constraints provided that the maps $\boldsymbol{\psi}_x$ and $\boldsymbol{\psi}_u$ are polynomial.

3 B-spline parameterized solutions

Let $\boldsymbol{\kappa} = (\kappa_0, \ldots, \kappa_{m+1})$ be a strictly increasing vector of points, k be a positive integer, and $\boldsymbol{v} = (v_1, \ldots, v_m)$ be a vector of integers with $0 \leq v_i \leq k-1$. Then, s is a polynomial spline of order k with break points $\boldsymbol{\kappa}$ and continuity conditions $\boldsymbol{v}$ if there exist polynomials $p_0, \ldots, p_l$ of order k such that

$$s(t) = p_i(t), \text{ for } \kappa_i \le t < \kappa_{i+1},\ i = 0, 1, \ldots, m-1$$
$$s(t) = p_m(t), \text{ for } \kappa_m \le t \le \kappa_{m+1},$$

and

$$p_{i-1}^{(j-1)}(\kappa_i) = p_i^{(j-1)}(\kappa_i) \text{ for } j = 1, \ldots, \nu_i,\ i = 1, \ldots, m.$$

The vector space of polynomial splines with given k, $\boldsymbol{\kappa}$ and $\boldsymbol{\nu}$ is denoted by $\Pi_{k,\kappa,\nu}$ and has dimension $n = (m+1)k - \sum_{i=1}^{m} \nu_i$. The normalized B-spline basis of order k, defined over the knot vector

$$\boldsymbol{t} = \left(\underbrace{\kappa_0, \ldots, \kappa_0}_{k}, \underbrace{\kappa_1, \ldots, \kappa_1}_{k-\nu_1}, \ldots, \underbrace{\kappa_m, \ldots, \kappa_m}_{k-\nu_m}, \underbrace{\kappa_{m+1}, \ldots, \kappa_{m+1}}_{k} \right)$$

is commonly used as a basis for this vector space as it has various useful properties: the basis functions are nonnegative, sum up to one (partition of unity) and have local (minimal) support (de Boor, 2001). It yields a stable evaluation of the functions and its derivatives. A spline $s \in \Pi_{k,\kappa,\nu}$ with B-spline basis $\boldsymbol{b}_s = (b_1, \ldots, b_n)$ and (B-spline) coefficients $\boldsymbol{\sigma} = (\sigma_1, \ldots, \sigma_n)$ is represented as

$$s(t) = \sum_{i}^{n} \sigma_i b_i(t) = \langle \boldsymbol{\sigma}, \boldsymbol{b}_s(t) \rangle.$$

The control polygon of the spline is the broken line with

$$c_i = \left(t_i^*, \sigma_i\right), i = 1, \ldots, n$$

as vertex sequence, where

$$t_i^* = \frac{t_{i+1} + \ldots + t_{i+k-1}}{k-1}, \forall i.$$

Figure 1 illustrates a fourth order spline and its control polygon, which can be regarded as an exaggerated version of the spline itself (de Boor, 2001).

The convex hull property of splines is essential for the further course of this paper and is repeated from de Boor (2001) for completeness:

Property 1 (Convex hull) *Let s be a polynomial spline of order k with knot vector* $\boldsymbol{t}$*. From the nonnegativity, partition of unity and local support property of the B-spline basis it follows immediately that the segment* $s(t)$, $t \in [t_i, t_{i+1}]$ *lies within the convex hull of its control points* $c_{i-k+1}, \ldots, c_i$.

The convex hull property is illustrated in Fig. 1. It follows immediately from Property 1 that for constants a and b.

$$a \le \boldsymbol{\sigma} \le b \Rightarrow a \le s(t) \le b, \forall t \in \left[\kappa_0, \kappa_{m+1}\right].$$

Thus, by constraining the spline's coefficients, semi-infinite bounds on the spline can easily be imposed. Furthermore, it is trivial to see that any polynomial function of splines is itself a spline. Moreover, its B-spline coefficients can be determined from the B-spline coefficients of its constituents using the sum and product properties detailed in Appendix A.

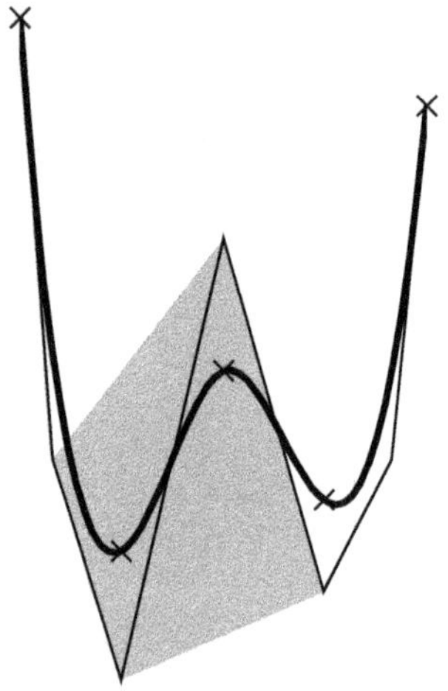

Figure 1. A continuous fourth order spline with five breaks indicated by the crosses. The spline's control polygon is the broken thin line. The gray area illustrates the convex hull property for points between the second and third break.

Now, let us apply Properties 1–3 to the optimization problem Eq. (3). Let $\boldsymbol{b}_{y_i}, i = 1, \ldots, n_u$ denote the B-spline basis for the ith flat output and $\boldsymbol{\gamma}_i$ the corresponding coefficients. Since $\boldsymbol{\psi}_x$, $\boldsymbol{\psi}_u$ and $\boldsymbol{h}$ are polynomial, we can determine the B-spline coefficients $\boldsymbol{\eta}_i(\boldsymbol{\gamma}_1, \ldots, \boldsymbol{\gamma}_{n_u})$ of the ith component of $\boldsymbol{h}(\boldsymbol{\psi}_x(\boldsymbol{y}, \boldsymbol{y}^{(1)}, \ldots, \boldsymbol{y}^{(r-1)}), \boldsymbol{\psi}_u(\boldsymbol{y}, \boldsymbol{y}^{(1)}, \ldots, \boldsymbol{y}^{(r)}))$. Then, an approximate solution for Eq. (3) is determined by solving

$$\begin{array}{lll} \underset{\boldsymbol{\gamma}_1, \ldots, \boldsymbol{\gamma}_{n_u}}{\text{minimize}} & \tilde{g}(\boldsymbol{\gamma}_1, \ldots, \boldsymbol{\gamma}_{n_u}, t_f) & \\ \text{subject to} & \langle \boldsymbol{\gamma}_i, \boldsymbol{b}_{y_i}^{(j)}(0) \rangle = \left(\boldsymbol{y}_0^{(j)}\right)_i, & j = 0, \ldots, r-1,\ i = 1, \ldots, n_u \\ & \langle \boldsymbol{\gamma}_i, \boldsymbol{b}_{y_i}^{(j)}(t_f) \rangle = \left(\boldsymbol{y}_{t_f}^{(j)}\right)_i, & j = 0, \ldots, r-1,\ i = 1, \ldots, n_u \\ & \boldsymbol{\eta}_i(\boldsymbol{\gamma}_1, \ldots, \boldsymbol{\gamma}_{n_u}) \ge 0, & i = 1, \ldots, n_h \end{array} \quad (4)$$

where, $\tilde{g}$ denotes the result of the substitution of the spline parameterization in the objective function of Eq. (3).

4 Discussion

4.1 Reducing conservatism

Imposing a semi-infinite constraint on a polynomial spline through constraints on its B-spline coefficients yields only sufficient conditions and hence, the optimal value of Eq. (4) is an upper bound on the optimal value of Eq. (3). This conservatism is due to the distance between the control polygon of the spline and the spline itself. By representing the spline in a higher dimensional basis that includes the original one, the control polygon can be brought closer to the spline. Such

```
basis = BSplineBasis([0, tf], 4, 11); % Basis of degree 4 with 11 knots
y = BSpline.sdpvar(basis, [1, 1]);    % scalar (1x1) spline variable
dy = y.derivative(1);
ddy = y.derivative(2);
dddy = y.derivative(3);

pu = c0 + c1 * y;                     % Upper bound (l=1)
pl = -c0 + c1 * y;                    % Lower bound (l=1)

obj = y^2;
con = [y.f(0) == 0.8, y.f(tf) == -0.8, % z.f evaluates the spline
       dy.f(0) == 0, dy.f(tf) == -0.8,
       ddy.f(0) == -0.2, ddy.f(tf) == 0.2,
       dddy.f(0) == 0, dddy.f(tf) == 0,
       M*g*L/k * pu + I1/k * ddy + y <= pi/4, % semi-infinite constraints
       M*g*L/k * pl + I1/k * ddy + y >= -pi/4,
       M*g*L/k * pu + I1/k * ddy <= pi/16,
       M*g*L/k * pl + I1/k * ddy >= -pi/16];
sol = optimize(con, obj.integral());

y_opt = value(y);                      % Retrieve numerical solution for z
```

Listing 1. Example code for solving Eq. (6) with $l = 1$.

a basis can be derived by inserting knots, increasing the order[1] or a combination of both.

More precisely, let $s \in \Pi_{k,\kappa,\nu}$ with B-spline coefficients $\boldsymbol{\sigma}$. Let $\Pi_{k,\kappa,\nu} \subset \Pi_{\hat{k},\hat{\kappa},\hat{\nu}}$ with $\hat{k} \geq k$, and $\hat{\boldsymbol{\kappa}}$ and $\hat{\boldsymbol{\nu}}$ the refined break and continuity vectors such that $\boldsymbol{\kappa} \subseteq \hat{\boldsymbol{\kappa}}$ and $\nu_i \geq \hat{\nu}_j$, $i = 1, \ldots, m$ with j: $\kappa_i = \hat{\kappa}_j$. Then $\hat{s} \in \Pi_{\hat{k},\hat{\kappa},\hat{\nu}}$ with B-spline coefficients

$$\hat{\boldsymbol{\sigma}} = \mathbf{T}_{\hat{s}}^{s} \boldsymbol{\sigma},$$

where $\mathbf{T}_{\hat{s}}^{s}$ denotes the linear mapping from $\boldsymbol{b}_{\hat{s}}$ to $\boldsymbol{b}_s$, equals s and it can be shown that the control polygon of $\hat{s}$ will lie closer to the graph than that of s (de Boor, 2001).

A refinement of the break and/or continuity vectors acts locally on the spline and can target specific regions where conservatism is high. Order elevation is a global approach and changes the entire shape of the control polygon. This is illustrated in Fig. 2. In both cases it is clear that the new control polygon lies closer to the spline than the original one and hence conservatism is reduced. Note that order elevation increases the number of coefficients, and hence also the number of constraints, by $m+1$. Inserting a single knot only amounts to one additional coefficient. Moreover, it can be shown that the convergence towards the spline for subsequent knot insertions is faster compared to order elevation (Prautzsch et al., 2002). For these reasons knot insertion is generally favored.

[1] Order elevation borrows from the idea of Polya's relaxation for polynomials.

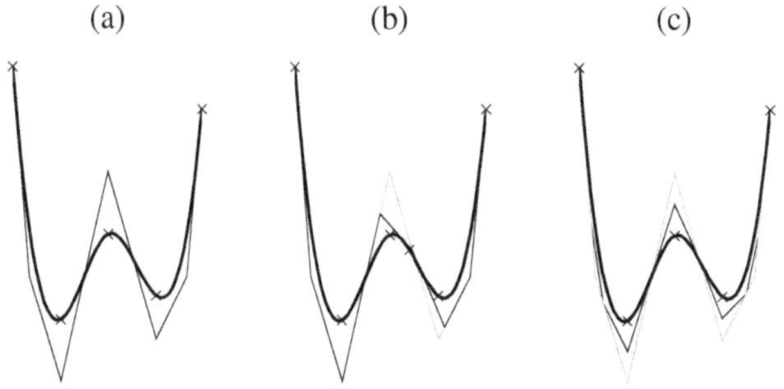

Figure 2. Refining the control polygon brings the control polygon closer to the spline: **(a)** the original spline, **(b)** one knot insertion and **(c)** elevate order by one. Note that knot insertion acts locally, while order elevation changes the control polygon globally.

4.2 Free end-time problems

For optimization problems where the final time t_f is a variable, a classical time scaling is applied to Eq. (3). The pseudo time $\tau = \frac{t}{t_f}$ is used as free variable in the parameterization for the flat output instead of the time t. Consequently, the derivatives must be scaled by t_f and for free end-time problem Eq. (3) can be formulated as follows:

$$\begin{array}{ll}
\underset{y(\cdot),t_f}{\text{minimize}} & g\left(\boldsymbol{\psi}_x\left(y,\ldots,t_f^{r-1}y^{(r-1)}\right),\boldsymbol{\psi}_u\left(y,\ldots,t_f^{r}y^{(r)}\right),t_f\right) \\
\text{subject to} & y^{(j)}(0) = t_f^{j}\, y_0^{(j)},\ j=0,\ldots,r-1 \\
& y^{(j)}(1) = t_f^{j}\, y_{t_f}^{(j)},\ j=0,\ldots,r-1 \\
& \boldsymbol{h}\left(\boldsymbol{\psi}_x\left(y,\ldots,t_f^{r-1}y^{(r-1)}\right),\boldsymbol{\psi}_u\left(y,\ldots,t_f^{r}y^{(r)}\right)\right) \geq 0,\ \forall\tau\in[0,1]
\end{array} \quad (5)$$

Therefore, the proposed approach remains applicable to free end-time problems as well.

4.3 Software

To facilitate computations with splines and the translation of problem Eq. (3) into Eq. (4), a Matlab toolbox is made available at http://gitlab.mech.kuleuven.be/meco/splines-m. The aim is to be able to model optimization problems as easily as Yalmip (Löfberg, 2004), but with the variables being polynomial spline functions. An example listing is discussed in the following section.

5 Numerical validation

This section validates the proposed approach on the motion planning problem of a flexible link manipulator, introduced by Faiz (1999) and subsequently treated by Louembet et al. (2010). Numerical values and the constraints are taken from Louembet et al. (2010). Both a convex and nonconvex problem formulation Eq. (4) are compared to a classical sampling based and a sum-of-squares approach.

The dynamics of the manipulator are described by the equations

$$I_1\ddot{q}_1 + MgL\sin q_1 + k(q_1 - q_2) = 0,$$
$$I_2\ddot{q}_2 - k(q_1 - q_2) = u.$$

The system's state is given by $\boldsymbol{x} = (q_1, \dot{q}_1, q_2, \dot{q}_2)^\top$. The goal is to steer the system from the initial state $\boldsymbol{x}_0 = (0.8\,\text{rad}, 0\,\text{rad}\,\text{s}^{-1}, 0.67\,\text{rad}, 0\,\text{rad}\,\text{s}^{-1})^\top$ to the final state $\boldsymbol{x}_{t_f} = (-0.8\,\text{rad}, 0\,\text{rad}\,\text{s}^{-1}, -0.67\,\text{rad}, 0\,\text{rad}\,\text{s}^{-1})^\top$ at $t_f = 5.35\,\text{s}$ with minimal deflection of the link, i.e. $g = \int_0^{t_f} q_1^2(t)\mathrm{d}t$, while obeying the constraint on the joint positions

$$-\frac{\pi}{3} \le q_1 \le \frac{\pi}{3}, -\frac{\pi}{4} \le q_2 \le \frac{\pi}{4}, -\frac{\pi}{16} \le q_2 - q_1 \le \frac{\pi}{16}.$$

Note that the constraints on q_2 and $q_2 - q_1$ already imply the constraint on q_1. The state vector is described by the flat output, $y = q_1$, as:

$$\boldsymbol{x} = \boldsymbol{\psi}_x(y, \dot{y}, \ddot{y}, \dddot{y}) = \left(y, \dot{y}, \frac{I_1}{k}\ddot{y} + \frac{MgL}{k}\sin y + y, \frac{I_1}{k}\dddot{y} + \frac{MgL}{k}\dot{y}\cos y + \dot{y}\right)^\top.$$

Obviously, the mapping $\boldsymbol{\psi}_x$ is not polynomial. Therefore, in order to use the proposed approach, the constraints must be approximated or manipulated into polynomial expressions. To this end, we search for polynomial lower and upper bounds $\underline{p}(y)$ and $\overline{p}(y)$ such that

$$\underline{p}(y) \le \sin y \le \overline{p}(y), \forall y \in \left[-\frac{\pi}{3}, \frac{\pi}{3}\right].$$

The feasible set can then be replaced by the semi-algebraic inner approximation

$$\begin{array}{ll} \frac{I_1}{k}\ddot{y} + \frac{MgL}{k}\overline{p}(y) + y \le \frac{\pi}{4}, & \frac{I_1}{k}\ddot{y} + \frac{MgL}{k}\overline{p}(y) \le \frac{\pi}{16} \\ \frac{I_1}{k}\ddot{y} + \frac{MgL}{k}\underline{p}(y) + y \ge -\frac{\pi}{4}, & \frac{I_1}{k}\ddot{y} + \frac{MgL}{k}\underline{p}(y) \ge -\frac{\pi}{16} \end{array}.$$

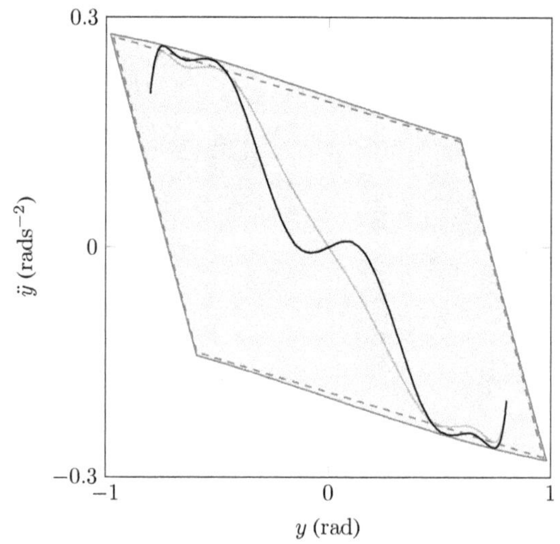

Figure 3. Solutions for the flexible manipulator problem with a polytopic approximation (gray) and semi-algebraic approximation (black).

The following polynomial bounds are used in this example:

$$\overline{p}(y), \underline{p}(y) = \pm c_0 + \sum_{i=1}^{l} c_i y^{2i-1},$$

where c_i, $i > 1$ are determined from a least-squares fit on $[-\frac{\pi}{3}, \frac{\pi}{3}]$ and the offset c_0 is determined such that $\underline{p}(y) \le \sin y \le \overline{p}(y)$. Note that for linear $\underline{p}(y)(\cdot)$, $\overline{p}(y)(\cdot)$, i.e. $l = 1$, we would get at a polytopic (convex) feasible set similar to the one suggested in Louembet et al. (2010). Note that in this example a semi-algebraic formulation is easily found. For more involved systems such as a six-dof robot, determining such a semi-algebraic approximation is a crucial step. In order to keep computation time low, it is key to limit the degree of the approximating polynomials.

We can now write the optimization problem as

$$\begin{array}{ll} \underset{y(\cdot)}{\text{minimize}} & \int_0^{t_f} y(t)^2\mathrm{d}t \\ \text{subject to} & y(0) = 0.8,\ y(t_f) = -0.8 \\ & \dot{y}(0) = 0,\ \dot{y}(t_f) = 0 \\ & \ddot{y}(0) = -0.2,\ \ddot{y}(t_f) = 0.2 \\ & \dddot{y}(0) = 0,\ \dddot{y}(t_f) = 0 \\ & \frac{I_1}{k}\ddot{y}(t) + \frac{MgL}{k}\overline{p}(y(t)) + y(t) \le \frac{\pi}{4},\ \forall t \in [0, t_f] \\ & \frac{I_1}{k}\ddot{y}(t) + \frac{MgL}{k}\underline{p}(y(t)) + y(t) \ge -\frac{\pi}{4},\ \forall t \in [0, t_f] \\ & \frac{I_1}{k}\ddot{y}(t) + \frac{MgL}{k}\overline{p}(y(t)) \le \frac{\pi}{16},\ \forall t \in [0, t_f] \\ & \frac{I_1}{k}\ddot{y}(t) + \frac{MgL}{k}\underline{p}(y(t)) \ge -\frac{\pi}{16},\ \forall t \in [0, t_f] \end{array} \quad (6)$$

The flat output is parameterized by a polynomial spline with 11 equidistant knots. Using the proposed approach the above optimization problem is cast in terms of its spline coefficients as in Eq. (4) using the accompanying software tool from Sect. 4.3. Listing 1 shows the code used for solving the problem with $l = 1$.

Figure 3 illustrates the solution for polynomial bounds of degree one ($l = 1$) (gray) and three ($l = 2$) (black). Clearly,

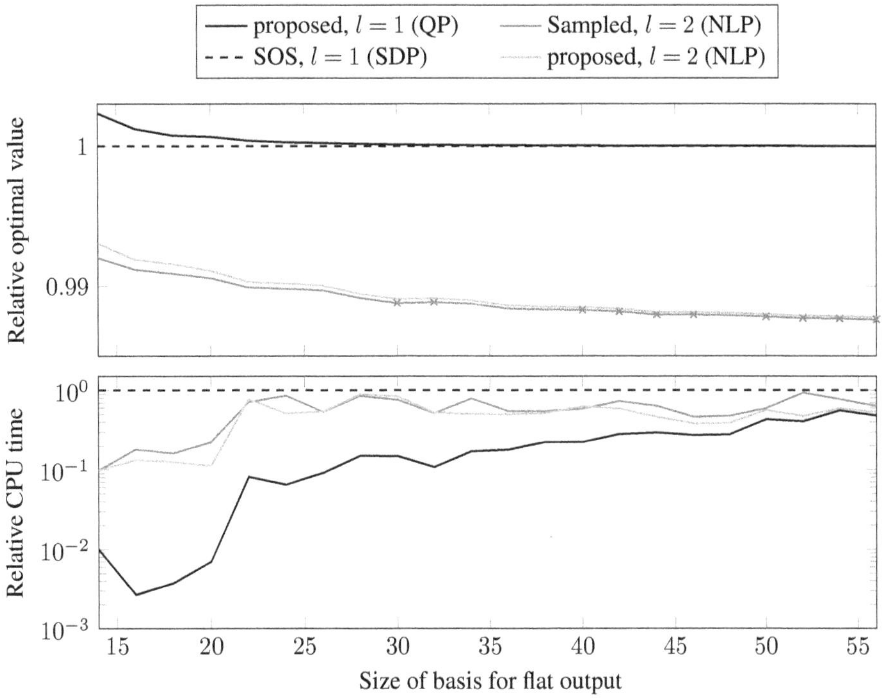

Figure 4. The optimal value and cpu time relative to the SDP for increasing size of the basis for the flat output. The crosses indicate where the sampled approach is feasible.

the latter solution is less conservative. Note, however, that the former problem is a convex quadratic program (QP) whereas the latter is nonconvex. Being able to use nonpolytopic sets is a clear advantage of our method over previous results (Louembet et al., 2010; Suryawan et al., 2012).

In a following numerical experiment, we compute the solution for increasing number of knots and compare the following cases:

1. Our proposed approach for $l = 1$, which is comparable to that of Suryawan et al. (2012). The resulting optimization problem is a convex QP and is solved using qpOASES (Ferreau et al., 2014).
2. Our proposed approach for $l = 2$. To the best of our knowledge, our method is unique in that it can guarantee these nonpolytopic constraints. The resulting optimization problen is a nonlinear progam (NLP). It is solved with Ipopt (Wächter and Biegler, 2006) with exact Hessians obtained using CasADi (Andersson, 2013).
3. A sampled solution for $l = 2$ as in Milam et al. (2000). The number of equidistant time samples is taken equal to the number of constraints in our approach. The resulting optimization problem is a NLP and similarly solved with Ipopt.
4. A globally optimal sum-of-squares (SOS) approach on each segment for $l = 1$ as in Henrion and Lasserre (2006) and Louembet et al. (2010). MOSEK ApS (2015) is used for solving the resulting convex semidefinite program (SDP). Note that this approach for $l > 1$ would require solving a nonconvex optimization problem with polynomial matrix inequalities for which to date no reliable solver exists.

Figure 4 shows the optimal value of the objective function and the CPU-time relative to that of the SDP as a function of n, the size of the basis for the flat output. The SDP is chosen as reference as it is the current state-of-the-art with respect to guaranteed constraint satisfaction. For the sampled approach, solutions that do not violate the constraints are indicated by crosses. A number of observations can be made.

1. For growing n, the optimal value of the QP converges to that of the SDP. This also illustrates that basis refinements by knot insertion or order elevation effectively reduce conservatism.
2. Solving a QP is significantly cheaper compared to the other approaches, especially for small n.
3. It is somewhat surprising to see that solving the convex SDP is more expensive than solving the NLP, illustrating great potential for NLP solvers. However, a global minimum cannot be guaranteed for the nonconvex programs.
4. The difference in optimal value between our approach and the sampled approach with $l = 2$ is small while the

former is guaranteed to be feasible. This is especially important in critical cases where constraint violation is not tolerated.

5. For $l = 2$ our approach is slightly cheaper to solve for larger bases compared to the sampled approach, offering a numerical advantage over traditional methods.

6 Conclusions

This paper focuses on optimal motion planning for systems that admit a polynomial description through differential flatness. The optimization problem is cast in terms of the flat output and a polynomial spline parameterization is proposed that allows us to guarantee state and input constraints by means of simple constraints on the B-spline coefficients. An intuitive relaxation of the constraints is achieved by representing the spline in a higher dimensional basis. Furthermore, a supporting software package is released. Numerical experiments show superior performance to existing approaches in the literature both in terms of computational time and optimality. For systems that do not admit a polynomial representation through differential flatness, an approximation is sought for as illustrated in the numerical validation.

Appendix A: Spline properties

In this Appendix we detail the sum and product properties of splines.

Property 2 (Summation) *Let* $p \in \Pi_{k,\kappa,\mu}$ *and* $r \in \Pi_{l,\lambda,\nu}$. *Then* $s = p + r \in \Pi_{\max(k,l),\xi,\omega}$, *where* $\boldsymbol{\xi} = \boldsymbol{\kappa} \underset{<}{\cup} \boldsymbol{\lambda}$ *the sorted, strictly increasing union of* $\boldsymbol{\kappa}$ *and* $\boldsymbol{\lambda}$, *and*

$$\omega_i = \begin{cases} \min(\mu_m, \nu_n) & \text{if } \xi_i = \kappa_m = \nu_n \text{ for some } m, n \\ \mu_m & \text{if } \xi_i = \kappa_m \text{ for some } m \\ \nu_n & \text{if } \xi_i = \lambda_n \text{ for some } n \end{cases}.$$

The spline coefficients, $\boldsymbol{\sigma}$, *of s are determined through a linear transformation of* $\boldsymbol{\pi}$ *and* $\boldsymbol{\rho}$:

$$\boldsymbol{\sigma} = \mathbf{T}_s^p \boldsymbol{\pi} + \mathbf{T}_s^r \boldsymbol{\rho},$$

where $\mathbf{T}_s^p$ *denotes the linear mapping from the B-spline basis* $\boldsymbol{b}_p$ *to* $\boldsymbol{b}_s$:

$$\boldsymbol{b}_p = \left(\mathbf{T}_s^p\right)^\mathsf{T} \boldsymbol{b}_s, \tag{A1}$$

and similarly for $\mathbf{T}_s^r$.

Property 3 (Multiplication) *Let* $p \in \Pi_{k,\kappa,\mu}$ and $r \in \Pi_{l,\lambda,\nu}$. Then $s = pr \in \Pi_{k+l,\xi,\omega}$, *where* $\boldsymbol{\xi}$ *and* $\boldsymbol{\omega}$ *are determined as in Property 2. Note, however, that the continuity over a given knot could also be higher. The spline coefficients,* $\boldsymbol{\sigma}$, *of s are determined through*:

$$\boldsymbol{\sigma} = \mathbf{T}_s^{p\otimes r}(\boldsymbol{\pi} \otimes \boldsymbol{\rho}),$$

where $\otimes$ *denotes the Kronecker product and* $\mathbf{T}_s^{p\otimes r}$ *is the linear mapping from* $\boldsymbol{b}_p \otimes \boldsymbol{b}_r$ *to* $\boldsymbol{b}_s$:

$$\boldsymbol{b}_p \otimes \boldsymbol{b}_r = \left(\mathbf{T}_s^{p\otimes r}\right)^\mathsf{T} \boldsymbol{b}_s. \tag{A2}$$

These linear mappings are easily found by solving a set of linear equations given by Eqs. (A1) or (A2), or by using dedicated algorithms that are readily available in the literature (e.g. Piegl and Tiller, 1997).

Acknowledgements. This research was supported by IWT ICON project Sitcontrol: Control with Situational Information, IWT SBO project MBSE4Mechatronics: Model-based Systems Engineering for Mechatronics, FWO project G0C4515N: Optimal control of mechatronic systems: a differential flatness based approach. This work also benefits from KU Leuven-BOF PFV/10/002 Center-of-Excellence Optimization in Engineering (OPTEC), from the Belgian Programme on Interuniversity Attraction Poles (DYSCO), initiated by the Belgian Federal Science Policy Office. Goele Pipeleers is partially supported by the Research Foundation Flanders (FWO Vlaanderen).

References

Andersson, J.: A General-Purpose Software Framework for Dynamic Optimization, PhD thesis, Arenberg Doctoral School, KU Leuven, Department of Electrical Engineering (ESAT/SCD) and Optimization in Engineering Center, Heverlee, Belgium, 2013.

de Boor, C.: A practical guide to splines, revised Edn.,Springer-Verlag, New York, 2001.

de Boor, C. and Daniel, J. W.: Splines with nonnegative *b*-spline coefficients, Math. Comput., 28, 565–568, 1974.

Faiz, T. N.: Real time and optimal trajectory generation for nonlinear systems, PhD thesis, University of Delaware, Newark, DE, 1999.

Ferreau, H., Kirches, C., Potschka, A., Bock, H., and Diehl, M.: qpOASES: A parametric active-set algorithm for quadratic programming, Math. Program. Comput., 6, 327–363, 2014.

Fliess, M., Lévine, J., Martin, P., and Rouchon, P.: Flatness and defect of nonlinear systems: Introductory theory and examples, Int. J. Control, 61, 1327–1361, doi:10.1020/00207179508921959, 1995.

Henrion, D. and Lasserre, J. B.: LMIs for constrained polynomial interpolation with application in trajectory planning, Syst. Control Lett., 55, 473–477, doi:10.1016/j.sysconle.2005.09.011, 2006.

Lévine, J.: Analysis and control of nonlinear systems: a flatness-based approach, in: Mathematical Engineering, Springer-Verlag, Berlin, Heidelberg, 2010.

Löfberg, J.: YALMIP : A Toolbox for Modeling and Optimization in MATLAB, in: Proceedings of the CACSD Conference, Taipei, Taiwan, http://users.isy.liu.se/johanl/yalmip (last access: 31 July 2015, 2004.

Louembet, C., Cazaurang, F., Zolghadri, A., Charbonnel, C., and Pittet, C.: Path planning for satellite slew manoeuvres: a combined flatness and collocation-based approach, Control Theory Appl., 3, 481–491, doi:10.1049/iet-cta.2008.0054, 2009.

Louembet, C., Cazaurang, F., and Zolghadri, A.: Motion planning for flat systems using positive *b*-splines: An LMI approach, Automatica, 46, 1305–1309, 2010.

Martin, P., Murray, R. M., and Rouchon, P.: Flat systems, equivalence and trajectory generation, unpublished technical report from Caltech, CDS Tech. Rep., CDS 2003-008, 2003.

Milam, M. B., Mushambi, K., and Murray, R. M.: A new computational approach to real-time trajectory generation for constrained mechanical systems, in: vol. 1, Proceedings of the IEEE Conference on Decision and Control, Sydney, Australia, 845–851, doi:10.1109/CDC.2000.912875, 2000.

MOSEK ApS: The MOSEK optimization toolbox for MATLAB manual, Version 7.1 (Revision 28), http://docs.mosek.com/7.1/toolbox/index.html, last access: 31 July 2015.

Piegl, L. and Tiller, W.: Symbolic operators for {NURBS}, Comput.-Aided Design, 29, 361–368, doi:10.1016/S0010-4485(96)00074-7, 1997.

Prautzsch, H., Boehm, W., and Paluszny, M.: Bezier and B-Spline Techniques, Springer-Verlag, New York, Inc., Secaucus, NJ, USA, 2002.

Suryawan, F., De Doná, J., and Seron, M.: Splines and polynomial tools for flatness-based constrained motion planning, Int. J. Syst. Sci., 43, 1396–1411, doi:10.1080/00207721.2010.549592, 2012.

Wächter, A. and Biegler, L. T.: On the implementation of an interior-point filter line-search algorithm for large-scale nonlinear programming, Math. Program., 106, 25–57, doi:10.1007/s10107-004-0559-y, 2006.

17

Remarks on the classification of wheeled mobile robots

Christoph Gruber[1] **and Michael Hofbaur**[2]
[1]eurofunk Kappacher GmbH, St. Johann im Pongau, Austria
[2]Institute of Robotics and Mechatronics, JOANNEUM RESEARCH, Klagenfurt, Austria
Correspondence to: Christoph Gruber (christoph.gruber@gmx.at)

Abstract. The subject of this work is modeling and classification of single-bodied wheeled mobile robots (WMRs). In the past, it was shown that the kinematics of each such robot can be modeled by one out of only five different generic models. However, the precise conditions under which a model is the proper description of the kinematic capabilities of a robot were not clear. These shortcomings are eliminated in this work, leading to a simple procedure for model selection. Additionally, a thorough analysis of the kinematic models and a classification of their singularities are presented.

1 Introduction

WMR are typically developed with a specific application in mind so that the resulting design provides the level of mobility that is appropriate for the robot's operation. The design of the WMR implies its specific kinematics that is then used to derive and program the controller for the robot, which translates a desired movement into the appropriate actuation of the individual wheels (steering angles and rotational speeds). Once in operation, the control law and also higher level control layers, such as the path planner, will use the (inverse) kinematics implicitly through the implemented control algorithms. It is thus often impossible for such controllers to adapt their control laws once the kinematics of the drive changes significantly.

Such a change can occur in many different scenarios: one example is the case of a fault in the drive – e.g. an impaired steering actuator – which obviously has big influence on the kinematic capabilities of the robot. A fault-tolerant controller for a wheeled robot could be able to adapt to this new situation. Another example would be a mobile robot pushing a (passive) roll container. If the container has less degrees of freedom (DoF) than the robot, then the kinematic and dynamic model of the whole system need to be considered by the controller. For control of modular WMR like those presented in Hofbaur et al. (2010), automatic model selection is a prerequisite for choosing the proper controller for a specific wheel configuration. Furthermore, online reconfiguration might lead to significant changes in the kinematics. A similar issue arises for teams of mobile robots which are linked by holonomic constraints, possibly virtual or real ones, for example in collaborative transportation. Despite the fact that all these problems could be solved through a custom-designed controller that can account for specific and pre-defined operational situations, it would be desirable to handle such situations in a more general way. When applying model based control, the controller is either required to deduce a model for an operational situation online or select the proper model from a pre-defined ensemble and parametrize it (Gruber and Hofbaur, 2012). For wheeled mobile robots, the latter method is preferable, because Campion et al. (1996) have shown that there is a *minimal* model, still sufficient to describe the posture kinematics of a WMR, which can only take five different forms for all kinds of single-bodied wheeled mobile robots. They obtain this result by introducing conditions for non-degeneracy of WMR. The contribution of this work is a controllability-study that also results in such a classification, making the notion of non-degeneracy unnecessary. In existing works, the precise conditions for using one of these minimal models for some specific wheel configuration are not made clear. This work eliminates these shortcomings and also investigates the consequences of choosing a specific model.

Modeling of systems defined by wheel-like constraints is a well studied field, and most of the pioneering work was conducted in the last two decades of the last century. Early

works are Alexander and Maddocks (1989) and Muir and Neuman (1987). A large step was the work of Campion et al. (1996), which showed – in addition to the existence of the above-mentioned "minimal" models – that WMR models are controllable, differentially flat and that robots with restricted mobility are non-holonomic. Furthermore, they can be transformed to chained form. These model properties were the onset for control design methodologies for tracking and stabilization controllers for WMR, and WMR are popular examples and benchmarks for controllers of non-holonomic systems (Morin and Samson, 2008). Also modeling of WMR is still an issue, and especially the growing interest in mobile manipulators gave rise to some issues related to the modeling and control of mobile robots with steering wheels only (Giordano et al., 2009; De Luca et al., 2010). One of the first publications related solely to the topic of modeling this kind of pseudo-omnidirectional robots was Betourne et al. (1996) and this work was supplemented by Thuilot et al. (1996). A differential geometric perspective on undulatory locomotion of wheeled robots is given in Ostrowski and Burdick (1998) or Bullo and Lewis (2005), who use the snakeboard as an example.

The classification of WMR models found by Campion et al. (1996) is reviewed in Sect. 2. Section 3 shows the demand for a precise understanding of the implications of choosing a certain model. In Sect. 4, we analyze the accessibility properties of the associated control system. Based on these structural properties, we are able to find conditions under which a kinematic model is the proper description for a robot. These conditions lead to different *modes of operation* for wheeled mobile robots, presented in Sect. 5. In this section, we further discuss the effects of neglecting sliding constraints in modeling and introduce a classification of WMR. A classification of singularities is suggested in Sect. 6. We show how our analysis is supported by results from differential geometry in Sect. 7.

2 Review

In this section, the classes of posture kinematic models of WMR are reviewed, as they are developed in Campion et al. (1996); Campion and Chung (2008). Only single-body-robots are considered, where wheels that roll without slipping are mounted over hinges to a single rigid body (the chassis). Types of wheels are Swedish wheels, casters, steerable standard and fixed standard wheels. Configuration variables of an unconstrained robot are elements of an l-dimensional configuration manifold Q, given by

$$Q = G \times M \tag{1}$$

$$= \mathrm{SE}(2) \times \mathbb{T}^{N_c} \tag{2}$$

where G is the Special Euclidean Group SE(2) describing rigid motions (rotation and translations) in two-dimensional Euclidean space and M the manifold of all possible wheel orientations of center steerable wheels, with N_c the number of such wheels. An element of G is a robot pose $\boldsymbol{\xi}$, an element of M a steering angle configuration $\boldsymbol{\beta}_c$. Accordingly, $\boldsymbol{q} = (\boldsymbol{\xi}^T \, \boldsymbol{\beta}_c^T)^T \in Q$. We assume that standard wheels may not slide in a direction perpendicular to their rolling plane and each wheel touches the ground in a single point. This restriction is modeled by so-called sliding constraints, which introduce a dependence between the configuration variables of the robot. Sliding constraints for standard wheels are in the form of Pfaffian equations

$$\omega_i(\boldsymbol{q})\dot{\boldsymbol{q}} = 0. \tag{3}$$

In coordinates, for example, the sliding constraint of a standard wheel i is

$$[\cos(\alpha_i + \beta_i) \;\sin(\alpha_i + \beta_i) \;l_i \sin(\beta_i)]\mathbf{R}^T(\theta)\dot{\boldsymbol{\xi}} = 0$$

$$\boldsymbol{c}_i(\beta_i)\mathbf{R}^T(\theta)\dot{\boldsymbol{\xi}} = 0, \tag{4}$$

see Fig. 1. There, (l_i, α_i) are the polar coordinates of the contact point of wheel i in a robot-fixed reference frame $\mathbb{X}_1, \mathbb{X}_2$. The origin of the robot-fixed frame is located at the Cartesian coordinates (x, y) in the inertial frame $\mathbb{I}_1, \mathbb{I}_2$. The rotation between the robot-fixed and the inertial frame is described by the matrix

$$\mathbf{R}(\theta) = \begin{bmatrix} \cos(\theta) & -\sin(\theta) & 0 \\ \sin(\theta) & \cos(\theta) & 0 \\ 0 & 0 & 1 \end{bmatrix}. \tag{5}$$

For steered wheels, the co-vector fields ω_i depend on β_i and therefore on $\boldsymbol{q}$. The sliding constraints of center-steerable and fixed standard wheels, as formulated in Eq. (3), are Pfaffian equations, since they equate to zero. In contrast to that, the rolling constraints of all kinds of wheels and the sliding constraints of off-centered orientable wheels are equations with a right-hand-side that may be set via an input (steering or wheel drive). For this reason, we will assume that the sliding constraints of off-centered orientable wheels and rolling constraints in general do not impose restrictions of mobility and will therefore be neglected in the remainder.

Let r be the number of center-steerable and fixed standard wheels. The restriction of mobility for WMR originates from their corresponding r sliding constraints (Eq. 3). A motion $\dot{\boldsymbol{q}}$ is consistent with these sliding constraints, if it is an element of the annihilator (null-space) of their corresponding Pfaffians $\omega_i, i \in \{1 \ldots r\}$. A basis of this annihilator spans a subspace of the tangential space $T_q Q$. This subspace is called a distribution $\Delta_c(\boldsymbol{q})$, which can be constructed in the following way:

Arrange the co-vectors $\boldsymbol{c}_i(\boldsymbol{q})$ from Eq. (4) in so-called sliding constraint matrices $\mathbf{C}_{1f}$ and $\mathbf{C}_{1c}$, corresponding to fixed standard wheels and steerable standard wheels, respectively, such that

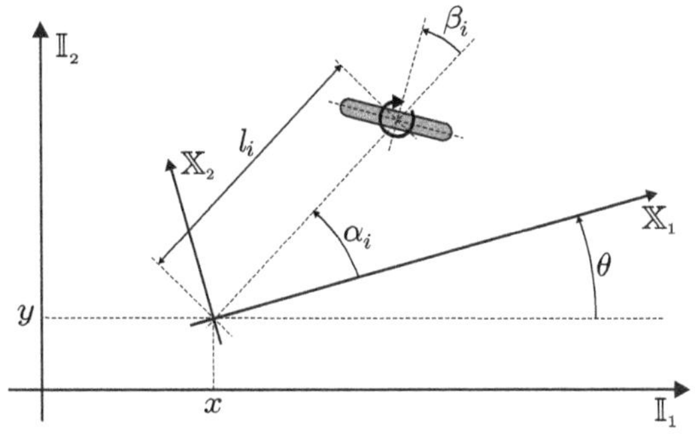

Figure 1. A center-steerable standard wheel.

$$\mathbf{C}_{1f}\mathbf{R}^T(\theta)\dot{\boldsymbol{\xi}} = 0$$
$$\mathbf{C}_{1c}\left(\boldsymbol{\beta}_c\right)\mathbf{R}^T(\theta)\dot{\boldsymbol{\xi}} = 0 \tag{6}$$

and

$$\mathbf{C}_1^*\left(\boldsymbol{\beta}_c\right) := \begin{pmatrix} \mathbf{C}_{1f} \\ \mathbf{C}_{1c}\left(\boldsymbol{\beta}_c\right) \end{pmatrix}. \tag{7}$$

A basis of the null-space of this matrix $\mathbf{C}_1^*(\boldsymbol{\beta}_c)$ spans the distribution Δ_c:

$$\operatorname{span}\left(\operatorname{col}\boldsymbol{\Sigma}\left(\boldsymbol{\beta}_c\right)\right) := \operatorname{null}\left(\mathbf{C}_1^*\left(\boldsymbol{\beta}_c\right)\right) \tag{8}$$
$$\Delta_c(\boldsymbol{q}) = \operatorname{span}\left(\operatorname{col}\mathbf{R}(\theta)\boldsymbol{\Sigma}\left(\boldsymbol{\beta}_c\right)\right). \tag{9}$$

This definition of $\Delta_c(\boldsymbol{q})$ follows the notation of Campion et al. (1996); Campion and Chung (2008) to illustrate that the null-space of $\mathbf{C}_1^*(\boldsymbol{\beta}_c)$ is the object that defines the dimension of the distribution. This is actually the case, since $\mathbf{R}(\theta)$ is of constant rank 3 and therefore has no influence on the dimension of $\Delta_c(\boldsymbol{q})$. Instead, this dimension depends on the steering angles $\boldsymbol{\beta}_c$ and is, by the rank-nullity theorem, given by

$$\delta_m\left(\boldsymbol{\beta}_c\right) = 3 - \operatorname{rank}\left(\mathbf{C}_1^*\left(\boldsymbol{\beta}_c\right)\right), \tag{10}$$

which is called *degree of mobility*. The *degree of steerability* is defined to be

$$\delta_s\left(\boldsymbol{\beta}_c\right) = \operatorname{rank}\left(\mathbf{C}_{1c}\left(\boldsymbol{\beta}_c\right)\right), \tag{11}$$

which is the number of independent steering actuators that have influence on the distribution.

According to the definition of Campion et al. (1996), a robot is *non-degenerate*, if

$$\operatorname{rank}\left(\mathbf{C}_{1f}\right) \leq 1 \tag{12}$$
$$\forall\boldsymbol{\beta}_c \in M: \ \operatorname{rank}\left(\mathbf{C}_1^*\left(\boldsymbol{\beta}_c\right)\right) = \operatorname{rank}\left(\mathbf{C}_{1f}\right) + \operatorname{rank}\left(\mathbf{C}_{1c}\left(\boldsymbol{\beta}_c\right)\right) \tag{13}$$
$$\exists\boldsymbol{\beta}_c^*: \ \operatorname{rank}\left(\mathbf{C}_1^*\left(\boldsymbol{\beta}_c^*\right)\right) \leq 2. \tag{14}$$

The numbers $\delta_m(\boldsymbol{\beta}_c)$, $\delta_s(\boldsymbol{\beta}_c)$ and the property of being either degenerate or non-degenerate allow the classification of *models* of WMR into five non-degenerate classes: $(3,0)$, $(2,0)$, $(2,1)$, $(1,1)$ and $(1,2)$ where each non-degenerate model class is identified by the notation (δ_m, δ_s). The sum of degree of mobility and degree of steerability is the number of overall DoF. Car-like robots $(1,1)$ and differential drive robots $(2,0)$ have 2 DoF, all other non-degenerate robots have 3 DoF. For every model class, Campion et al. provide a generic *posture kinematic model*, describing the posture kinematics of every model in a certain class. Each of these generic models is based on a specific choice of body-fixed frame (see Campion et al., 1996). These five posture kinematic models can be written in the form

$$\dot{\boldsymbol{x}} = \mathbf{G}(\boldsymbol{x})\boldsymbol{u} \tag{15}$$

with $\boldsymbol{x}$ the $3+\delta_s$-dimensional state vector, $\boldsymbol{u} = (\boldsymbol{\eta}^T \ \boldsymbol{\zeta}^T)^T$ the input vector and the matrix $\mathbf{G}(\boldsymbol{x})$ given by

$$\mathbf{G}(\boldsymbol{x}) = \mathbf{R}(\theta)\boldsymbol{\Sigma}\left(\boldsymbol{\beta}_c\right)$$
$$= \left(\boldsymbol{g}_{\eta_1}\cdots\boldsymbol{g}_{\eta_{\delta_m}}\boldsymbol{g}_{\zeta_1}\cdots\boldsymbol{g}_{\zeta_{\delta_s}}\right) \tag{16}$$

where $\boldsymbol{\eta}$ defines the δ_m wheel-speed-related inputs and $\boldsymbol{\zeta}$ the δ_s steering inputs.

3 Problem description

The models of WMR presented so far have some special features which are not addressed in popular publications like Campion et al. (1996), Canudas-de Wit et al. (1996), Siegwart and Nourbakhsh (2004), Campion and Chung (2008), etc. In contrast to these publications, the dependence of $\delta_m(\boldsymbol{\beta}_c)$, $\delta_s(\boldsymbol{\beta}_c)$ and $\Delta_c(\boldsymbol{q})$ on the steering angles $\boldsymbol{\beta}_c$ was explicitly pointed out in the previous section. This dependence already indicates that the mapping of a WMR to a certain model class is a *local* one. This means, that by the dependence of $\delta_m(\boldsymbol{\beta}_c)$ and $\delta_s(\boldsymbol{\beta}_c)$ on the steering angles $\boldsymbol{\beta}_c$, a specific robot can be mapped to multiple different classes according to its operational situation.

Recall that $\delta_m(\boldsymbol{\beta}_c)$ is the dimension of the constraint distribution Δ_c. Since this dimension changes with $\boldsymbol{\beta}_c$, Δ_c is no regular distribution on all of Q for all WMR. The mechanism which leads to a change in $\delta_m(\boldsymbol{\beta}_c)$ is the following: by setting specific steering angles $\boldsymbol{\beta}_c$, rows in $\mathbf{C}_1^*(\boldsymbol{\beta}_c)$ become linear dependent. As a result, the dimension of the null-space of $\mathbf{C}_1^*(\boldsymbol{\beta}_c)$ increases. Note that the set D of steering angles decreasing the rank of $\mathbf{C}_1^*(\boldsymbol{\beta}_c)$

$$D = \left\{\boldsymbol{\beta}_c \in M \middle| \operatorname{rank}\left(\mathbf{C}_1^*\left(\boldsymbol{\beta}_c\right)\right) < \max\left(\operatorname{rank}\left(\mathbf{C}_1^*\left(\boldsymbol{\beta}_c\right)\right)\right)\right\} \tag{17}$$

is nowhere dense in M. The occurrence of such nowhere dense sets and their relation to the model types is illustrated in the following example:

Example 1 *Consider a WMR with two center-orientable standard wheels as shown in Fig. 2a and b. In the configuration shown in Fig. 2a, the wheel axes intersect in a*

point defining the Instantaneous Center of Rotation (ICR). Another steering configuration is shown in Fig. 2b. In this configuration, the wheel axes coincide and the rank of the annihilator of the constraint distribution drops from two to one: the constraint distribution Δ_c has constant dimension one on the open dense subset

$$Q_1 = Q \setminus Q_2, \tag{18}$$

consisting of configurations similar to the one shown in Fig. 2a, and dimension two on

$$Q_2 = \{(x, y, \theta, \beta_1, \beta_2) \,|\, \beta_1 \in \{0, \pi\} \text{ and } \beta_2 \in \{0, \pi\}\}, \tag{19}$$

which contains four disjoint submanifolds representing steering configurations similar to those shown in Fig. 2b. These special configurations are instances of what will be called singularity of type A in the remainder. Following the procedure from Campion et al. (1996), reviewed in Sect. 2, one obtains

$$\delta_m = 1,\ \delta_s = 2 \text{ on } Q_1 \tag{20}$$

$$\delta_m = 2,\ \delta_s = 0 \text{ on } Q_2. \tag{21}$$

This illustrates the configuration dependency of the numbers (δ_m, δ_s) *and* Δ_c *is not a regular distribution.*

Example 1 shows that the dimension of the distribution Δ_c is not constant on all of Q. Intuitively, this is in contradiction with the configuration-independent number of DoFs. However, this is not the case. The point is that on Q_1 (wheel axes not aligned), the ICR is fixed by the intersection of the wheels axes. The location of this ICR is fixed via *two* steering inputs, while the tangential velocity is set via *one* wheel speed. The speed of the other wheel cannot be freely chosen but needs to be consistent with its rolling constraint. However, on Q_2, the sliding constraints only restrict the ICR to lie on the common axis through both wheels. The *two* wheel speeds define the location of the ICR on this axis via the rolling constraints. As soon as a steering angle is changed, the configuration leaves Q_2.

In the set Q_2, kinematic constraints of the robot become dependent, increasing its number of directly accessible degrees of freedom δ_m. This fact can be asserted in two simple thought experiments: think of a passive robot with two center-steerable standard wheels. In the first experiment, wheels are oriented as in Fig. 2a and the robot is pushed manually in a direction while the steering angles are held constant. Then the ICR remains at a fixed point and the motion of the robot describes a circle around this ICR. For the second experiment, change the steering angles such that the wheel axes coincide (the rolling planes are parallel, see Fig. 2b). In this situation, the robot can be pushed to any desired position without steering the wheels but rotating the whole robot.

To the best of our knowledge, such a property was not mentioned in literature and shows a demand for a deeper understanding of the relation between robots and models.

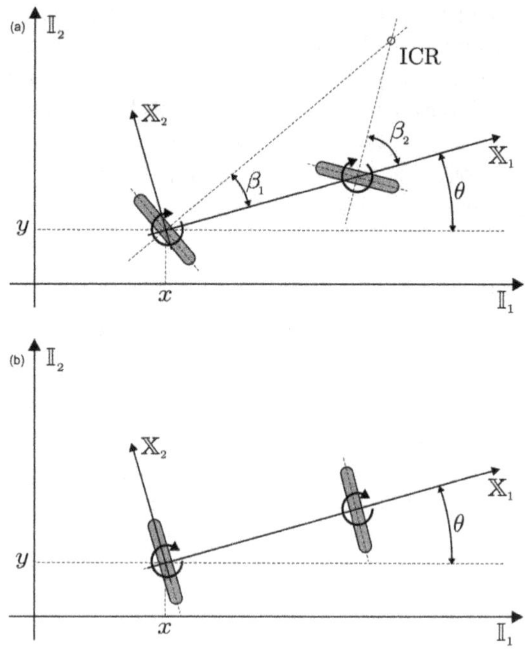

Figure 2. **(a)** A WMR with two center-steerable standard wheels in a non-singular configuration. **(b)** A WMR with two center-steerable standard wheels in a singular configuration.

4 Accessibility

Any robot which is at least capable of reaching any pose in the plane is able to fulfill most of the typical tasks of WMR. From a control engineering point of view, this formulation already sounds familiar, since accessibility is a well-known concept in the field of controllability analysis of nonlinear control systems.

Recall from Sect. 2, that the vector fields that span the nullspace of $\mathbf{C}_1^*(\boldsymbol{\beta}_c)$ span the distribution $\Delta_c(\boldsymbol{q})$, and a robot is able to move along these vector fields. To account for steering motions, this distribution $\Delta_c(\boldsymbol{q})$ is extended by the corresponding number of dimensions and vector fields modeling these steering motions to obtain a distribution $\Delta(\boldsymbol{q})$. Even though it seems like $\Delta(\boldsymbol{q})$ should now contain all the motions consistent with the sliding constraints, this is actually not the case. Nonlinear systems with multiple inputs, like WMR, are not limited to move along these vector fields in $\Delta(\boldsymbol{\beta}_c)$. By consecutive execution of vector field motions (e.g. steer-drive-steer-drive), "new" motion directions might result. A motion along such a "new" direction, a so-called Lie bracket motion, is described by a vector field $\boldsymbol{g}_m$, obtained by applying the Lie bracket operation on two vector fields $\boldsymbol{g}_k, \boldsymbol{g}_l \in \Delta$:

$$\boldsymbol{g}_m = [\boldsymbol{g}_k, \boldsymbol{g}_l] = \frac{\partial \boldsymbol{g}_l}{\partial \boldsymbol{q}} \boldsymbol{g}_k - \frac{\partial \boldsymbol{g}_k}{\partial \boldsymbol{q}} \boldsymbol{g}_l. \tag{22}$$

The involutive closure of the distribution Δ is denoted inv(Δ). It is a distribution spanned by the vector fields in Δ and a minimal subset of Lie brackets of these vector fields,

such that all remaining Lie brackets lie in inv(Δ). This involutive closure of the distribution Δ is finally the object, that is truly capable of describing all the motions that are consistent with the sliding constraints of a WMR (this holds because the vector fields that span Δ are analytic, see e.g. Isidori, 1995).

The set of poses accessible by motions in inv(Δ) is found by applying Chow's Theorem (Choset et al., 2005; Bullo and Lewis, 2005): if

$$T_{\xi}G \subset \text{inv}(\Delta(\boldsymbol{q})) \tag{23}$$

over a trajectory (integral curve) $\boldsymbol{q}(t)$, $\forall t \in [0, T]$, then all poses $\boldsymbol{\xi}_T \in G$ are accessible from any initial configuration $\boldsymbol{q}_0$. Remarkably, this holds not just for the tangent space $T_{\xi}G$, but for the whole tangent bundle TG: the dependence of $\Delta(\boldsymbol{q})$ on $\boldsymbol{\xi}$ originates from the rotation matrix in Eq. (9). However, such a rotation does not influence the dimension of the space that is spanned by the distribution. As a result, condition Eq. (23) can be given in the more general, pose-independent form:

$$TG \subset \text{inv}\left(\Delta\left(\boldsymbol{\beta}_c\right)\right). \tag{24}$$

Accordingly, in order to show that every pose $\boldsymbol{\xi} \in G$ is accessible, one only has to find a single steering configuration for which inv($\Delta(\boldsymbol{\beta}_c)$) spans TG. However, this is still a somewhat abstract condition. For this reason, we give an equivalent condition in the following theorem:

Theorem 1 *A WMR is able to reach any pose in the obstacle-free plane, if and only if all its fixed wheels are co-aligned, that is,*

$$\text{rank}\left(\mathbf{C}_{1f}\right) \leq 1. \tag{25}$$

Proof:
Necessity: if this condition is violated, then Δ contains only one single vector field (rank($\mathbf{C}_{1f}$) = 2) or is empty (rank($\mathbf{C}_{1f}$) = 3). In the latter case, the robot is not able to move at all. In the former case, the distribution is trivially involutive and thus not able to span a space of dimension larger than 1, but dim($T_{\xi}G$) = 3. This means that when rank($\mathbf{C}_{1f}$) = 2, the robot is only capable of moving around a fixed ICR, that is, on a straight line or circle and therefore not able to reach any desired pose in the plane.

For sufficiency, we have to show that every robot with rank($\mathbf{C}_{1f}$) $\leq$ 1 is able to reach any desired pose. The rank of the sliding constraint matrix rank($\mathbf{C}_1^*(\boldsymbol{\beta}_c)$) can take values from zero to three. Let us analyze these valuations case-by-case:

1. rank($\mathbf{C}_1^*(\boldsymbol{\beta}_c)$) = 0: the distribution spanned by the nullspace of $\mathbf{C}_1^*(\boldsymbol{\beta}_c)$ is by itself three-dimensional, involutive and spans $T_{\xi}G$. Due to Eq. (24), every pose in the plane is reachable.

2. $\exists \boldsymbol{\beta}_c^*$: rank($\mathbf{C}_1^*(\boldsymbol{\beta}_c^*)$) = 1: the nullspace of $\mathbf{C}_1^*(\boldsymbol{\beta}_c^*)$ has dimension two. Let the two vector fields, that span this nullspace be $\boldsymbol{g}_{\eta_1}(\boldsymbol{\beta}_c^*)$ and $\boldsymbol{g}_{\eta_2}(\boldsymbol{\beta}_c^*)$, i.e. $\Delta_c = \{\boldsymbol{g}_{\eta_1}, \boldsymbol{g}_{\eta_2}\}$ and dim(Δ_c) = 2. The Lie bracket $[\boldsymbol{g}_{\eta_1}(\boldsymbol{\beta}_c^*), \boldsymbol{g}_{\eta_2}(\boldsymbol{\beta}_c^*)]$ does not lie in the space spanned by $\Delta_c(\boldsymbol{\beta}_c^*)$ and thereby gives the third direction, such that $TG \subset \text{inv}(\Delta_c(\boldsymbol{\beta}_c^*))$ and the robot is able to reach any pose.

   Note that for rank($\mathbf{C}_1^*(\boldsymbol{\beta}_c^*)$) to be one, the axes of all wheels – if there are more than one – must be co-aligned.

3. $\exists \boldsymbol{\beta}_c^*$: rank($\mathbf{C}_1^*(\boldsymbol{\beta}_c^*)$) = 2: the nullspace of $\mathbf{C}_1^*(\boldsymbol{\beta}_c^*)$ has dimension one. Let the vector field, that spans this nullspace be $\boldsymbol{g}_{\eta_1}(\boldsymbol{\beta}_c^*)$. At all but a nowhere dense set S of singularities (these singularities are treated in more detail in Sect. 7), this vector field depends at least on one steering angle, because otherwise rank($\mathbf{C}_{1f}$) would be two. Construct Δ by extending Δ_c about a dimension and a steering vector field $\boldsymbol{g}_{\zeta_1}$. Build the following Lie brackets to obtain inv(Δ): inv(Δ) = $\{\boldsymbol{g}_{\eta_1}, \boldsymbol{g}_{\zeta_1}, [\boldsymbol{g}_{\eta_1}, \boldsymbol{g}_{\zeta_1}], [\boldsymbol{g}_{\eta_1}, [\boldsymbol{g}_{\eta_1}, \boldsymbol{g}_{\zeta_1}]]\}$. Again, we get the result that $TG \subset \text{inv}(\Delta_c(\boldsymbol{\beta}_c^*))$, allowing to conclude that the robot is able to reach any pose.

   Note that for rank($\mathbf{C}_1^*(\boldsymbol{\beta}_c)$) to be two the axes of all wheels must intersect in one single point, the ICR.

4. $\exists \boldsymbol{\beta}_c^*$: rank($\mathbf{C}_1^*(\boldsymbol{\beta}_c^*)$) = 3: in this case, the nullspace of $\mathbf{C}_1^*(\boldsymbol{\beta}_c^*)$ is empty. However, the fixed wheels only contribute one to this rank of three. This means that the remaining rank of two must originate from the center-steered wheels. In other words, for steering angle configurations in

$$B_3 = \left\{\boldsymbol{\beta}_c \in M | \text{rank}\left(\mathbf{C}_1^*\left(\boldsymbol{\beta}_c\right)\right) = 3\right\}, \tag{26}$$

   the axes of the wheels do not intersect in one point. However, since all but the co-aligned fixed wheels are steerable, it must be possible to steer the wheels such that all axes intersect in one point (see Sect. 7 for details). Summing up, this means that every robot with a steering angle configuration $\boldsymbol{\beta}_c^*$ such that rank($\mathbf{C}_1^*(\boldsymbol{\beta}_c^*)$) = 3 must have another steering angle configuration $\boldsymbol{\beta}_c^+$ with rank($\mathbf{C}_1^*(\boldsymbol{\beta}_c^+)$) = 2 if condition (Eq. 25) holds. ∎

An immediate consequence of Theorem 1 is the classification discovered by Campion et al. (1996): rank($\mathbf{C}_1^*(\boldsymbol{\beta}_c)$) can either be 0, 1 or 2, giving degrees of mobility δ_m of 3, 2 or 1, respectively. The number of independent steering degrees of freedom δ_s is given by rank($\mathbf{C}_1^*(\boldsymbol{\beta}_c)$) − rank($\mathbf{C}_{1f}$), which can also only take values 0, 1, 2. Table 1 shows inv(Δ) for all model classes, where vector fields that span TG are underlined. The Degree of Nonholonomy (DoN), also shown in the table, is the highest degree of the Lie-brackets in inv(Δ).

There is a slight difference between the set of robots considered here, and the set considered by Campion et al. (1996): in their work, some robots with pathological wheel distributions like for example the one shown in Fig. 3

Table 1. How inv(Δ) spans TG. Vector fields and Lie brackets that span TG are underlined.

(δ_m, δ_s)	inv(Δ)	DoN	DoF
(3, 0)	$\underline{\boldsymbol{g}_{\eta_1}}, \underline{\boldsymbol{g}_{\eta_2}}, \underline{\boldsymbol{g}_{\eta_3}}$	1	3
(2 ,1)	$\underline{\boldsymbol{g}_{\eta_1}}, \underline{\boldsymbol{g}_{\eta_2}}, \boldsymbol{g}_{\zeta_1}, \underline{[\boldsymbol{g}_{\eta_1}, \boldsymbol{g}_{\eta_2}]}$	2	3
(1, 2)	$\underline{\boldsymbol{g}_{\eta_1}}, \boldsymbol{g}_{\zeta_1}, \boldsymbol{g}_{\zeta_2}, \underline{[\boldsymbol{g}_{\eta_1}, \boldsymbol{g}_{\zeta_1}]}, \underline{[\boldsymbol{g}_{\eta_1}, \boldsymbol{g}_{\zeta_2}]}$	2	3
(2, 0)	$\underline{\boldsymbol{g}_{\eta_1}}, \underline{\boldsymbol{g}_{\eta_2}}, \underline{[\boldsymbol{g}_{\eta_1}, \boldsymbol{g}_{\eta_2}]}$	2	2
(1, 1)	$\underline{\boldsymbol{g}_{\eta_1}}, \boldsymbol{g}_{\zeta_1}, \underline{[\boldsymbol{g}_{\eta_1}, \boldsymbol{g}_{\zeta_1}]}, \underline{[\boldsymbol{g}_{\eta_1}, [\boldsymbol{g}_{\eta_1}, \boldsymbol{g}_{\zeta_1}]]}$	3	2

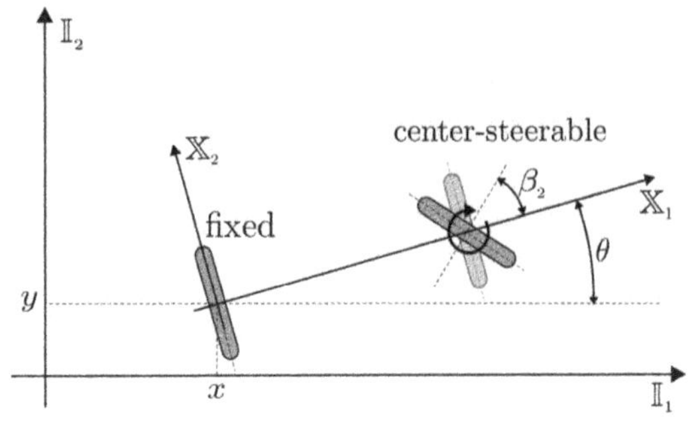

Figure 3. A robot with one fixed and one center-steerable wheel, that would be degenerate in the sense of Campion et al. (1996), although it is still able to reach any posture in the plane.

are defined to be degenerate. This is seen in Eq. (13): the robot has one fixed wheel, so $\text{rank}(\mathbf{C}_{1f}) = 1$ and one steerable wheel, giving $\text{rank}(\mathbf{C}_{1c}(\boldsymbol{\beta}_c)) = 1 \forall \boldsymbol{\beta}_c$. However, there exists a steering angle configuration $\hat{\boldsymbol{\beta}}_c$ such that $\text{rank}(\mathbf{C}_1^*(\hat{\boldsymbol{\beta}}_c)) = 1 \neq \text{rank}(\mathbf{C}_{1f}) + \text{rank}(\mathbf{C}_{1c}(\hat{\boldsymbol{\beta}}_c))$. Equation (13) does therefore not hold for all $\boldsymbol{\beta}_c$. In our work, we want to include such robots as they may appear as a result of reconfiguration of modular WMRs (Mutambara, 1998; Hofbaur et al., 2010) or faults and are still able to reach any pose in the plane.

There is another difference in the approach of Campion et al. (1996) and Theorem 1: Campion et al. (1996) first define conditions for non-degeneracy, from which they obtain the classification. Then they show that non-degenerate robots are controllable. In the approach taken in this work, the classification follows from a controllability argument

We will therefore avoid using the notation of degeneracy and instead simply distinguish between wheel configurations that are controllable or not.

5 Classification

When talking about wheel (hardware) configurations and model classes, it is impractical to use the same notation (δ_m, δ_s) for both. Therefore, in contrast to Campion et al. (1996), we do not associate sets of wheel (hardware) configurations to the pair (δ_m, δ_s). Instead, we interpret (δ_m, δ_s) as a model class or "mode of operation" (MoO) and introduce a separate notation for hardware types in this section.

Table 2. The relation of hardware designs and modes of operation.

MoO↓ HW-type→	I	III	II	Va	Vb	IV
(3, 0)	r	n	–	n	n	–
(2, 1)	v	r	–	n	n	–
(2, 0)	v	v	r	n	r	n
(1, 2)	v	v	–	r	r	–
(1, 1)	v	v	v	v	v	r
Number of Swedish wheels	≥ 0	≥ 0	≥ 0	≥ 0	≥ 0	≥ 0
Number of oc-steerable wheels	≥ 0	≥ 0	≥ 0	≥ 0	≥ 0	≥ 0
Number of c-steerable wheels	0	1	0	≥ 3	≥ 2[a]	≥ 1
Number of fixed[b] wheels	0	0	≥ 1	0	0	≥ 1
Singularities in MoO (1, 2)	–	–	–	B, C	A, C	–

[a] The contact points of all center-steerable wheels lie on a straight line. [b] fixed standard wheels have to share a common axis. If steered standard wheels are placed on this axis, they must be counted as fixed wheels with similar orientation as the fixed wheels.

Controllable robots may have multiple structurally different controllable wheel configurations. This fact was illustrated in Example 1, where the robot was found to have two disjoint sets of controllable wheel configurations. The property of the singular set Q_2 of being *nowhere dense* suggests the following method to find a configuration-independent classification: (1) partition Q into sets Q_i where $\delta_m(q_i)$ and $\delta_s(q_i)$ is constant $\forall q_i \in Q_i$. (2) Choose the only set Q_i which is dense in Q. However, this approach fails for robots with more than two center-steerable standard wheels: in this case the only dense set Q_i corresponds to degenerate configurations, since more than two arbitrarily oriented standard wheels will block any motion.

Previous works Thuilot et al. (1996), Betourne et al. (1996) and Giordano et al. (2009) already deal with robots with more than two center-steerable or fixed standard wheels. However, they treat this topic in substantially different ways: Thuilot et al. (1996) and Betourne et al. (1996) represent all these robots with (1, 2) models, whereas Giordano et al. (2009) use (2, 1).

To resolve this ambiguity the following definition is made:

Definition 1 *A WMR is said to operate in mode* $(\overline{\delta}_m, \overline{\delta}_s)$ *in the time interval* $[t_0, t_0 + T]$, *if it moves along a trajectory (integral curve)* $\boldsymbol{q}(t)$ *for which* $\delta_m(\boldsymbol{q}) = \overline{\delta}_m = const.$ *and* $\delta_s(\boldsymbol{q}) = \overline{\delta}_s = const.$ $\forall t \in [t_0, t_0 + T]$.

When a robot operates in a mode $(\overline{\delta}_m, \overline{\delta}_s)$ at some time instant t, then the corresponding model is the proper description of its posture kinematics at that time. In this context, the following question comes to mind: which WMR hardware designs are able to operate in which modes of operation? The answer to this question is given in Table 2, which is read in the following way: sum up the number of wheels of a specific type, then search the column that matches. Watch the coalignment conditions from footnotes a and b. These wheel counts determine the hardware type. Individual Swedish and off-center steerable wheels in drives that combine different

types of wheels do not lead to a restriction of mobility. Accordingly, they have no influence on the hardware type. The upper part of the table then shows the possible modes of operation for the hardware type. Cells containing an "r" mark a Mode of Operation (MoO) resulting from the restriction due to physical sliding constraints. We call these the *physical* modes. The modes marked by "v" can be enforced by additional *virtual* constraints. If a mode is labeled with "n", this means that by driving the robot in this mode, physical sliding constraints of steerable wheels are neglected. With *neglected*, we mean that not all physical sliding constraints are represented in the model. The effect of such a concept is that for sufficiently fast actuators and motions with sufficiently low accelerations all sliding constraints are respected, while for motions with higher accelerations or slower actuators one or more sliding constraints will be violated and the robot will slip. Details are discussed in Sect. 5.3.

The hardware design-types that we introduce in Table 2 are equivalence classes of hardware designs under the following equivalence relation: $R_1 \sim R_2 \Leftrightarrow$ Robot R_1 and robot R_2 are able to operate in the same set of modes of operation without neglecting sliding constraints.

Example 2 *Let R_1 be a robot with two center-steerable wheels and R_2 a robot with two co-aligned fixed wheels. $R_1 \sim R_2$, because R_1 is able to operate in mode (1, 2), (2, 0) (cf. Example 1) and even in mode (1, 1), whereas R_2 is only able to operate in modes (2, 0) and (1, 1).*

Example 3 *Let R_1 be a car with four (two steered and two co-aligned fixed) wheels and R_2 a motorcycle. R_1 and R_2 do not belong to the same class, because both are able to operate in mode (1, 1) only.*
The robot type numbers were chosen to match the classification of two- and three-wheeled robot classification from Gracia and Tornero (2007).

The last row in Table 2 shows which kinds of singularities must be considered, when the corresponding hardware-type is operating in mode $(1, 2)$.

5.1 A HW-type Vb-robot operating in different modes

Exemplary, the following description shows how a robot with two or more coaligned center-steerable standard wheels (HW-type Vb) can operate in different modes. An example for a HW-type Vb robot is the snakeboard (Ostrowski and Burdick, 1998), typically operating in mode $(1, 2)$.

$(2, 0)$: a HW-type-Vb robot operates in mode $(2, 0)$, if the axes of all wheels are co-aligned. In such a configuration, the rank of $\mathbf{C}_{1c}(\boldsymbol{\beta}_c)$ is one, which gives $\delta_m = 2$. This is equivalent to forcing the ICR to lie on the common axis of all wheels. Due to this co-alignment-condition there is no freedom in choosing $\boldsymbol{\beta}_c$, resulting in $\delta_s = 0$.

$(1, 2)$: a robot is able to operate in this mode if the ICR is defined by a unique intersection of the wheel axes. If these axes coincide, then this condition is not satisfied and the robot is forced to operate in mode $(2, 0)$. This means that, in order to operate in mode $(1, 2)$, a robot must not reach a configuration that is consistent with the conditions of operating in mode $(2, 0)$. Thus, for HW-type-Vb robots, $Q_{(2,0)}$ appears as set of singularities in $Q_{(1,2)}$. This is a serious issue, as illustrated in the following example:

Example 4 *Figure 6a and b illustrates a HW-type Vb WMR with two center-orientable wheels moving along a circular path. In Fig. 6a, the orientation of the robot frame is constant relative to a Frenet frame. The steering angles $\boldsymbol{\beta}_c$ are constant and the robot is able to follow the circular path in mode of operation (1, 2). In Fig. 6b, the orientation of the robot frame is held constant relative to an inertial frame. In this case, the steering angles $\boldsymbol{\beta}_c$ are not constant and the robot is not able to follow the circular path in mode of operation (1, 2). At the time instant shown in the upper right, the ICR lies on the straight line through both wheel contact points. This corresponds to mode of operation (2, 0). Along every circular path with constant orientation, there will be two configurations where the ICR is required to lie on the line through the contact points of both wheels, which is a configuration in $Q_{(2,0)}$.*

$(1, 1)$: mode $(1, 1)$ can be enforced for a HW-type-Vb robot by introducing a virtual constraint that blocks the steering of exactly one wheel. The axis of this blocked wheel must not coincide with the wheel contact points of the other wheels, otherwise forcing mode $(2, 0)$.

$(3, 0)$: by driving a HW-type-Vb robot in this mode, the restrictions from the sliding constraints are neglected for all wheels.

$(2, 1)$: when a $(2, 1)$ model is used for a HW-type-Vb robot, the sliding constraints are neglected for all but one wheel.

Due to its *two* different "physical modes", HW-type Vb is the most complex hardware type. For the other hardware types, finding the physically enforced modes of operation is as simple as determining δ_m and δ_s for an arbitrary configuration that allows motion.

5.2 Virtually constrained modes

For the virtually constrained modes respecting all sliding constraints, the following conditions must be satisfied:

1. the number of DoF of the desired mode must be lower or equal to the number of DoF of the physical mode;

2. the DoN of the desired mode must be higher or equal to the DoN of the physical mode (see also Table 1).

While the necessity of the first condition is obvious, the second one might require explanation. In the previous section, Lie-bracket motions were introduced as directions reachable by consecutive motions along different vector fields (e.g. "drive-steer-drive"). So, while some robots are able to drive directly in a certain direction, others may require Lie-bracket motions to do so. For this reason, Lie-bracket motions can be considered slower than vector-field motions. The higher the order of the Lie-bracket, the slower the motion. A WMR is therefore only able to drive in a mode involving Lie-brackets of equal or higher order than its own physical mode.

5.3 The effects of neglecting constraints

The mappings from Table 2 marked with "r" or "v" provide accurate models for each hardware type. This is to be understood in the following way: if a model for a mode of operation is selected according to the classification from Table 2, then this model captures all the restrictions of mobility of the WMR originating from its sliding constraints. The modes marked with "n" in Table 2 do not contain all of these restrictions. Especially for hardware-types Va and Vb, operating in mode (1, 2), these accurate models come along with high complexity compared to other hardware-types and modes of operation. The generic model for mode of operation (2, 1) has the same number of inputs, but a less complex model than the one of MoO (1, 2) and, moreover, is free of singularities. Let us further analyze the differences between these models by bringing a (1,2) model to a form similar to a (2,1) model:

Place the origin O_R of the robot-fixed reference frame $\Sigma_R = \{O_R : \mathbb{X}_1, \mathbb{X}_2\}$ in the contact point of wheel i. Let the $\mathbb{X}_1$ axis point towards the contact point of wheel j. Then, $l_i = \alpha_i = \alpha_j = 0$. With this choice, the (1, 2) posture kinematic model is given by:

$$\dot{\boldsymbol{x}} = \mathbf{B}(\boldsymbol{x})\boldsymbol{u} \tag{27}$$

$$= \begin{pmatrix} l_j\left(\cos\left(\theta+\beta_i+\beta_j\right)-\cos\left(\theta+\beta_i-\beta_j\right)\right) & 0 & 0 \\ l_j\left(\sin\left(\theta+\beta_i+\beta_j\right)-\sin\left(\theta+\beta_i-\beta_j\right)\right) & 0 & 0 \\ 2\sin\left(\beta_i-\beta_j\right) & 0 & 0 \\ 0 & 1 & 0 \\ 0 & 0 & 1 \end{pmatrix} \begin{pmatrix} \eta_1 \\ \zeta_1 \\ \zeta_2 \end{pmatrix} \tag{28}$$

with l_j the distance between wheel i and wheel j. The input transformation

$$\eta_v = 2l_j \sin\left(\beta_j\right)\eta_1 \tag{29}$$

brings the system into the form

$$\begin{pmatrix} \dot{x} \\ \dot{y} \\ \dot{\theta} \\ \dot{\beta}_i \\ \dot{\beta}_j \end{pmatrix} = \begin{pmatrix} -\sin(\theta+\beta_i) & 0 & 0 \\ \cos(\theta+\beta_i) & 0 & 0 \\ \dfrac{\sin\left(\beta_i-\beta_j\right)}{l_j \sin\beta_j} & 0 & 0 \\ 0 & 1 & 0 \\ 0 & 0 & 1 \end{pmatrix} \begin{pmatrix} \eta_v \\ \zeta_1 \\ \zeta_2 \end{pmatrix}. \tag{30}$$

In this representation, the only expression depending on β_j is the one corresponding to $\dot{\theta}$. Taking

$$\eta_\theta = \frac{\sin\left(\beta_i-\beta_j\right)}{l_j \sin\beta_j}\eta_v \tag{31}$$

as new input gives

$$\dot{\theta} = \eta_\theta. \tag{32}$$

Equation (31) is a transformation that decouples the equation for $\dot{\theta}$ from the other equations. However, β_j needs to be controlled such that Eq. (31) is satisfied. Solving for β_j gives

$$\beta_j = \arctan\left(\frac{\sin(\beta_i)\eta_v}{l_j\eta_\theta + \cos(\beta_i)\eta_v}\right) + k\pi \tag{33}$$

with $k \in \mathbb{Z}$ chosen such that β_j is continuous. If a controller was able to make this equation invariant, then the posture kinematic model would simplify to

$$\begin{pmatrix} \dot{x} \\ \dot{y} \\ \dot{\theta} \\ \dot{\beta}_i \end{pmatrix} = \begin{pmatrix} -\sin(\theta+\beta_i) & 0 & 0 \\ \cos(\theta+\beta_i) & 0 & 0 \\ 0 & 1 & 0 \\ 0 & 0 & 1 \end{pmatrix} \begin{pmatrix} \eta_v \\ \eta_\theta \\ \zeta_1 \end{pmatrix}, \tag{34}$$

which is similar to the model for MoO (2, 1).

The reason why a (2, 1) model is not as accurate as a (1, 2) model for a robot operating in mode (1, 2) is found in Table 1: for models in class (3, 0), (2, 1) and (2, 0) no steering-angle-related vector field is involved in spanning TG. A steering-angle-related vector field can only be involved in spanning TG via Lie-bracket motions. As afore mentioned, Lie-bracket motions can be considered to be slower than pure vector field motions. For (1, 2) and (1, 1) models, such Lie bracket motions are required for spanning TG, that is, for reaching any pose in the plane.

An intuitive interpretation of these arguments is found by considering actuation: in a (2, 1) model, there is no state corresponding to β_j. It is thus not possible to introduce a model of the steering actuator into the model. Making Eq. (33) invariant would require a steering actuator that takes no time for reorientation of wheel j. If such an actuator would be available, then there would be no restriction of mobility due to the sliding constraints at all. This shows that a (2, 1) model does not capture the possible delay that originates from reorientation of wheels. However, such reorientations are necessary for the robot to reach any pose in the plane.

One could argue that also for a (1, 2) model there are unmodeled steering actuators. If a robot has more than two center-steered wheels, then only two of them appear in the model. However, since the ICR is 2-DoF, at most two steering angles can be restricting the motion of the robot at a time. This can be accounted for using a switching strategy (Betourne et al., 1996; Bak et al., 2003) or a clever choice for a coordinate chart of the steering angle manifold M (Thuilot et al., 1996).

6 Singularities

By singularities we mean configurations, at which sliding constraints become linear dependent. Note that this definition is not bound to a rank loss of $\mathbf{C}_1^*$, like for example the definition in Gracia and Tornero (2007). The reason for that is found in the models of class $(1,2)$. Although a robot that is described by such a model might have more than two steered wheels, only two steering angles appear in the state vector. Linear dependence of the corresponding sliding constraints causes a situation similar to the singularity in the two-wheel case, illustrated in Example 1. We therefore also call it singularity in the multi-wheel case.

Singularities of type A are those, where the wheel axes of all wheels of a robot with coaligned center-steerable wheels (HW-type Vb) coincide. This singularity was introduced in Example 1 as a model-switch, leading to a hybrid WMR model. From a mathematical point of view, for hybrid systems, the meaning of *solution* and its existence and uniqueness are not per se defined (Filippov, 1988). In this special case, however, the meaning of *solution* is rather clear and indeed equal to the meaning of *solution* in the continuous case. From a physical point of view also existence and uniqueness of solutions are given, as a real robot indeed executes a concrete motion in an experiment. The latter could however just be the result of dynamical effects that are not contained in our kinematic model.

The mathematical theory that allows to conclude about the properties of existence and uniqueness is the theory of Carathèodory differential equations and Lebesgue integration. A generalization of the notion of *solution* is called a solution in an extended sense $\boldsymbol{x}(t)$. Such a solution is absolutely continuous and satisfies the differential equation $\dot{\boldsymbol{x}} = \boldsymbol{f}(\boldsymbol{x},t)$ almost everywhere. For existence, it is required that the right hand side $\boldsymbol{f}(\boldsymbol{x},t)$ of the equation satisfies the following conditions:

1. $\boldsymbol{f}(\boldsymbol{x},t)$ must be defined and continuous in x for almost all t;
2. $\boldsymbol{f}(\boldsymbol{x},t)$ must be measurable in t for each x;
3. there must exist a function $m(t) \geq |\boldsymbol{f}(\boldsymbol{x},t)|$ that must be summable on every finite interval $\mathcal{I} \ni t$.

Uniqueness is given if there exists a summable function $\boldsymbol{l}(t)$ such that $\boldsymbol{f}(\boldsymbol{x},t)$ satisfies the following Lipschitz-like condition:

$$||\boldsymbol{f}(\boldsymbol{x}_1,t) - \boldsymbol{f}(\boldsymbol{x}_2,t)|| \leq \boldsymbol{l}(t)||\boldsymbol{x}_1 - \boldsymbol{x}_2||. \quad (35)$$

For a type Vb robot operating in mode $(1,2)$, the vector field $\boldsymbol{f}(\boldsymbol{x},t)$ is defined by Eq. (15) with $\boldsymbol{u}$ to be set by a controller. Since the set of configurations Q_2 where the mode switches to $(2,0)$ is nowhere dense in the configuration manifold, the robot operates in mode $(1,2)$ for almost all t if this controller does not contain any feedback and is fully actuated also in MoO $(2,0)$. In such a case, it is not required to consider the $(2,0)$ model at all. The vector field $\boldsymbol{f}(\boldsymbol{x},t)$ corresponding to mode $(1,2)$ indeed satisfies the conditions for existence and uniqueness of a solution in the extended sense. Type A singularities can therefore be ignored in feedforward control, at least from an idealized point of view also ignoring dynamical effects.

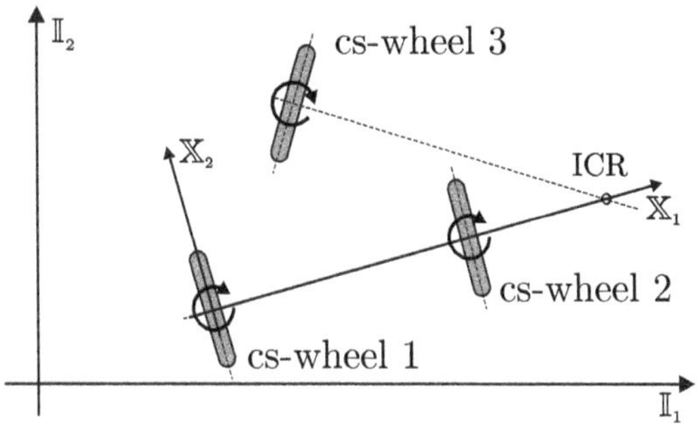

Figure 4. An example for a singularity of type B for a robot with three cs-wheels that are not coaligned: no unique ICR is defined by wheels one and two, but one and three.

In the feedback-case, the situation is more challenging: suppose a reference trajectory is designed in a way that requires the robot to stay in a singular configuration for longer time. In the feedforward-case this does not compromise the Carathèodory conditions, as the robot will still be in $Q \setminus Q_2$ almost always. However, if a feedback controller is active, the robot is forced to enter Q_2 again and again (by the controller), leading to high-frequency switching. Continuity of $\boldsymbol{f}(\boldsymbol{x},t)$ in x for almost all t is therefore questionable, as is the use of a controller that is based on a $(1,2)$-model in such a case. In practical experiments with controllers based solely on $(1,2)$-models, a chattering of the steering actuators is observed on such a "singular" trajectory (Gruber and Hofbaur, 2014). The reason for this chattering is that a $(1,2)$-controller uses steering actuators to move the ICR and correct small orientation errors of the robot. Since the wheel axes are parallel in a type-A singularity, this problem is ill conditioned. The proper actuators for this task would be the wheel speeds, cf. Example 1.

A singularity of type A is only present for robots with coaligned wheel contact points (HW-type Vb). When the wheel contact points are not on a straight line (HW-type Va), then one or more (but not all) wheel axes may be coaligned. Such a situation is a singularity of type B, shown in Fig. 4. Since only two steering angles appear in the posture kinematic model of the robot, singularities of type B appear similar to singularities of type A if the corresponding wheels are coaligned. The actual difference is that in type B singularities, the number of immediately accessible DoF does not increase because there is at least one more wheel that still restricts motion. Type B singularities can therefore be avoided

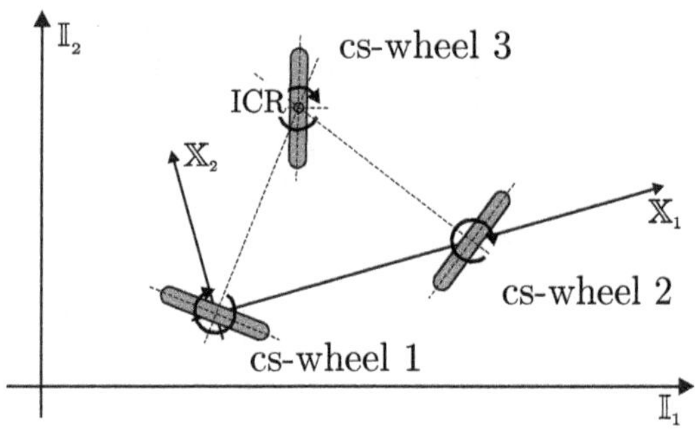

Figure 5. An example for a singularity of type C for a robot with three cs-wheels that are not coaligned: the ICR, defined by wheels one and two, coincides with the contact point of wheel three.

by using a hybrid system model (Bak et al., 2003) or a clever choice of coordinates for M.

A robot is in a type C singularity when the ICR coincides with the contact point of a center-steerable standard wheel, see Fig. 5. The challenges associated with type C singularities are two-fold: on one hand, once the ICR lies exactly in the contact point of a center-steerable wheel then its steering angle cannot be computed from the ICR-relation. The practically more relevant problem is that steering speeds get high when the ICR is required to get close to a wheel contact point. Some works exist that deal with this singularity problem and avoidance strategies (D'Andrea-Novel et al., 1995; Connette et al., 2009; Thuilot et al., 1996; Dietrich et al., 2011).

7 The geometry of wheeled mobile robots

The previous sections tried to require a minimum of preknowledge in differential geometry to make this article easily accessible to a broader audience. A crash-course for control engineers is found in Schlacher and Schöberl (2014). An early, comprehensive treatment of nonholonomy in dynamic systems is found in Neimark and Fufaev (1972). In this section, we analyze wheeled mobile robots from the point of view of differential geometry. This allows us to precisely reason about state manifolds, singularities and conditions for non-degeneracy.

Let us again start at the root cause for restriction of motion, the sliding constraints from Eq. (3). For each standard wheel i, the sliding constraint is defined via an analytic 1-form $\omega_i(\boldsymbol{q})$. Such a constraint reduces the number of degrees of freedom of a robot only if it is integrable, that is, holonomic. The Frobenius theorem provides a criterion to determine if a set of Pfaffian equations is holonomic. Let $\wedge$ be the exterior (wedge) product on forms and define

$$\Omega(\boldsymbol{q}) = \omega_1(\boldsymbol{q}) \wedge \cdots \wedge \omega_{N_c}(\boldsymbol{q}) \tag{36}$$

as an N_c-form (Chen et al., 1997). A set of constraints is holonomic, if at every point $\boldsymbol{q}$, for all $r = 1, \ldots, N_c$,

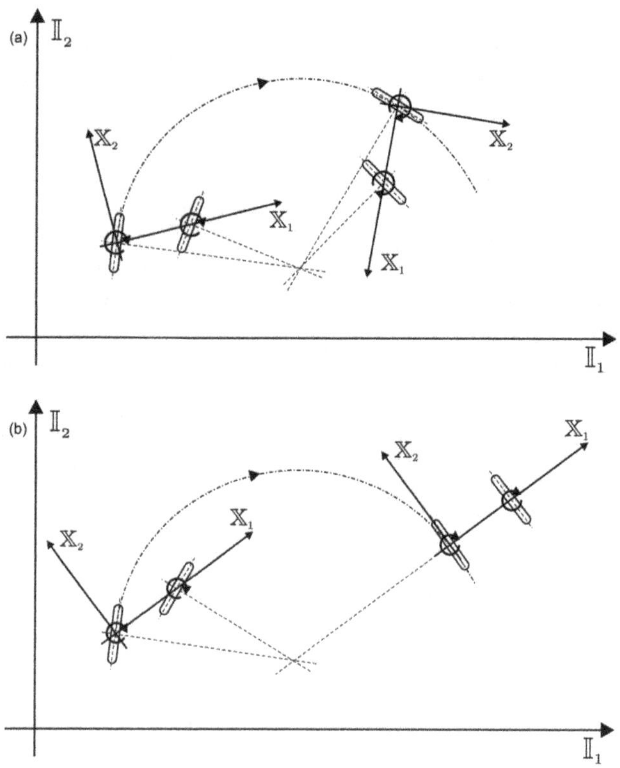

Figure 6. **(a)** Circular motion of a HW-type Vb WMR with constant steering angles. **(b)** Circular motion of a HW-type Vb WMR with constant orientation.

$$F_r(\boldsymbol{q}) = (\mathrm{d}\omega_r \wedge \Omega)(\boldsymbol{q}) = 0, \tag{37}$$

where $\mathrm{d}\omega_r$ denotes the exterior derivative of ω_r. However, we do not just want to reason whether all constraints are integrable or not, but also consider partial integrability. To do so, let us analyze the space spanned by the 1-forms ω in more detail: the collection of 1-forms ω locally spans a linear subspace of the cotangent space

$$D(\boldsymbol{q}) = \operatorname{span}\{\omega_1(\boldsymbol{q}), \ldots, \omega_{N_c}(\boldsymbol{q})\} \subset T_{\boldsymbol{q}}^* Q. \tag{38}$$

A family of such subspaces is called a *generalized codistribution* (Cortés et al., 2001). If the dimension of this codistribution is constant on an open neighborhood of $\boldsymbol{q}_0$, then $\boldsymbol{q}_0$ is called a regular point of D. Otherwise, it is a singular point. If the dimension of D is constant on all of Q, then D is called a regular codistribution. Given a generalized codistribution D, let us define its coannihilator D^0 as the *generalized distribution* Δ (in fact, this is exactly the object we were dealing with in Sect. 4).

We are looking for the "largest" connected submanifold X of Q on which the dimension of Δ is constant. On this submanifold, also δ_{m} – which is the dimension of Δ – and δ_{s}, together defining the mode of operation, are constant. Given the generalized codistribution D, let $\operatorname{inv}(\Delta)$ be the involutive closure of Δ. Since $\operatorname{inv}(\Delta)$ is involutive and spanned by a

locally finitely generated (see Table 1) set of vector fields, it has the maximal integral manifolds property (Isidori, 1995, Theorem 2.1.5). This guarantees the existence of such manifolds X. Now let us introduce these formally:

The submanifold X of Q is an integral manifold of inv(Δ) if $T_x X$ is spanned by inv(Δ)($\boldsymbol{x}$) at each $\boldsymbol{x} \in X$. An open submanifold X of Q is a maximal integral manifold (leaf) if every connected integral manifold of inv(Δ) which intersects X is an open submanifold of X. These leaves thereby are the "largest" connected subsets in Q, on which δ_m and δ_s are constant. A distribution has the maximal integral manifolds property, if there exists a leaf passing through every point on the manifold. Since inv(Δ) has this property, there exists a leaf passing through every point $\boldsymbol{q} \in Q$. These submanifolds X therefore perfectly qualify as state manifolds for the associated control systems.

Since inv(Δ) has the maximal integral manifolds property but may be singular, WMR are simple examples for an interesting class of systems: depending on the initial state, one may obtain different control systems that evolve on manifolds of different dimensions, cf. (Isidori, 1995, Remark 2.2.4). Bullo and Lewis (2005, Chapter 4.5.2) find "work to be done" in the field of singular codistributions.

Let us now apply this theory to wheeled mobile robots. Consider a robot with only one center-steerable or fixed standard wheel. In this case, there is only one 1-form $\omega_1(\boldsymbol{q})$ due to the single sliding constraint. In this special case, the integrability conditions from Frobenius theorem, Eq. (37), are violated globally. The codistribution D is regular, and, as a result, δ_m and δ_s are constant on all of Q. This allows to draw the following conclusions: WMR with only one center-steerable wheel or one fixed standard wheel are (i) nonholonomic, (ii) globally controllable, since the LARC (Isidori, 1995) holds globally, and (iii) the state manifold X of the associated control system is the whole configuration manifold Q.

However, for two sliding constraints, the situation already gets more complicated. Consider the robot with two center-steerable wheels from Example 1. Without loss of generality, place the origin of the robot fixed frame in the contact point of wheel 1 and let the $\mathbb{X}_1$ axis point towards the contact point of wheel 2 (see Fig. 2a). This specific choice of coordinates is just to simplify expressions ($\alpha_1 = 0$, $\alpha_2 = 0$, $l_1 = 0$): the 1-forms are

$$\omega_1 = \cos(\beta_1 + \theta)\mathrm{d}x + \sin(\beta_1 + \theta)\mathrm{d}y$$
$$\omega_2 = \cos(\beta_2 + \theta)\mathrm{d}x + \sin(\beta_2 + \theta)\mathrm{d}y + l_2 \sin(\beta_2)\mathrm{d}\theta$$

and Ω evaluates to:

$$\begin{aligned}\Omega = & -\sin(\beta_1 - \beta_2)\mathrm{d}x \wedge \mathrm{d}y \\ & + \cos(\beta_1 + \theta) l_2 \sin(\beta_2)\mathrm{d}x \wedge \mathrm{d}\theta \\ & + \sin(\beta_1 + \theta) l_2 \sin(\beta_2)\mathrm{d}y \wedge \mathrm{d}\theta. \end{aligned} \tag{39}$$

For the integrability conditions one obtains

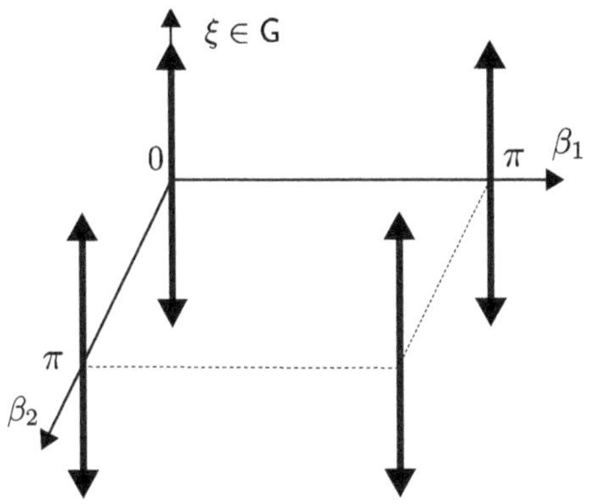

Figure 7. For two sliding constraints, the involutive closure of the coannihilator of the generalized codistribution D foliates Q into five separate leaves. The four leaves indicated by the bold rays correspond to the leaves at the four singular points of D. The fifth leaf fills the remaining open space (schematic).

$$F_1 = l_2 \sin(\beta_2)\mathrm{d}x \wedge \mathrm{d}y \wedge \mathrm{d}\theta \wedge \mathrm{d}\beta_2 = 0 \tag{40}$$
$$F_2 = l_2 \sin(\beta_1)\mathrm{d}x \wedge \mathrm{d}y \wedge \mathrm{d}\theta \wedge \mathrm{d}\beta_1 = 0, \tag{41}$$

which both hold only when the axes of both wheels coincide, that is, both conditions hold on

$$Q_2 = \{(x, y, \theta, \beta_1, \beta_2) | \beta_1 \in \{0, \pi\} \text{ and } \beta_2 = \{0, \pi\}\} \tag{42}$$

which contains four disjoint submanifolds, each of them corresponding to a connected set of singular points of D. This generalized codistribution D foliates Q into five separate leaves, see Fig. 7. At every regular point $\boldsymbol{q} \in Q_1 = Q \setminus Q_2$, the dimension of inv($\Delta$) is equal to the dimension of Q, that is, 5. On Q_2, ω_1 and ω_2 become linear dependent and take the form

$$\omega = \cos(\theta)\mathrm{d}x + \sin(\theta)\mathrm{d}y \tag{43}$$

and the dimension of inv(Δ) is reduced to 3. These singular points of the codistribution D are those that were introduced as *singularities of type A* (cf. Table 2). An example for a robot with two steered wheels in a type-A singularity is shown in Fig. 2b. Summing up, this gives the following results: WMR with two center-steerable or fixed standard wheels are (i) nonholonomic, since sliding constraints are at most partly integrable; (ii) the state manifold X of the currently active model is the leaf determined by the steering angle configuration; and (iii) globally controllable, since the LARC holds at each $\boldsymbol{x} \in X$. Note, however, that both, the dimension of X and the rank of the Lie algebra are not constant on all of Q.

For three sliding constraints, the integrability condition (Eq. 37) holds globally. In similar coordinates as in the case of two sliding constraints, the condition $\Omega = 0$ from Eq. (36) can be solved for β_3. Thereby, one obtains

$$\beta_3 = -\arctan\left(\frac{-l_2\cos(\alpha_3-\beta_1+\beta_2)+l_2\cos(\alpha_3-\beta_1-\beta_2)}{l_2\sin(\alpha_3-\beta_1+\beta_2)-l_2\sin(\alpha_3-\beta_1-\beta_2)+2l_3\sin(\beta_1-\beta_2)}\right) + k\pi \quad (44)$$

with $k \in \mathbb{Z}$ chosen such that β_3 is continuous. This condition ensures the existence of a unique ICR. On the surface N defined by this equation, two of the three 1-forms (sliding constraints) are linear dependent and similar conclusions as for the two-sliding-constraint-case can be drawn.

Constraint (Eq. 44) is singular on $\beta_1, \beta_2 \in Q_2$, in which case no unique ICR is defined by wheel 1 and 2. This is the singularity introduced in Sect. 6 as *singularity of type B*. On this singularity, 1-forms ω_1 and ω_2 become linear dependent. However, if $\alpha_3 \neq k_3\pi, k_3 \in \mathbb{Z}$ (wheels $1-3$ are not colinear), then ω_1 and ω_3 are independent. The ICR is therefore defined by the axes of wheel 1 and wheel 3. An example for such a singularity is shown in Fig. 4. To avoid singularities of type B, it is required to either switch (Betourne et al., 1996) to coordinates β_1, β_3 or β_2, β_3 or choose a completely different chart on N, see Thuilot et al. (1996) for a clever choice. Note that singularities of type B are no singularities of the codistribution D.

A *singularity of type C* appears when only the denominator of the fraction in the arctangent in Eq. (44) is zero. This is the case when the ICR is placed in the contact point of wheel 3. An example for such a singularity is shown in Fig. 5.

If $\alpha_3 = k_3\pi, k_3 \in \mathbb{Z}$ and $\boldsymbol{q} \in Q_3$,

$$Q_3 = \{(x, y, \theta, \beta_1, \beta_2, \beta_3) \,|\, \beta_1 \in \{0, \pi\},\ \beta_2 = \{0, \pi\},\ \beta_3 = \{0, \pi\}\}, \quad (45)$$

then all 1-forms become linear dependent and the dimension of D reduces to 1 at these singular points. This is the case when all wheel axes are co-aligned and reduces to the same situation as for co-aligned wheels in the two-wheel case. Consistently, these singularities are also classified as *singularities of type A*.

The foliation of Q for three sliding constraints is illustrated in Fig. 8. As in the two-wheel case, inv(Δ) foliates Q into several structurally different leaves: (i) the dashed horizontal planes are the leaves that correspond to a robot in rest (ξ = const.). Only a nowhere dense subset of steering angle configurations allow motion. One of those is (ii) the dash-dotted surface defined by Eq. (44). Another one is the dash-dotted plane $\beta_1, \beta_2 = 0$. Only if $\alpha_3 = k_3\pi, k_3 \in \mathbb{Z}$, then also the (iv) bold vertical arrows are leaves, corresponding to the case when all wheel axes are co-aligned.

It can be concluded that WMR with three center-steerable wheels are (i) nonholonomic, (ii) the state manifold X of the currently active model is the leaf determined by the steering angle configuration; and (iii) globally controllable, since the LARC holds at each $\boldsymbol{x} \in X$.

The case of three sliding constraints can be analogously extended to an arbitrary number of wheels: for every further wheel, either its axis must be co-aligned to at least one of the existing wheels or coordinated by a coordination function asserting the existence of a unique ICR.

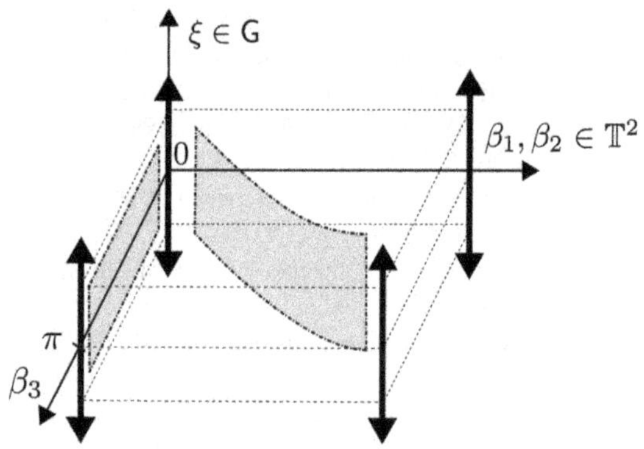

Figure 8. For three sliding constraints, the involutive closure of the coannihilator of the generalized codistribution D foliates Q into several leaves (schematic).

8 Conclusions

This article presented a detailed view on the kinematics of single-bodied wheeled mobile robots.

One suggestion is to drop the notion of degeneracy of wheeled mobile robots and replace it with controllability. The usage of this term is consistent with the standard definition in control theory. Furthermore, a simple condition for controllability was presented.

Based on this controllability study, a classification of wheeled mobile robots into six hardware types is introduced. This classification is solely based on the type, location and number of wheels. Each hardware type is able to operate in one or more modes of operation. A mode of operation corresponds to a set of configurations, in which a specific model is an accurate representation of the kinematic capabilities of the robot.

Moreover, we provide a detailed analysis of the geometry of wheeled mobile robots by which we are able to give a general view on state manifolds and singularities.

Acknowledgements. This work was supported by the Austrian Science Fund (FWF) under Grant PN 20041.

References

Alexander, J. C. and Maddocks, J. H.: On The Kinematics of Wheeled Mobile Robots, Int. J. Robot. Res., 8, 15–27, 1989.

Bak, T., Bendtsen, J., and Ravn, A.: Hybrid control design for a wheeled mobile robot, in: Hybrid Systems: Computation and Control, Springer-Verlag, 50–65, http://www.springerlink.com/index/e7a3gw8umlkyuggl.pdf (last access: 5 April 2016), 2003.

Betourne, A., Campion, G., and Bètournè, A.: Kinematic modelling of a class of omnidirectional mobile robots, in: Proceedings of IEEE International Conference on Robotics and Automation, 22–28 April 1996, Minneapolis, 3631–3636, doi:10.1109/ROBOT.1996.509266, 1996.

Bullo, F. and Lewis, A. D.: Geometric Control of Mechanical Systems, Springer-Verlag, New York, XXIV, 727, 2005.

Campion, G. and Chung, W.: Wheeled Robots, in: Springer Handbook of Robotics, chap. 17, edited by: Siciliano, B. and Kathib, O., Springer-Verlag, Berlin, Heidelberg, 391–410, 2008.

Campion, G., Bastin, G., and D'Andrèa-Novel, B.: Structural properties and classification of kinematic and dynamic models of wheeled mobile robots, IEEE T. Robot. Automat., 12, 47–62, doi:10.1109/70.481750, 1996.

Canudas-de Wit, C., Siciliano, B., and Bastin, G.: Theory of Robot Control, Springer-Verlag, London, XVI, 392, 1996.

Chen, B., Wang, L.-S., Chu, S.-S., and Chou, W.-T.: A new classification of non-holonomic constraints, P. Roy. Soc. A, 453, 631–642, doi:10.1098/rspa.1997.0035, 1997.

Choset, H., Lynch, K. M., Hutchinson, S., Kantor, G. A., Burgard, W., Kavraki, L. E., and Thrun, S.: Principles of Robot Motion, MIT Press, Boston, 632, 2005.

Connette, C. P., Parlitz, C., Hagele, M., Verl, A., and Hägele, M.: Singularity avoidance for over-actuated, pseudo-omnidirectional, wheeled mobile robots, in: International Conference on Robotics and Automation, 2009, ICRA'09, 12–17 May 2009, Kobe, 4124–4130, 2009.

Cortés, J., Deleón, M., de Diego, D. M., and Martínez, S.: Mechanical systems subjected to generalized non-holonomic constraints, P. Roy. Soc. Lond. A, 457, 651–670, doi:10.1098/rspa.2000.0686, 2001.

D'Andrea-Novel, B., Campion, G., and Bastin, G.: Control of Nonholonomic Wheeled Mobile Robots by State Feedback Linearization, Int. J. Robot. Res., 14, 543–559, doi:10.1177/027836499501400602, 1995.

De Luca, A., Oriolo, G., and Giordano, P. R.: Kinematic Control of Nonholonomic Mobile Manipulators in the Presence of Steering Wheels, in: IEEE International Conference on Robotics and Automation, 3–7 May 2010, Anchorage, 1792–1798, 2010.

Dietrich, A., Wimböck, T., Albu-Schäffer, A., and Hirzinger, G.: Singularity Avoidance for Nonholonomic, Omnidirectional Wheeled Mobile Platforms with Variable Footprint, ICRA, 9–13 May 2011, Shanghai, 6136–6142, 2011.

Filippov, A.: Differential Equations with Discontinuous Right Hand Sides, Springer Netherlands, X, 304, 1988.

Giordano, P. R., Fuchs, M., Albu-Schäffer, A., and Hirzinger, G.: On the kinematic modeling and control of a mobile platform equipped with steering wheels and movable legs, in: 2009 IEEE International Conference on Robotics and Automation, 12–17 May 2009, Kobe, 4080–4087, doi:10.1109/ROBOT.2009.5152625, 2009.

Gracia, L. and Tornero, J.: A new geometric approach to characterize the singularity of wheeled mobile robots, Robotica, 25, 627–638, doi:10.1017/S0263574707003578, 2007.

Gruber, C. and Hofbaur, M.: Distributed Configuration Discovery for Modular Wheeled Mobile Robots, in: IFAC Symposium on Robot Control (SYROCO), 5–7 September 2012, Dubrovnik, 689–696, doi:10.3182/20120905-3-HR-2030.00042, 2012.

Gruber, C. and Hofbaur, M.: Practically Stabilizing Motion Control of Mobile Robots with Steering Wheels, in: IEEE Multi-Conference on Systems and Control (MSC), 8–10 October 2014, Juan Les Pins, 1312–1317, 2014.

Hofbaur, M., Brandstötter, M., Jantscher, S., and Schörghuber, C.: Modular re-configurable robot drives, in: International Conference on Robotics and Automation and Mechatronics (RAM 2010), 28–30 June 2010, Singapore, 150–155, 2010.

Isidori, A.: Nonlinear Control Systems, 3rd Edn., Springer, Springer-Verlag, London, XV, 549, 1995.

Morin, P. and Samson, C.: Motion control of wheeled mobile robots, in: Springer Handbook of Robotics, chap. 34, edited by: Siciliano, B. and Kathib, O., Springer, Springer-Verlag, Berlin, Heidelberg, 799–826, 2008.

Muir, P. and Neuman, C.: Kinematic modeling for feedback control of an omnidirectional wheeled mobile robot, in: IEEE International Conference on Robotics and Automation, March 1987, Raleigh, 1772–1778, 1987.

Mutambara, A. G.: Decentralized Estimation and Control for Mulitsensor Systems, CRC Press LLC, Boca Raton, 256 pp., 1998.

Neimark, I. I. and Fufaev, N. A.: Dynamics of Nonholonomic Systems, Translations of Mathematical Monographs, American Mathematical Society, 1972.

Ostrowski, J. and Burdick, J.: The Geometric Mechanics of Undulatory Robotic Locomotion, Int. J. Robot. Res., 17, 683–701, 1998.

Schlacher, K. and Schöberl, M.: Geometrische Darstellung nichtlinearer Systeme, At-Automatisierungstechnik, 62, 452–462, doi:10.1515/auto-2014-1090, 2014.

Siegwart, R. and Nourbakhsh, I. R.: Introduction to Autonomous Mobile Robots, vol. 23, MIT Press, 2004.

Thuilot, B., D'Aandrea-Novel, B., Micaelli, A., and D'Andrea-Novel, B.: Modeling and feedback control of mobile robots equipped with several steering wheels, IEEE T. Robot. Automat., 12, 375–390, doi:10.1109/70.499820, 1996.

18

Solving the double-banana rigidity problem: a loop-based approach

Florian Simroth[1], Huafeng Ding[2], and Andrés Kecskeméthy[1]

[1]University of Duisburg-Essen, Duisburg, Germany

[2]China University of Geosciences Wuhan, Wuhan, China

Correspondence to: Florian Simroth (florian.simroth@uni-due.de)

Abstract. Rigidity detection is an important tool for structural synthesis of mechanisms, as it helps to unveil possible sources of inconsistency in Grübler's count of degrees of freedom (DOFs) and thus to generate consistent kinematical models of complex mechanisms. One case that has puzzled researchers over many decades is the famous "double-banana" problem, which is a representative counter-example of Laman's rigidity condition formula for which existing standard DOF counting formulas fail. The reason for this is the body-by-body and joint-by-joint decomposition of the interconnection structure in classical algorithms, which does not unveil structural isotropy groups for example when whole substructures rotate about an "implied hinge" according to Streinu. In this paper, a completely new approach for rigidity detection for cases as the "double-banana" counterexample in which bars are connected by spherical joints is presented. The novelty of the approach consists in regarding the structure not as a set of joint-connected bodies but as a set of interconnected loops. By tracking isolated DOFs such as those arising between pairs of spherical joints, rigidity/mobility subspaces can be easily identified and thus the "double-banana" paradox can be resolved. Although the paper focuses on the solution of the double-banana mechanism as a special case of paradox bar-and-joint frameworks, the procedure is valid for general body-and-joint mechanisms, as is shown by the decomposition of spherical joints into a series of revolute joints and their rigid-link interconnections.

1 Introduction

Rigidity detection has attracted researchers in the field of robotics and mechanisms since many decades, as it is a necessary component in the creative automatic synthesis of mechanisms (Ding et al., 2012), the processing of CAD-generated models (Lee, 2008; Moinet et al., 2014), the analysis of flexibility and dynamics of proteins (Jacobs et al., 2002), and others. The problem is (1) to identify whether a mechanism contains substructures with negative DOF (so-called "degenerate chains"), and if so, (2) to detect the parts forming those (over-) rigid substructures, such that their negative DOF do not illicitly cancel out positive DOF in other parts of the structure (Fig. 1).

Based on Laman's theorem (Laman, 1970), many algorithms were proposed to detect the rigidity for both planar and simple spatial mechanisms. Tay addressed the detection of rigidity for rigid bodies connected by bars represented by multi-graphs using screw-theory based constraints (Tay, 1984). Agrawal and Rao analyzed and determined the mobility of kinematic chains with fractionated DOFs (Agrawal and Rao, 1987). Based on the SE(3) displacement groups, Fanghella and Galletti analyzed the mobility properties of single-loop kinematic chains by regarding the connectivity between any two links in a chain and the invariant properties of the displacement group of their relative motion (Fanghella and Galletti, 1994). Hwang and Hwang presented a loop-decreasing method for the detection of rigid subchains of rigid planar kinematic chains (Hwang and Hwang, 1991). But according to Sunkari and Schmidt, it fails to detect some of the degenerate chains (Sunkari and Schmidt, 2005). Lee and Yoon proposed a method that involved deleting binary chains in turn to simplify the chain for rigid subchain detection, which is applicable to planar kinematic chains whose corresponding graph representations are planar or nonpla-

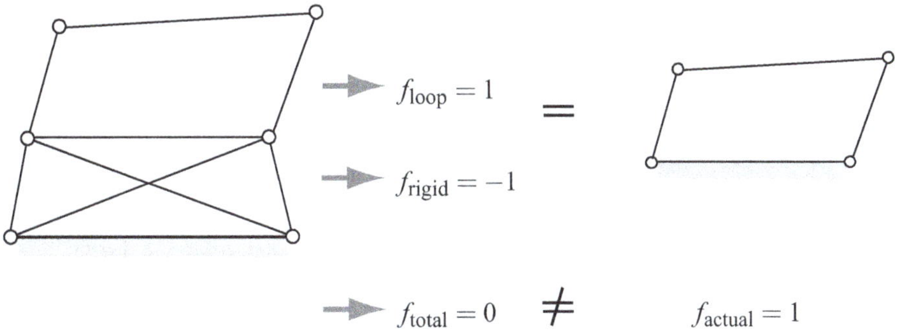

Figure 1. Simple example for an overconstrained mechanism.

nar (Lee and Yoon, 1992). Sunkari adopted Lee and Yoon's method to detect rigid subchains in the structural analysis and synthesis of planar and spatial mechanisms satisfying Gruebler's DOF equation (Sunkari, 2006). Tuttle obtained basic rigid chains with 7 or fewer links, and then attempted to use these chains to identify rigid subchains in systems with more links (Tuttle, 1996). But with increasing complexity, this method can hardly be applied to mechanisms with more than 10 links. Moukarzel adopted the description of a system as a collection of rigid bodies connected by bars to improve efficiency of rigidity detection (Moukarzel, 1996). In the context of material sciences, Jacobs and Hendrickson developed a graph-based algorithm for detecting overconstrained regions and rigid clusters of two-dimensional networks (Jacobs and Hendrickson, 1997). Another method used for structural analysis is the ear decomposition which originates from the work of Robbins and Whitney (Robbins, 1939; Whitney, 1932) for which a mechanism is decomposed by sequentially removing a sequence of "ears", i.e. serial chains with internal binary bodies connected at their endpoints to bodies of at least degree two, from the graph. Franzblau generalized this approach to chains including cycles and bridges, and by successively analyzing the subsystems using theorems based on the rigidity matrix, he was able to determine the minimal and maximal boundaries for the degrees of freedom for a given bar-and-joint framework (Franzblau, 1999, 2000), however without providing an explicit DOF count formula. A further approach is the use of matroids for determining globally rigid substructures (Servatius and Servatius, 2010). This method is very powerful for planar systems but has not been fully extended yet to spatial mechanisms. Moreover, the methodology is based on bar-frameworks whose generalization to kinematical joints is still a pending problem. Lee, Streinu and Theran as well as Chubynsky and Thorpe used combinatorial approaches like the pebble game algorithm to analyze rigidity (Lee et al., 2005, 2008; Chubynsky and Thorpe, 2007; Streinu and Theran, 2008). Shai, Servatius, Sljoka, Whiteley, and Müller focused on the decomposition of mechanisms into Assur components for mobility analysis (Servatius et al., 2010; Sljoka et al., 2011; Shai et al., 2013; Shai and Müller, 2013). Michelucci and Foufou proposed the

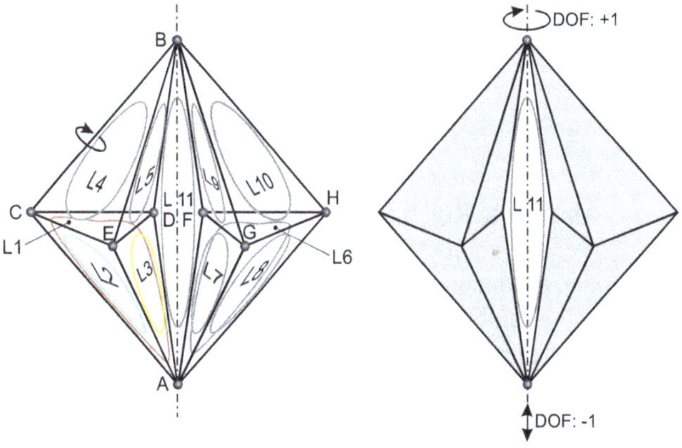

Figure 2. The "double-banana" mechanism.

"witness" method in which infinitesimal mobility is analyzed by evaluating the rank of the Jacobian at a given position of a given parametrization of the mechanism (Michelucci and Foufou, 2009). A method based on an independent loop set was presented by Ding, Huang, and Mu (Ding et al., 2008). The method was used to detect and then delete kinematic chains containing rigid subchains in the structural synthesis of planar simple and multiple-joint kinematic chains, but is not able to localize the minimal rigid substructures. More recently, Xia, Ding and Kecskeméthy presented a novel loop-based method for identifying rigid or over-rigid planar, spherical, spatial or mixed subsystems based on the kinematical network (Xia et al., 2012). The advantage of this method is that the mobility counts can be made separately within each loop and then assembled together at the level of relative kinematics (i.e. joint motions), such that mobility of hybrid systems containing substructures moving relatively to each other in different subspaces of SE(3) can be determined appropriately (for example when a planar four-bar mechanism is embedded in a spatial structure). The loop-based approach was verified in Xia et al. (2012) for planar systems, and is extended here to 3-D systems, including spherical-spherical bars. This shows that the loop-based method is suitable also for revolving paradox 3-D bar-and-joint framework cases.

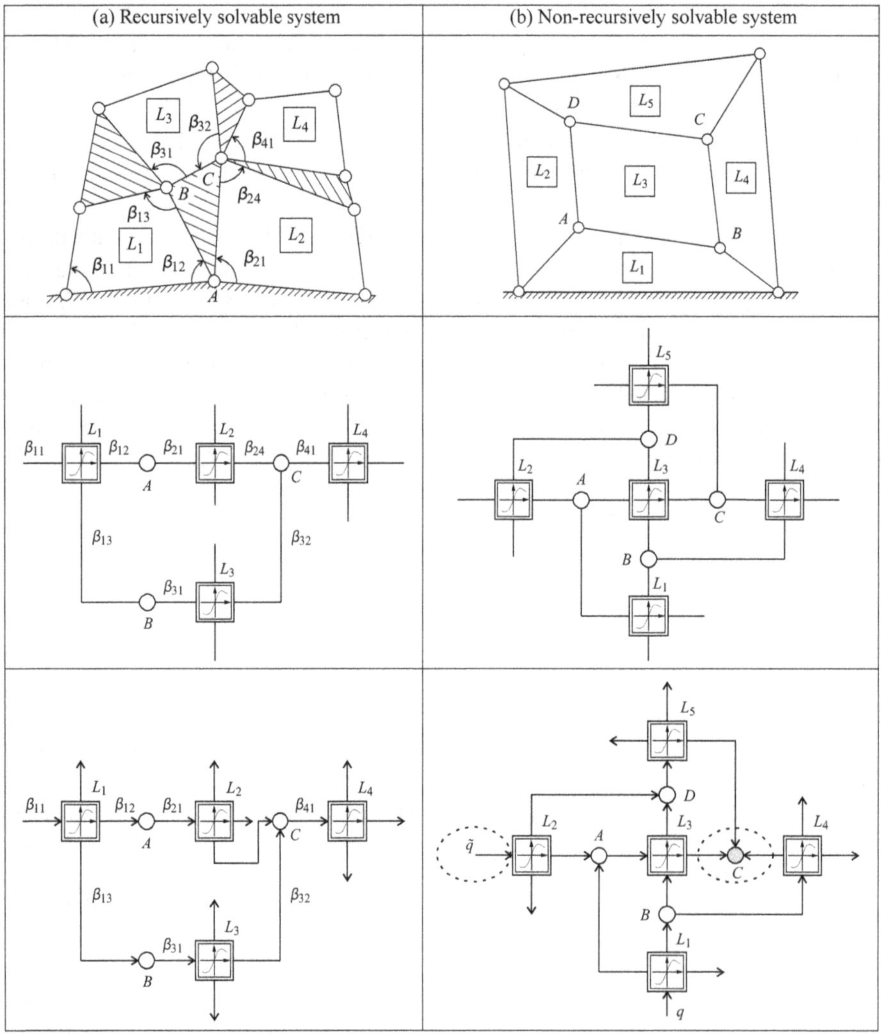

Figure 3. Mapping of mechanisms to kinematical networks.

2 The "double-banana" case

Although many rigidity detection algorithms have been provided to date, an automatic finite-mobility rigidity detection algorithm for systems involving "implied hinges" (see below) such as the famous "double-banana" case shown in Fig. 2, to the knowledge of the authors has not been reported so far. The "double-banana" consists of 18 bars connected at 8 nodes which can be interpreted as (multiple) spherical joints. If we give each node 3 DOFs and subtract one constraint per bar, then the Grübler count gives DOF $=0$, thus suggesting that the structure is rigid. However, one can clearly "see" in Fig. 2 that the "double-banana" structure consists of two rigid bipyramids, one at the left and one at the right, which are coupled together by two spherical joints (A and B). Thus, the segment $\overline{AB}$ is an "implied hinge" (Cheng et al., 2009) about which the two halves can rotate with respect to each other, while the longitudinal direction along this edge is over-constrained with DOF $= -1$. This implied DOF remains also undetected by the famous test equation for rigidity according to Maxwell's rule proposed in (Maxwell, 1864), which states that a statically and kinematically determined framework with b bars and j joints follows the equation

$$b = 3j - 6. \qquad (1)$$

It is obvious that a bipyramid (single banana), with 9 bars and 5 joints (correctly) satisfies the condition of the Maxwell's rule. However, also the "double-banana" structure obeys the Maxwell's rule, although it is obviously not rigid. Thus, this structure is a classical (non-trivial) example of a mechanism satisfying Maxwell's rule of a statically and kinematically determined framework, but which is nevertheless generically able to move, and which has thus attracted considerable attention in literature: Fowler analyzed the symmetry of the structure and presented the result of the symmetry of multiple banana mechanisms (Fowler and Guest, 2002); Cheng, Sitharam, and Streinu studied the rigidity of the 3-D struc-

ture with the "double-banana" as an example (Cheng et al., 2009), and Rojas proposed the "double-banana" closure conditions as a paradigm for position analysis in robots (Rojas and Thomas, 2013). The cause for the "double-banana" paradox is that there exist "isolated" DOFs in the mechanism (e.g. the bar spin between two spherical joints) which create an own level of transmission kinematics, and which thus need to be tracked in order to be able to detect which substructures are rigid and which are movable. We propose in this paper a novel procedure for tracking such isolated DOFs by using an alternative method for describing kinematical interdependencies, called the "kinematical network" (Kecskeméthy, 1993), in which the mechanism is regarded not as a system of connected bodies, but as a system of interconnected kinematical loops. A kinematical loop can be regarded as a closed chain of bodies which corresponds to an elementary cycle in a graph theoretical sense. If the underlying graph of the mechanism is not at least 2-edge-connected, i.e. it contains bridges or is separated, the graph first is dissected into its 2-edge-connected components, called clusters, for which the loop decomposition algorithm is applied separately. Of course, rigidity needs to be detected only within each cluster, as non-connected or 1-edge-connected clusters can never reduce the DOF in comparison with the sum of their internal DOFs. In the following, the method of "kinematical network" is briefly reviewed here for better understanding of the present paper, and then our novel rigidity-detection procedure for "double-banana" types of mechanisms is elucidated.

3 Description of mechanisms as kinematical networks

The concept of mechanism description using a network of linearly connected loops was introduced in (Kecskeméthy, 1993), and a fully automatic implementation in the symbolic formula manipulation software Mathematica for general planar, spherical, translational and spatial cases is described in (Kecskeméthy et al., 1997). The following planar examples are used for illustration of the method, but the concepts apply one-to-one also to spatial cases.

In this approach, as a first step, a minimal cycle basis comprising a set of smallest independent loops L_i of the mechanism is determined, where "length" is measured in terms of the number of involved joint variables. According to the Euler cyclomatic number, the number n_{L} of independent loops is given by

$$n_{\mathrm{L}} = n_{\mathrm{G}} - n_{\mathrm{B}} + 1, \tag{2}$$

where n_{B} is the number of bodies including the ground frame, and n_{G} is the number of binary joint connections. For example, for Fig. 3a one has $n_{\mathrm{B}} = 10$ and $n_{\mathrm{G}} = 13$ and thus $n_{\mathrm{L}} = 4$, while for Fig. 3b it holds $n_{\mathrm{B}} = 12$ and $n_{\mathrm{G}} = 16$, and thus $n_{\mathrm{L}} = 5$. Each loop is first regarded independently of the others, and its local degree of freedom f_{L_i} is determined, which is the number of loop joint coordinates minus the dimension of the subgroup of Euclidean motion in which the bodies of the loop locally move (spatial, planar, spherical, translational, etc.).

As the next step, the coupling conditions between the loops are determined. These arise exactly at those joints in which the number of incident loops is equal to or larger than the number of incident bodies and correspond to a balance of inner joint coordinates of the incident loops and some constants. More precisely, if a joint G_i connects $n_{\mathrm{B}}(G_i)$ bodies, then the bodies at the joint G_i can move with $[\ n_{\mathrm{B}}(G_i)-1\ ]\cdot k$ relative degrees of freedom with respect to each other, where k is the dimensionality of the joint G_i, e.g. $k = 1$ for a revolute joint or $k = 3$ for a spherical joint. Thus if $n_{\mathrm{L}}(G_i)$ loops are incident at joint G_i, the relative joint variables introduced independently of each other in each loop at that joint must fulfill

$$n_{\mathrm{C}}(G_i) = [\ n_{\mathrm{L}}(G_i) - n_{\mathrm{B}}(G_i) + 1\ ] \cdot k \tag{3}$$

balance equations, which therefore yields uniquely the number of coupling conditions at that joint (Kecskeméthy, 1993). For example, in joint A of Fig. 3a, there are two incident loops and two incident bodies, yielding one coupling equation. This corresponds to the condition that the sum of joint coordinates β_{12} and β_{21} plus a constant is equal to 360°, yielding a linear coupling between the loops L_1 and L_2. Note that for $n_{\mathrm{C}} > k$, while the number of coupling conditions is unique, the choice of the balance conditions is non-unique, as any set of combinations of admissible balance conditions gives again a suitable balance condition.

The thus connected loops form the so-called "undirected kinematical network" in which the loops represent local nonlinear transmission elements and the couplings between the loops represent the global interrelationships. From the kinematical network, the DOF can be obtained as the sum of local DOFs of all loops minus the sum of loop coupling conditions. In Fig. 3a, there are four four-bar loops with local DOF = 1 each, and three loop coupling conditions. Thus the overall DOF is 1. In Fig. 3b, there are five four-bar loops with local DOF = 1 each, and four loop coupling conditions, thus the global DOF is again 1. This shows that the DOF counting over the loops and their couplings is fully equivalent to the usual Grübler count, giving a new and consistent type of Grübler DOF formula. Note however that here, in contrast to the classical Grübler count, one can easily combine planar, spherical, and spatial loops.

After having established the undirected kinematical network, the next step is to orient the edges and to prescribe external inputs such that each loop can be computed as a function of outputs of previous loops and/or of the external inputs, and that no closed cycles occur. The thus ensuing oriented kinematical network is termed the "solution flow". The external inputs can be "true" DOFs, or auxiliary "pseudo" inputs are needed when the structure has no recursive solution flow and there will remain some implicit constraints to be solved

iteratively. An advantage of the kinematical network is that one can easily recognize recursively solvable substructures by the so-called *sink method* (Kecskeméthy, 1993): here, one searches iteratively for elements in the kinematical network for which the number of edges is less than or equal to the local degree of freedom of the element. After finding such an element, all edges are oriented into the element, the element together with all ingoing edges is removed, and the procedure is re-applied to the rest of the system. If the procedure covers all loops, one obtains a recursive solution flow, i.e. each loop can be solved as a function of a "true" external input and/or outputs of previously solved loops (example: Fig. 3a). If this is not the case, the network must be solved iteratively by choosing additional "pseudo" inputs and following the solution flow until an element is overconstrained, which yields the implicit constraint equation needed to be solved by an iterative root finding algorithm such as Newton's method (example: Fig. 3b) where a pseudo input $\widetilde{q}$ has been applied to loop L_2 and the implicit constraint equation results in loop coupling condition "C").

4 Types of isolated degrees of freedom

Isolated degrees of freedom are mobilities that leave some characteristic relative measurements between surrounding substructures locally invariant. In literature, several kinds of names and concepts are used in this respect. The different concepts and terminology might lead to confusion so that a short introduction is given.

Hunt introduced "superfluous spin-freedoms" which arise for the rotation of a bar in-between two spherical joints, which he suggests to remove for kinematic analysis (Hunt, 1978). Waldron uses the term "passive degree of freedom" for the same type of motion (Waldron, 1973). However, this term might lead to confusion as in parallel kinematics machines it refers to non-actuated chains (legs) for which the joint variables are dependent functions of actuated ones (Liu et al., 2014). This is also the meaning used in robotics by (Bennett and Hollerbach, 1991) when the joint variables of a virtual contact joint are a function of the actuated robot degrees of freedom. Also, in control theory, the term passive DOF refers to non-controlled degrees of freedom which however are properly transmitted throughout the mechanism (Shiriaev et al., 2010). For example, one four-bar mechanism could have one actuated DOF, while a second four-bar mechanism on top of it could be non-actuated (passive), thus the control of the first four-bar mechanism is sought which controls both stably. Furthermore, the term passive degree of freedom is sometimes used for underactuated manipulators and are accounted for structural flexibilities (Jain and Rodriguez, 1993; Casals and Amat, 1996). Another term one encounters in this setting is "redundant DOF" (Zhang et al., 2003), which however may lead to confusion as the term redundant DOF is used in robotics for robots featuring a greater

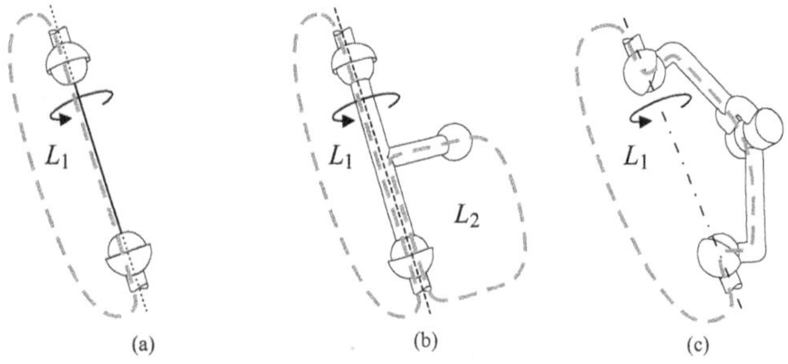

Figure 4. Different types of isolated DOFs: **(a)** "*Fully isolated*", **(b)** "*transmitted isolated*", **(c)** "*structurally isolated*".

number of actuators than necessary for a given task space which allows for special control movements (such as the human arm having seven relative joint motions for six DOFs at the hand). In some cases the term "parasitic DOF" is used for isolated DOFs (Rolland, 1999); however, the term "parasitic motion" (Carretero et al., 2000) or "parasitic DOF" is also (and more frequently) used to describe unwanted dependent motions in some direction such as the vertical axis of a lower mobility platform (Hsu et al., 2004; Isaksson et al., 2015; Merlet, 2005). As a further variant one finds the term parasitic degrees of freedom meaning additional degrees of freedom added to the principal DOFs by compliant joints (Togashi et al., 2014). Finally, in some cases the term "idle DOF" is used in this context (Razmara et al., 2000); however, the term "idle" would imply that the related motion of the isolated DOF is not changing, which is not the case in the present discussion.

In view of the plurality of notions and terms used in different contexts, we propose here to use the term "isolated" which most accurately describes the notion of degrees of freedom which have no influence on certain relative quantities at the ends of a serial chain, i.e. remain non-transmitted in a certain neighborhood of this chain, but may be transmitted to surrounding substructures. This is in close correspondence to the original term "superfluous spin-freedom" used by Hunt, and helps to convey the idea that the isolated DOFs might not be "superfluous" or pure "spinning" (meaning that they might be transmitted to overlayed structures or that there is a material axis of rotation). It might be noted that this term has been used in the past in the modeling of active meshes for grasping devices (Mazzone et al., 2003).

In order to track isolated DOFs, three types can be distinguished according to their scope of action (Fig. 4):

(a) "Fully isolated DOF": these are immaterial motions which can be completely removed from the mechanism without any effect; this is the case for example for the isolated spin of an infinitesimally thin spherical-spherical bar about its longitudinal axis in Fig. 4a. Fully isolated DOFs will be displayed below in the loop connection graphs as dotted lines.

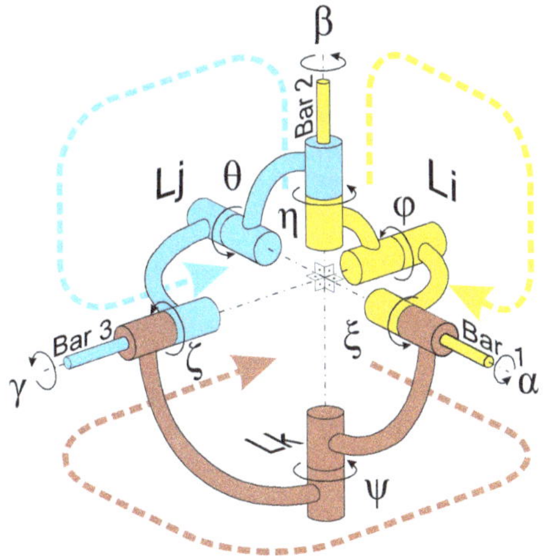

Figure 5. Loop coupling conditions at a multiple spherical joint.

(b) "Transmitted isolated DOF": these are isolated DOFs which have already been counted once as "fully isolated" in one loop, so that they become transmitted in a neighboring loop; an example is shown in Fig. 4b, where the spin of the rod has been counted as isolated in loop L_1, but becomes transmitted to loop L_2; this will be essential when regarding the transmission of isolated DOFs at loop coupling conditions of spherical joints. Transmitted isolated DOFs will be displayed below in the loop connection graphs as dashed lines.

(c) "Structurally isolated DOF": these are isolated DOFs that comprise whole substructures; they leave the internal motion of whole subchains within one loop invariant but operate on the absolute motion of the chain as proper DOFs; an example is shown in Fig. 4c, where the isolated DOF does not change the internal configuration of the revolute joint on the spun subchain, but the chain rotates as a whole about the implied spin axis connecting both spherical joints. Note that in this type of isolated DOFs the spin axis is not constant with respect to any body, but changes with the internal motion of the involved subchain. Structurally isolated DOFs will be displayed below in the loop connection graphs as dot-dashed lines.

5 Loop coupling conditions at a spherical joint

When multiple loops coincide at one spherical joint, the product of rotation matrices of the relative rotations over all incident loops must yield unity, thus producing 3 independent coupling conditions. Furthermore, when multiple bars coincide in a spherical joint, one can always decompose the relative rotation between the bars within a loop into two terminal rotations about the connecting bars, and one intermediate rotation about some arbitrary axis, which should however be warranted never to become parallel to one of the bars over the complete motion. In this way, one obtains loop couplings at a spherical joint between two isolated bar spin DOFs and one proper aperture angle for each loop incident to that joint. An example is given in Fig. 5. Here, the loop coupling equation between the three loops L_i, L_j and L_k can be expressed as

$$\mathbf{R}_k(\xi,\psi,\zeta) = \mathbf{R}_j(\eta,\theta,\gamma)\cdot\mathbf{R}_i(\alpha,\varphi,\beta), \tag{4}$$

where $\mathbf{R}(\ldots)$ denotes the rotation matrix in terms of (z-x-z) Euler angles in the order of the arguments. The coupling equation can be interpreted such that the three internal rotations ξ,ψ,ζ of loop L_k result numerically as a (nonlinear) function of the three internal rotations α,φ,β and η,θ,γ of loops L_i and L_j, respectively. Topologically, this coupling has particular properties due to the implicitly assumed isolated DOFs about the bars that are incident to the spherical joint, which can be described qualitatively without resorting to the explicit resolution of the spatial loop coupling conditions in terms of the output angles ξ,ψ,ζ, as discussed next.

Assuming that the other bar ends in Fig. 5 (not shown) of the bars are also attached to spherical joints, the bars will be allowed to spin about their longitudinal axes. However, only one rotation per bar can be regarded as truly fully isolated: if a loop "registers" one isolated spin as fully isolated, then for the neighboring loop this rotation becomes transmitted. This is shown by the chosen colors in Fig. 5: assuming that the spin of bars 1 and 2 have been registered as fully isolated in loop L_i, these can be regarded as immaterial spinnings of the pins within the sleeves of the joints of bars 1 and 2, denoted by angles α and β, respectively; however, the rotation within loop L_k about bar 1, denoted by ξ, is then not fully isolated anymore, and actually sways the brown chain of loop L_k as when unfolding two faces of an origami at the edge folded along the axis of bar 1. Thus the angle ξ is a transmitted isolated DOF. Similarly, the rotation within loop L_j about bar 2, denoted by η, rotates the blue chain within loop L_j about the axis of bar 2, and is thus again a transmitted isolated DOF. Finally, the rotation γ of the inner pin of the revolute joint along bar 3 can be regarded again as a fully isolated DOF once, which in the case of Fig. 5 has been arbitrarily assigned to loop L_j. This makes the rotation about bar 3 within loop L_k, denoted by ζ, again a transmitted isolated DOF.

By the loop coupling conditions, one can recognize that, if loop L_k is the "output" of the coupling conditions according to Eq. (4), then two of the incoming variables of loop L_k, namely ξ and ζ, are transmitted isolated DOFs that do not affect the inner kinematics of loop L_k, but the third one, namely ψ, is a proper transmission angle which regulates the relative orientation of bars 1 and 3 with respect to each other. This angle is a function of the opening angles φ and θ of loops L_i and L_j, respectively, but also of the transmitted isolated DOF η of loop L_j, which thus becomes material. If the three loops L_i, L_j, L_k are assumed to be triangles with spherical joint nodes (as will be the case for the regarded mechanism), then the angles φ, θ, ψ must remain

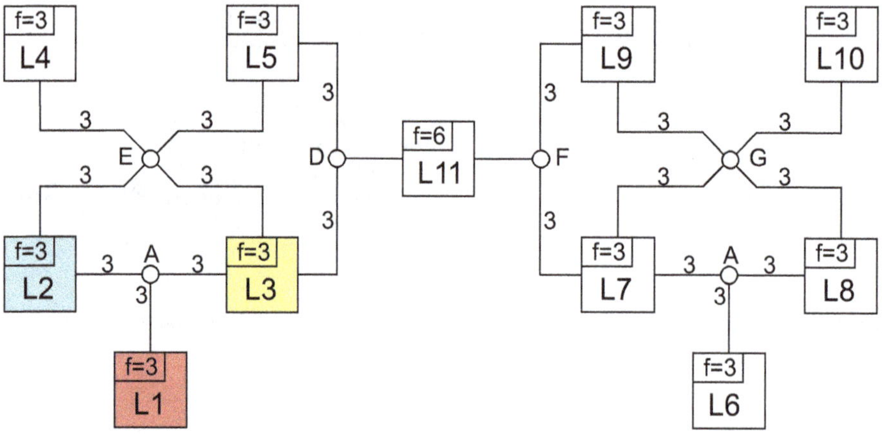

Figure 6. The undirected kinematical network of the "double-banana" mechanism.

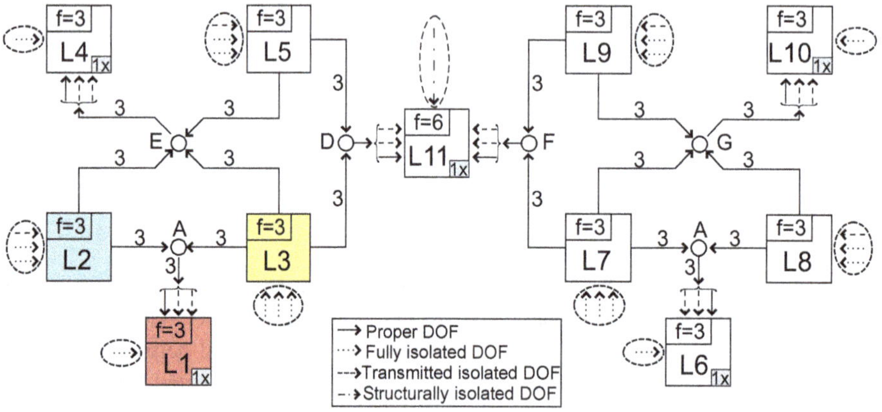

Figure 7. The directed kinematical network of the "double-banana" mechanism.

constant, which means that loop L_k induces an implicit constraint equation for the transmitted isolated DOF η of loop L_j. This kind of tracking of transmitted isolated DOFs up to locations of implicit constraint equations will prove useful for detecting rigid and movable substructures in multiple spherical-spherical bar mechanisms, as shown below.

6 Kinematical network of the "double-banana" mechanism

As described in section 3, initially the interconnected kinematical loops are chosen as a set of "smallest" independent loops, in which the sum of numbers of relative joint variables within each loop over all loops is smallest. While the number n_L of independent loops is unique, the choice of the independent loops can have many solutions (Kecskeméthy, 1993). In the present case, we have $n_L = 11$ for which we choose 10 triangles within the two bipyramid halves and the quadrilateral loop L_{11} between them shown in Fig. 2. At the beginning, each loop is regarded as an independent transmission element, i.e., all relative motions within a loop are regarded as independent. The loop coupling condition gives one spherical coupling condition at joints E, G, D and F and two spherical coupling conditions at joint A. Using the coupling conditions described in Section 5, one obtains for the "double-banana" example of Fig. 2 the undirected kinematical network shown in Fig. 6.

In the process of orienting the edges of the kinematical network, the following steps arise:

1. Each of the loops $L_1, \ldots, L_{10}$ displays 3 local DOFs, which are the three isolated DOFs about the three bars; likewise, loop L_{11}, which comprises four spherical joints, displays 6 local DOFs.

2. At each coupling joint, three coupling equations according to Eq. (4) are produced. Two of these coupling conditions go into isolated DOFs, while one propagates into a transmission angle, which is the aperture angle of the intermediate joint of the z-x-z decomposition.

3. By the sink method, one can start the orientation of the edges with the sink loops L_1, L_4, L_6, L_{10} and L_{11}. However, for each of these loops, the "intermediate" aperture coupling angle (θ in the case of loop L_1) actu-

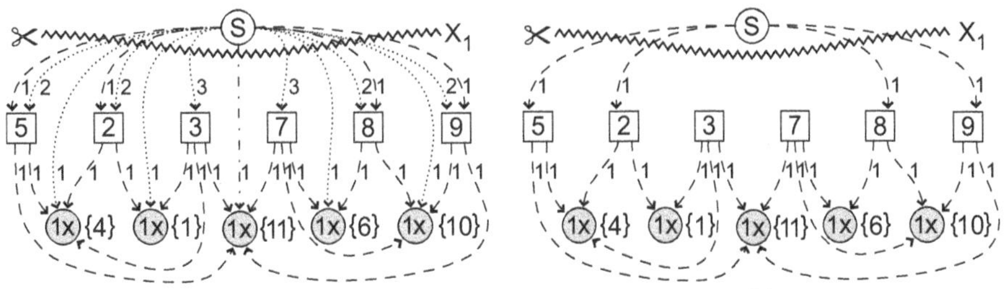

(a) (b)

Figure 8. Loop connection graph for the kinematical network of Fig. 7 **(a)** with and **(b)** without isolated DOFs.

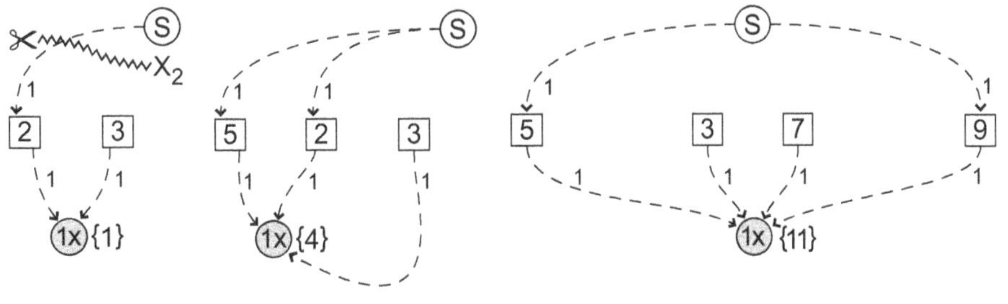

(a) (b) (c)

Figure 9. Pruned graphs for each of the four special shared joints.

ally transmits into the loop kinematics and is subject to a constraint condition there due to the triangle condition. Thus, the sink loop actually introduces an implicit equation (marked "1 ×" in a gray box within the loop) which must be fulfilled by the transmitted DOFs entering the loop. On the other hand, each sink loop features an additional fully isolated DOF, which is about the third edge that is not shared with any other loop. Thus, after orienting the edges into the sinks one obtains one implicit equation and one "true" DOF per sink (Fig. 7).

4. After having determined all sinks, the orientation of the edges of the remaining loops are completed by feeding enough inputs to each loop such that the local loop DOF count is obeyed. Here, it is important that, as discussed above, exactly (only) one fully isolated DOF is registered for each bar. In the example of Fig. 7, we registered all three isolated DOFs of loop L_3 as fully isolated; thus in loop L_2 we can only register two isolated DOFs as fully isolated, while the third (about the common bar with L_3) must be chosen as a transmitted isolated DOF. After receiving its inputs at joints D and F (which can be regarded as fixed), loop L_{11} can detect locally (internally) a fully isolated DOF about the axis connecting joints A and B, which is registered as a structurally additional fully isolated DOF operating on this loop.

The fully isolated DOFs can be applied irrespectively of the rest of the structure, thus from Fig. 7 one can see that there are in total 19 fully isolated DOFs, 18 representing the spins of the bars, and one in loop L_{11} representing the spinning rotation between the two bananas about the implied axis A-B. Thus the kinematical network already is able to detect that, apart from the 18 isolated bar spins, loop L_{11} features an additional DOF that is not canceled by the rest of the structure. However, it remains to detect which parts of the structure are rigid or over-rigid. This is done in the next section.

7 Loop connection graph and rigidity detection for the "double-banana" case

A kinematical network can be transformed into a "loop connection graph" describing the level of dependency of the individual loops and possible implicit conditions (Xia et al., 2012). We show here the loops as boxes and implicit conditions as gray disks, with the number n of implicit equations denoted by "$n\times$" within the disk, and the indices i, j, ... of loops L_i, L_j, ... embracing these implicit conditions in braces besides the disk. The connections between loops and circles are represented by weighted edges, where the weight represents the number of joint variables transferred through this edge. External inputs are depicted by edges from the source "S" to the corresponding loop, while the level of dependency is expressed by the distance (in rows) from the source to the node, as shown in Fig. 8a.

According to (Xia et al., 2012), rigidity detection can be performed on this acyclic graph by evaluating directed "cuts" such that the source is on one side, the sink(s) is on the other, and all edges through which the cut passes are di-

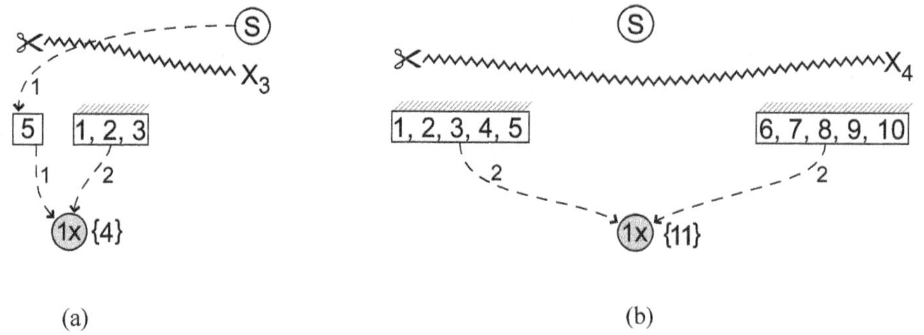

Figure 10. Minimal cuts and resulting DOFs for each substructure.

rected from the "upstream" to the "downstream" side. The sum of weights of the cut edges is termed the *weight* of the cut, while the sum of implicit equations in the downstream side of the cut is termed the *absorbing degree* of the cut. The DOF of the cut is equal to its weight minus its absorbing degree. This DOF is equal to the DOF of the nodes on the downstream side of the cut. Whenever the DOF of the cut is less or equal to zero, the downstream subsystem will represent a rigid or an overconstrained subsystem. The cut with the minimal DOF will determine the most overconstrained substructure. Such cuts can be easily determined using state-of-art graph-theoretic methods (Abel and Bicker, 1982).

For example, the cut X_1 in Fig. 8a displays a weight of 23 and an absorbing degree of 5, which results into a DOF of 18 that correspond to the 18 isolated bar rotations. But since the isolated degrees of freedom need not be taken into account for rigidity detection, only the transmitted isolated DOFs (dashed lines) need to be regarded, leading to the reduced loop connection graph in Fig. 8b, where the weight of cut X_1 now can be recognized as 4 with an absorbing degree of 5. Thus, the whole loop transmission structure without isolated DOFs is recognized as over-rigid with DOF $= -1$. However, as there is an additional structurally isolated DOF entering loop L_{11}, one can recognize a global proper DOF. The question of which parts are rigid and which parts are movable can be answered by regarding so-called pruned graphs containing subsets of implicit constraints and their predecessor nodes from the root of the loop connection graph, as explained below.

As a "pruned" sub-graph, one understands the sub-graph obtained when taking some sinks and removing all nodes that are not on any path from the sink to the source (Fig. 9). Clearly, only one such pruned sub-graph has a cut with non-positive DOF (X_2 in Fig. 9a). Thus the subsystem L_1, L_2, L_3 can be replaced by a rigid body. Restarting the algorithm with these loops replaced by a rigid body gives Fig. 10a, which again displays a cut X_3 with DOF $= 0$, hence the subsystem L_1, L_2, L_3, L_4, L_5 is again rigid. Replacing this again by a rigid body and carrying out the analogous steps for the right half of the "double-banana" graph gives the pruned graph of Fig. 10b. This graph has now a cut X_4 with transmitted weight zero (not counting the fully isolated DOF of loop L_{11}) and an absorbing degree of 1, hence the cut has DOF $= -1$. As a result, the algorithm is able to detect both halves of the "double-banana" as rigid, and also to detect that their assembly into loop L_{11} features one structurally isolated DOF.

8 Discussion

The proposed approach shows that by regarding loops as transmission elements and tracking the transmission of isolated DOFs through the mechanism one can detect non-trivial cases of rigidity involving implied hinges. Although the method may seem complicated, it can actually be broken down into simple steps for which graph-theoretic algorithms already exist. For example, steps for dissecting a mechanism into kinematical loops and setting up the kinematical network and loop connection graph have already been implemented in Mathematica by our group using well-known graph theoretical algorithms for finding minimal cycles or minimal cut sets (Kecskeméthy, 1993). The advantage of using loops for transmission evaluation lies evidently in the fact that one can combine substructures of different mobility spaces such as planar, spherical or spatial, and that structurally isolated DOFs can be detected within one loop by regarding the corresponding isotropy groups (Kecskeméthy, 1993). Thus, in our opinion this new methodology could prove useful for rigidity detection algorithms in other areas of applications. While the automatic implementation of the classification of different types of isolated DOFs is still to be completed, we believe that this is accomplishable in a general setting. Future research will be devoted to demonstrate that the presented algorithm also holds for so-called nucleation free mechanisms (Cheng et al., 2009), which comprise individual mobile structures that become rigid and form "implied hinges" once embedded in a certain mechanism.

9 Conclusions

Based on the concept of the kinematical network, a loop-based rigidity detection method is proposed to detect mo-

bility and rigidity for systems featuring spherical-spherical pairs, such as the "double-banana" case. The method consists in viewing the mechanism as an assembly of coupled independent kinematical loops instead of regarding it as a system of coupled bodies and joints (or bars and nodes). Due to this, it is possible to track isolated degrees of freedom separately from transmitted joint angles, and by this to detect rigid subsystems and "implied hinge" finite mobility for complex systems as the "double-banana" case. In this setting, it is interesting to note that this approach is implementable using standard tools from graph theory. Although we have not yet completed this automation, we believe that this is feasible, which is a topic of future research.

Acknowledgements. The financial support of the second author as a Humboldt Research Fellow by the Alexander von Humboldt Foundation is gratefully acknowledged.

References

Abel, U. and Bicker, R.: Determination of All Minimal Cut-Sets between a Vertex Pair in an Undirected Graph, IEEE T. Reliab., R-31, 167–171, 1982.

Agrawal, V. and Rao, J.: Fractionated Freedom Kinematic Chains and Mechanisms, Mech. Mach. Theory, 22, 125–130, 1987.

Bennett, D. J. and Hollerbach, J. M.: Autonomous calibration of single-loop closed kinematic chains formed by manipulators with passive endpoint constraints, IEEE T. Robotic. Autom. 7, 597–606, 1991.

Carretero, J., Podhorodeski, R., Nahon, M., and Gosselin, C. M.: Kinematic analysis and optimization of a new three degree-of-freedom spatial parallel manipulator, J. Mech. Design, 122, 17–24, 2000.

Casals, A. and Amat, J.: Automatic Guidance of an Assistant Robot in Laparoscopic Surgery, in: Proceedings of the 1996 IEEE International Conference on Robotics and Automation, Minneapolis, MN, USA, April 1996, 895–900, 1996.

Cheng, J., Sitharam, M., and Streinu, I.: Nucleationfree 3d rigidity, in: In Proceedings of the 21st Canadian Conference on Computational Geometry(CCCG2009), 71–74, 2009.

Chubynsky, M. V. and Thorpe, M. F.: Algorithms for three-dimensional rigidity analysis and a first-order percolation transition, Phys. Rev. E, 76, 041135, doi:10.1103/PhysRevE.76.041135, 2007.

Ding, H., Kecskeméthy, A., and Huang, Z.: Synthesis of the Whole Family of Planar 1-DOF Kinematic Chains and Creation of their Atlas Database, Mech. Mach. Theory, 47, 1–15, 2012.

Ding, H., Huang, Z., and Mu, D.: Computer-aided structure decomposition theory of kinematic chains and its applications, Mech. Mach. Theory, 43, 1596–1609, 2008.

Fanghella, P. and Galletti, C.: Mobility Analysis of Single-Loop Kinematic Chains: An Algorithmic Approach Based on Displacement Groups, Mech. Mach. Theory, 29, 1187–1204, 1994.

Fowler, P. W. and Guest, S. D.: Symmetry analysis of the double banana and related indeterminate structures, In New approaches to structural mechanics, shells and biological structures, edited by: Drew, H. R. and Pellegrino, S., 91–100, 2002.

Franzblau, D. S.: Ear decomposition with bounds on ear length, Inform. Process. Lett., 70, 245–249, 1999.

Franzblau, D. S.: Generic rigidity of molecular graphs via ear decomposition, Discrete Appl. Math., 101, 131–155, 2000.

Hsu, K., Karkoub, M., Tsai, M.-C., and Her, M.: Modelling and index analysis of a Delta-type mechanism, P. I. Mech. Eng. K-J. Mul., 218, 121–132, 2004.

Hunt, K.: Kinematic Geometry of Mechanisms, Clarendon Press, Oxford, 334, 1978.

Hwang, W.-M. and Hwang, Y.-W.: An algorithm for the detection of degenerate kinematic chains, Math. Comput Model., 15, 9–15, 1991.

Isaksson, M., Eriksson, A., Watson, M., Brogårdh, T., and Nahavandi, S.: A method for extending planar axis-symmetric parallel manipulators to spatial mechanisms, Mech. Mach. Theory, 83, 1–13, 2015.

Jacobs, D. and Hendrickson, B.: An Algorithm for Two Dimensional Rigidity Percolation: The Pebble Game, J. Comput. Phys., 137, 346–365, 1997.

Jacobs, D. J., Kuhn, L. A., and Thorpe, M. F.: Flexible and Rigid Regions in Proteins, in: Rigidity Theory and Applications, edited by: Thorpe, M. and Duxbury, P., Fundamental Materials Research, 357–384, Springer USA, 2002.

Jain, A. and Rodriguez, G.: An Analysis of the Kinematics and Dynamics of Underactuated Manipulators, IEEE T. Robot., 9, 411–422, 1993.

Kecskeméthy, A.: Objektorientierte Modellierung der Dynamik von Mehrkörpersystemen mit Hilfe von Übertragungselementen, Fortschrittberichte VDI, Reihe 20, Nr. 88, VDI-Verlag, Düsseldorf, 1993.

Kecskeméthy, A., Krupp, T., and Hiller, M.: Symbolic Processing of Multiloop Mechanism Dynamics Using Closed-Form Kinematics Solutions, Multibody Syst. Dyn., 1, 23–45, 1997.

Laman, G.: On graphs and rigidity of plane skeletal structures, J. Eng. Math., 4, 331–340, 1970.

Lee, A.: Geometric constraint systems with applications in CAD and biology, ph.D. Thesis, 2008.

Lee, A., Streinu, I., and Theran, L.: Finding and Maintaining Rigid Components, in: CCCG'05, 219–222, 2005.

Lee, A., Streinu, I., and Theran, L.: Analyzing rigidity with pebble games, in: Proceedings of the Twenty-Fourth Annual Symposium on Computational Geometry, SCG '08, 226–227, ACM, New York, NY, USA, 2008.

Lee, H.-J. and Yoon, Y.-S.: Detection of Rigid Structure in Enumerating Basic Kinematic Chain by Sequential Removal of Binary Link String, JSME international journal. Ser. 3, Vibration, control engineering, engineering for industry, 35, 647–651, 1992.

Liu, H., Huang, T., Kecskeméthy, A., and Chetwynd, D. G.: A generalized approach for computing the transmission index of parallel mechanisms, J. Mech. Mach. Theory, 245–256, 2014.

Maxwell, J. C.: On reciprocal figures, frames and diagrams of forces, Philos. Mag., 427, 250–261, 1864.

Mazzone, A., Spagno, C., and Kunz, A.: A haptic feedback device based on an active mesh, in: Proceedings of the ACM sympo-

sium on Virtual reality software and technology, 188–195, ACM, 2003.
Merlet, J.-P.: Kinematics and synthesis of cams-coupled parallel robots, in: International Workshop on Computational Kinematics, Cassino, Italy, May 2005.
Michelucci, D. and Foufou, S.: Interrogating Witnesses for Geometric Constraint Solving, in: 2009 SIAM/ACM Joint Conference on Geometric and Physical Modeling, SPM '09, 343–348, ACM, New York, NY, USA, 2009.
Moinet, M., Mandil, G., and Serre, P.: Defining Tools to Address Over-constrained Geometric Problems in Computer Aided Design, Comput. Aided Des., 48, 42–52, 2014.
Moukarzel, C.: An efficient algorithm for testing the generic rigidity of graphs in the plane, J. Phys. A, 29, 8079–8098, 1996.
Razmara, N., Kohli, D., and Dhingra, A. K.: On the degrees of freedom of motion of planar-spatial mechanisms, Volume 7, in: Proceedings of the 2000 ASME Design Engineering Technical Conferences and Computers and Information in Engineering Conference, Baltimore, MD, USA, 2000
Robbins, H.: A theorem on graphs, with an application to a problem of traffic control, Am. Math. Mon., 281–283, 1939.
Rojas, N. and Thomas, F.: The closure condition of the double banana and its application to robot position analysis, in: IEEE International Conference on Robotics and Automation (ICRA), 6–10, Karlsruhe, Germany, 2013.
Rolland, L.: The manta and the kanuk: Novel 4-dof parallel mechanisms for industrial handling, Proc. of ASME Dynamic Systems and Control Division IMECE, 99, 14–19, 1999.
Servatius, B. and Servatius, H.: Rigidity, global rigidity, and graph decomposition, Eur. J. Comb., 31, 1121–1135, 2010.
Servatius, B., Shai, O., and Whiteley, W.: Combinatorial characterization of the Assur graphs from engineering, Eur. J. Comb., 31, 1091–1104, 2010.
Shai, O. and Müller, A.: A novel combinatorial algorithm for determining the generic/topological mobility of planar and spherical mechanisms, in: ASME 2013 International Design Engineering Technical Conferences and Computers and Information in Engineering Conference, V06BT07A073–V06BT07A073, American Society of Mechanical Engineers, 2013.
Shai, O., Sljoka, A., and Whiteley, W.: Directed graphs, decompositions, and spatial linkages, Discrete Appl. Math., 161, 3028–3047, 2013.
Shiriaev, A. S., Freidovich, L. B., and Gusev, S. V.: Transverse linearization for controlled mechanical systems with several passive degrees of freedom, IEEE T. Automat. Contr., 55, 893–906, 2010.
Sljoka, A., Shai, O., and Whiteley, W.: Checking mobility and decomposition of linkages via pebble game algorithm, in: ASME 2011 International Design Engineering Technical Conferences and Computers and Information in Engineering Conference, 493–502, American Society of Mechanical Engineers, 2011.
Streinu, I. and Theran, L.: Combinatorial genericity and minimal rigidity, in: Proceedings of the Twenty-Fourth Annual Symposium on Computational Geometry, SCG '08, 365–374, ACM, New York, NY, USA, 2008.
Sunkari, R. P.: Structural synthesis and analysis of planar and spatial mechanisms satisfying Gruebler's degrees of freedom equation, PhD thesis, University of Maryland, College Park, Maryland, 2006.
Sunkari, R. P. and Schmidt, L. C.: Critical Review of Existing Degeneracy Testing and Mobility Type Identification Algorithms for Kinematic Chains, ASME Conference Proceedings 2005, 255–263, 2005.
Tay, T.-S.: Rigidity of multi-graphs. I. Linking rigid bodies in n-space, J. Combin. Theory B, 36, 95–112, 1984.
Togashi, J., Matsuda, T., and Mitobe, K.: A low cost and lightweight wire driven robot arm by using elastic strings, in: System Integration (SII), 2014 IEEE/SICE International Symposium, 436–440, IEEE, 2014.
Tuttle, E.: Generation of planar kinematic chains, Mech. Mach. Theory, 31, 729–748, 1996.
Waldron, K.: A Study of Overconstrained Linkage Geometry by Solution of Closure Equations – Part II. Four-Bar Linkages With Lower Pair Joints Other Than Screw Joints, Mech. Mach. Theory, 8, 233–247, 1973.
Whitney, H.: Non-Separable and Planar Graphs, Trans. Am. Math. Soc., 34, 339–362, 1932.
Xia, S., Ding, H., and Kecskeméthy, A.: A Loop-Based Approach for Rigid Subchain Identification in General Mechanisms, in: Proceedings of the ARK, Innsbruck, Austria, 19–26, 2012.
Zhang, Y., Finger, S., and Behrens, S.: Introduction to mechanisms, Carnegie Mellon University, Sec. 4.4.1, 2003.

Multi-objective optimization of a type of ellipse-parabola shaped superelastic flexure hinge

Zhijiang Du, Miao Yang, and Wei Dong

State Key Laboratory of Robotics and System, Harbin Institute of Technology, 2 Yikuang Street, Harbin, 150080, China

Correspondence to: Wei Dong (dongwei@hit.edu.cn)

Abstract. Flexure hinges made of superelastic materials is a promising candidate to enhance the movability of compliant mechanisms. In this paper, we focus on the multi-objective optimization of a type of ellipse-parabola shaped superelastic flexure hinge. The objective is to determine a set of optimal geometric parameters that maximizes the motion range and the relative compliance of the flexure hinge and minimizes the relative rotation error during the deformation as well. Firstly, the paper presents a new type of ellipse-parabola shaped flexure hinge which is constructed by an ellipse arc and a parabola curve. Then, the static responses of superelastic flexure hinges are solved via non-prismatic beam elements derived by the co-rotational approach. Finite element analysis (FEA) and experiment tests are performed to verify the modeling method. Finally, a multi-objective optimization is performed and the Pareto frontier is found via the NSGA-II algorithm.

1 Introduction

Flexure hinges is a substitution of conventional rotational joint in precision engineering. The monolithic structure brings flexure hinge various superior properties, e.g. high resolution, fine precision, compact structure, and the elimination of friction and lubrication (Paros and Weisbord, 1965). Therefore, flexure hinges are widely employed in micro positioning stages, precision alignment devices, micro manipulators, scanning electron microscopy and antennas where high precision and high resolution are required (Pham and Chen, 2005; Yong et al., 2008; Du et al., 2014). However, despite all the advantages aforementioned, flexure hinges also have some inherent shortcomings, for example the limited motion range restricted by the allowable stress and strain of the material. In addition, the rotational center shift also reduces the absolute accuracy of flexure hinge based compliance mechanisms.

The movability of a flexure hinge comes from the deformation of its structure. The 3-D sketch of a flexure hinge is shown in Fig. 1, it can be seen that the deformation almost exclusively takes place in the weakest region. In order to avoid plastic deformation the movability of the flexure hinges are severely restricted (Friedrich et al., 2014). A promising approach to enhance their motion capacity is to use superelastic materials, e.g. shape memory alloy (SMA), to fabricate flexure hinges since they can provide much larger allowable strains. The maximum recoverable strain of SMA is about 8 %, while for the most frequently used materials to fabricate flexure hinges, e.g. steel or aluminum, the maximum recoverable strain is 0.2–0.4 % (Desroches and Delemont, 2002). As it is mentioned in Hesselbach and Raatz (2000), the motion range of a flexure hinge made of SMA can reach as much as 30°, which can satisfy most of the application requirements for revolution joints in parallel mechanisms.

However, formulating a static deformation model for the superelastic flexure hinge is a challenging work. Although the flexure hinges can be modeled as Euler beams since the minimum thickness of a planar flexure hinge is much less than the hinge length (Chen et al., 2011), nonlinearities introduced by large deformation and the constitutive relationship of superelastic material made it extremely difficult to obtain analytic solutions to describe the static response of the flexure hinges. Previous modeling methods, like the integration of beam theory (Wu and Zhou, 2002), Castigliano's second theorem (Lobontiu et al., 2002a), and pseudo rigid body method (Kim et al., 2013) are also infeasible to solve

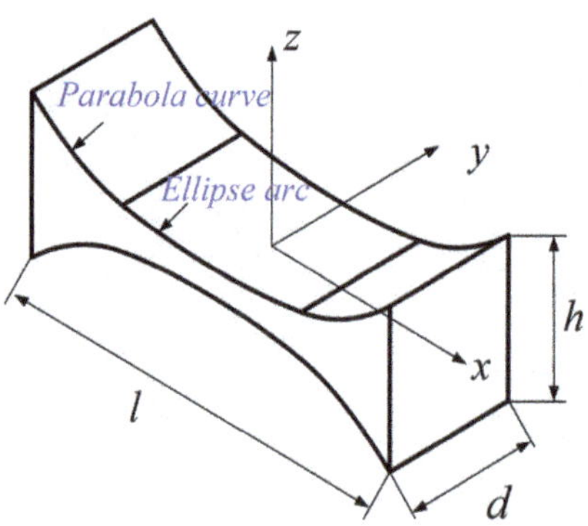

Figure 1. The 3-D model of the proposed ellipse-parabola shaped flexure hinge.

this problem. In this paper, we will use nonlinear finite element method (FEM) to formulate the relationship between the deformation and loads of superelastic flexure hinges.

The performance of a flexure hinge is completely determined by its notch shape, once the material of the flexure hinge is chosen. Various types of flexure hinges with different shapes have been investigated in the past years, e.g. V-shaped flexure hinges (Tian et al., 2010), parabolic and hyperbolic flexure hinges (Lobontiu et al., 2002b; Chen et al., 2009), elliptical-arc-fillet flexure hinges (Chen et al., 2008), flexure hinges that considered the machining errors (Shi et al., 2013), and even the exponent-sine-shaped flexure hinges (Wang et al., 2013). But these flexure hinges defined by only few parameters that restrict the degree of freedom to choose optimal shapes of flexure hinges. De Bona and Gh Munteanu (2005) illustrated a flexure hinge constructed by a group of cubic-spline curves which maximized the probability to find out the optimal flexure hinges, however it needs 9 parameters to formulate the shape, which will increase the computational complexity of the optimization dramatically, especially for flexure hinges made of superelastic materials.

To obtain a balance between the diversity of the optimization and the computational complexity, this paper proposed a new type of ellipse-parabola shaped flexure hinge which is constructed by an ellipse arc and a parabola curve. The parameters that defining the profile of the ellipse-parabola shaped flexure hinge are the notch length, the minimum thickness, the ellipticity of the ellipse arc and the terminated eccentric angle. It can describe a large range of flexure hinges by change the values of the parameters and all the parameters have certain meanings in geometry.

The multi-objective optimization is performed for the ellipse-parabola shaped superelastic flexure hinge to obtain excellent comprehensive performances. The remainder of this paper is organized as follows. Section 2 gives a description of the proposed ellipse-parabola shaped flexure hinge. Section 3 establishes the static deformation model of the flexure hinge by using nonlinear beam elements derived by the

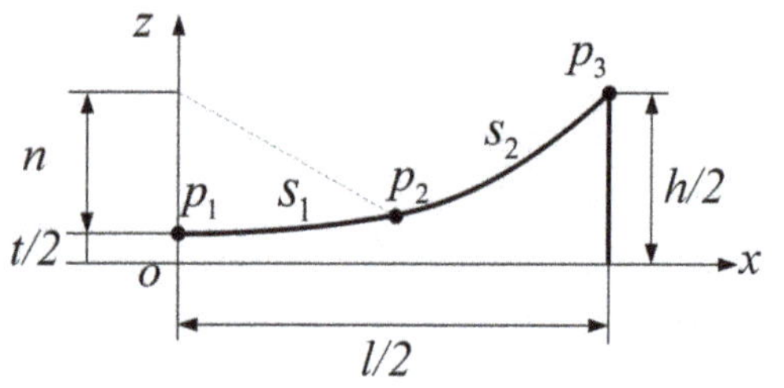

Figure 2. The 2-D schematic of the ellipse-parabola shaped flexure hinge.

co-rotational approach. The deformation model has been verified by FEA and experiment test in Sect. 4. In Sect. 5, the multi-objective optimization of the ellipse-parabola shaped superelastic flexure hinge is formulated and the Pareto frontier is found via the NSGA-II algorithm. Finally, the conclusions of this paper is presented in Sect. 6.

2 Structure of the ellipse-parabola shaped flexure hinge

The elliptic flexure hinge has relative larger motion range than the parabolic flexure hinge and the hyperbolic flexure hinge, and the parabolic flexure hinge has a higher transmission accuracy than the elliptic flexure hinge (Lobontiu et al., 2002a). Better comprehensive performances of a flexure hinge can be obtained by combining the two profiles together. The proposed ellipse-parabola shaped flexure hinge is shown in Fig. 1.

It can be seen that the ellipse-parabola shaped flexure hinge is fully symmetric and *o-xyz* is the corresponding Cartesian coordinate, the origin is located at the center of the structure, x axis follows with the longitudinal axis of symmetry and z axis coincides with the transverse axis. A 2-D sketch of the flexure hinge is illustrated in Fig. 2. The profile consists of two parts, i.e. an ellipse arc and a parabola curve, the two curves are smoothly connected at the intersection p_2. Moreover, l denotes the notch length of the flexure hinge, t denotes the minimum thickness, and h and d are the height and width of the flexure hinge respectively.

Curve s_1 is an ellipse arc with its center is located at $(0, \frac{h}{2})$, and it can be represented in a parametric form as below,

$$\begin{cases} x(\varphi) = ne\cos\varphi \\ z(\varphi) = n\sin\varphi + \dfrac{h}{2} \end{cases} \quad \varphi_t \leq \varphi \leq 0 \text{ and } -\frac{\pi}{2} \leq \varphi_t \leq 0, \tag{1}$$

where $n = \frac{h-t_s}{2}$ is the vertical semi-axe, e is the ellipticity of the ellipse arc which is the quotient of the horizontal semi-axe and the vertical semi-axe, the parameter φ is the eccentric angle, and φ_t is the terminated eccentric angle at p_2.

The parabola curve s_2 passes through $p_2 = \left(ne\cos\varphi_t, n\sin\varphi_t + \frac{h}{2}\right)$ and $p_3 = \left(\frac{l}{2}, \frac{h}{2}\right)$, and have the same slope with s_1 at point p_2. Then, the thickness of the flexure hinge in s_2 can be formulated by applying the

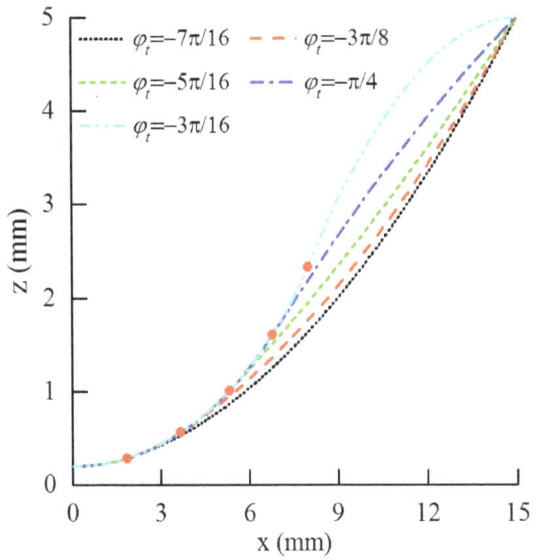

Figure 3. Notch shapes with different φ_t ($e = 2$ and φ_t ranges from $-\frac{7\pi}{16}$ to $-\frac{3\pi}{16}$).

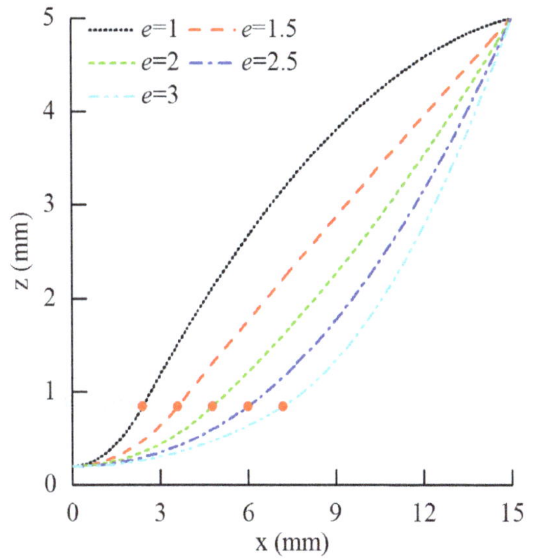

Figure 4. Notch shapes with different e ($\varphi_t = -\frac{\pi}{3}$ and e ranges from 12 to 20).

Hermite interpolation.

$$z(x) = a(x - ne\cos\varphi_t)^2 + b(x - ne\cos\varphi_t) + c. \tag{2}$$

The coefficients in Eq. (2) are given by

$$a = \frac{2l\cos\varphi_t - 4ne}{e\sin\varphi_t(l - 2ne\cos\varphi_t)^2}, \quad b = -\frac{1}{e\tan\varphi}, \quad \text{and}$$
$$c = \left(\frac{h}{2} + n\sin\varphi_t\right).$$

To further illustrate the shape of the proposed ellipse-parabola shaped flexure hinge, suppose that $l = 30$ mm, $t_s = 0.4$ mm, and $h = 10$ mm, then the profile is fully determined by e and φ_t. Figure 3 shows the profiles of the flexure hinges with different φ_t, where the ellipticity $e = 2$, φ_t ranges from $-\frac{7\pi}{16}$ to $-\frac{3\pi}{16}$, and the red dots in the figure are the intersections, i.e. p_2. It can be seen that the percentage of s_1 in the whole profile increases with the increase of φ_t. On the other hand, for a given $\varphi_t = -\frac{\pi}{3}$, and e ranges from 1 to 3, the profiles are shown in Fig. 4, it can be seen that the flexure hinges become more flat with the increase of e, thus the effective deform length increases.

3 Modeling of the superelastic flexure hinges

The ellipse-parabola shaped superelastic flexure hinge can be considered as a cantilever beam with variable cross section, geometric and material nonlinearities. In this paper, we use the nonlinear FEM to model deformation of the flexure hinge. The co-rotational approach provides an effective way to derive beam elements with geometric nonlinearity. By introducing a local coordinate system (LCS) attached on the element, the displacement and the deformation of the element is separated, and then a simple form of the local tangent stiffness matrix and internal force vector of the element are obtained. The co-rotational method used here is similar to those in Pacoste and Eriksson (1997) and Criesfield (1991), and briefly described in the following discussion.

Figure 5 shows a two-node beam element in the initial and current configurations. The coordinate of the two nodes in the global coordinate system (GCS) xoz are (x_1, y_1) and (x_2, y_2) initially, and the global nodal displacement vector of the element is given by

$$\boldsymbol{q}_g = \{u_1, w_1, \theta_1, u_2, w_2, \theta_2\}^T, \tag{3}$$

where u and w are nodal displacements in the x and z direction respectively; θ is the nodal rotation angle, and the subscripts 1 and 2 refer to the nodes 1 and 2.

The LCS $x_l o_l z_l$ is assigned as that its origin is located at node 1 and x axis directs to node 2. Thus, the local displacement vector of the element contains only 3 components shown as below

$$\boldsymbol{q}_l = \{\bar{u}, \bar{\theta}_1, \bar{\theta}_2\}^T. \tag{4}$$

Based on the kinematics depicted in Fig. 5, the axis displacement and rotation angles of the two nodes in LCS can be obtained

$$\begin{aligned} \bar{u} &= L - L_0 \\ \bar{\theta}_1 &= \theta_1 - \alpha \\ \bar{\theta}_2 &= \theta_2 - \alpha \end{aligned} \tag{5}$$

where L_0 and L are the initial and current length of the element respectively, and α is the rigid rotational displacement

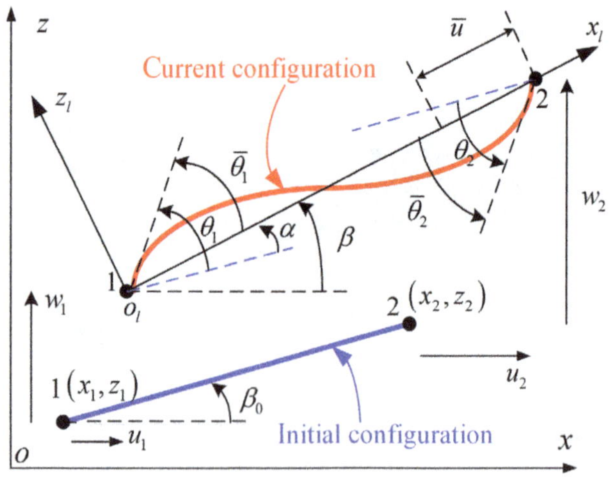

Figure 5. Beam kinematics.

of the element and formulated as below,

$$L_0 = \sqrt{(x_2 - x_1)^2 + (z_2 - z_1)^2}$$

$$L = \sqrt{(x_2 + u_2 - x_1 - u_1)^2 + (z_2 + w_2 - z_1 - w_1)^2} \tag{6}$$

$$\alpha = \beta - \beta_0 = \arctan\left(\frac{z_2 + w_2 - z_1 - w_1}{x_2 + u_2 - x_1 - u_1}\right) - \arctan\left(\frac{z_2 - z_1}{x_2 - x_1}\right)$$

Taking a differentiation of the Eq. (5) on both side, the relationship between the local virtual displacement $\delta \boldsymbol{q}_1$ and the global virtual displacement $\delta \boldsymbol{q}_g$ can be expressed as

$$\delta \boldsymbol{q}_1 = \mathbf{B} \delta \boldsymbol{q}_g, \tag{7}$$

where $\mathbf{B}$ is the transformation matrix.

Since the virtual work of the element is equivalent in both LCS and GCS, the global internal nodal force vector $\boldsymbol{f}_g$ can be formulated as

$$\boldsymbol{f}_g = \mathbf{B}^T \boldsymbol{f}_1, \tag{8}$$

where $\boldsymbol{f}_1 = \{\overline{N}, \overline{M}_1, \overline{M}_2\}$ is the local internal nodal force vector.

By taking the differentiation of Eq. (8) with respect to the global nodal displacement, the global tangent stiffness matrix is obtained

$$\mathbf{K}_g = \frac{\partial \boldsymbol{f}_g}{\partial \boldsymbol{q}_g} = \mathbf{B}^T \mathbf{K}_1 \mathbf{B} + \frac{\overline{N}}{L} \mathbf{z}\mathbf{z}^T + \frac{\overline{M}_1 + \overline{M}_2}{L^2} \left(\mathbf{z}\mathbf{r}^T + \mathbf{r}\mathbf{z}^T\right), \tag{9}$$

where $\mathbf{K}_1$ is the local stiffness matrix and defined by $\delta \boldsymbol{f}_1 = \mathbf{K}_1 \delta \boldsymbol{q}_1$, $\mathbf{r}$ and $\mathbf{z}$ are transformation matrices $\mathbf{r} = \mathbf{b}_1^T$, $\mathbf{z} = \frac{\partial \mathbf{r}}{\partial \beta}$, and $\mathbf{b}_1$ is the first row of matrix $\mathbf{B}$.

Based on the Euler–Bernoulli beam theory, the axial strain in a beam element for large deformation is given by

$$\varepsilon(\bar{x}, \bar{z}) = \frac{\partial u_0(\bar{x})}{\partial \bar{x}} + \frac{1}{2}\left(\frac{\partial w_0(\bar{x})}{\partial \bar{x}}\right)^2 - \bar{z}\frac{\partial^2 w_0(\bar{x})}{\partial \bar{x}^2}, \tag{10}$$

where $u_0(\bar{x})$ and $w_0(\bar{x})$ are the axial and transverse displacement on the beam mid-plane respectively. Applying the classical linear and cubic interpolation for the axial displacement $u_0(\bar{x})$ and transverse displacement $w_0(\bar{x})$ respectively, the axial displacement $u_0(\bar{x})$ can be then rewritten as

$$u_0 = \frac{\bar{x}}{L}\bar{u} \tag{11}$$

and the transverse displacement $w_0(\bar{x})$ is given as

$$w_0 = \left(\bar{x} - 2\frac{\bar{x}^2}{L} + \frac{\bar{x}^3}{L^2}\right)\bar{\theta}_1 + \left(-\frac{\bar{x}^2}{L} + \frac{\bar{x}^3}{L^2}\right)\bar{\theta}_2. \tag{12}$$

Considering the membrane locking involved in this problem, the first two items in Eq. (10) are replaced by the average axial strain which is defined as following

$$\varepsilon_{ma} = \frac{1}{L}\int_0^L \left[\frac{\partial u_0}{\partial \bar{x}} + \frac{1}{2}\left(\frac{\partial w_0}{\partial \bar{x}}\right)^2\right] \mathrm{d}\bar{x}. \tag{13}$$

Computing the integration in Eq. (13) and combine with Eq. (10), the axial strain ε can be rewritten as

$$\varepsilon = \frac{\bar{u}}{L} + z\left(\left(\frac{4}{L} - \frac{6x}{L^2}\right)\bar{\theta}_1 + \left(\frac{2}{L} - \frac{6x}{L^2}\right)\bar{\theta}_2\right) + \frac{1}{15}\bar{\theta}_1^2 + \frac{1}{30}\bar{\theta}_1\bar{\theta}_2 + \frac{1}{15}\bar{\theta}_2^2 \tag{14}$$

The local internal forces are then calculated by applying principle of virtual work.

$$V = \int_{\mathrm{v}} \sigma \delta \varepsilon \mathrm{d}v = N\delta\bar{u} + M_1 \delta\bar{\theta}_1 + M_2 \delta\bar{\theta}_2. \tag{15}$$

Superelasticity of SMA refers to that the material can undergo very large non-elastic strains and get fully recovered by unloading because of the phase transportation of the material's microstructure (Liew et al., 2004). The uniaxial isothermal stress-strain curve of Nitinol (the most frequently used SMA) measured by experiment is presented in Fig. 6. To describe the nonlinear relationship between the stress and strain of the material, a bilinear constitutive model is adopted in this paper.

$$\sigma = \begin{cases} E\varepsilon & (\varepsilon \leq \varepsilon_s) \\ \sigma_s + E_1(\varepsilon - \varepsilon_s) & (\varepsilon > \varepsilon_s) \end{cases} \tag{16}$$

where E is the Young's modulus of the material in the elastic stage, E_1 is the modulus in phase transformation, ε_s and σ_s are the strain and stress at the start point of transformation respectively, ε_f and σ_f are the strain and stress at the finial point of transformation.

Substituting Eqs. (14) and (16) into Eq. (15) and separating variables yield the local internal forces of the element.

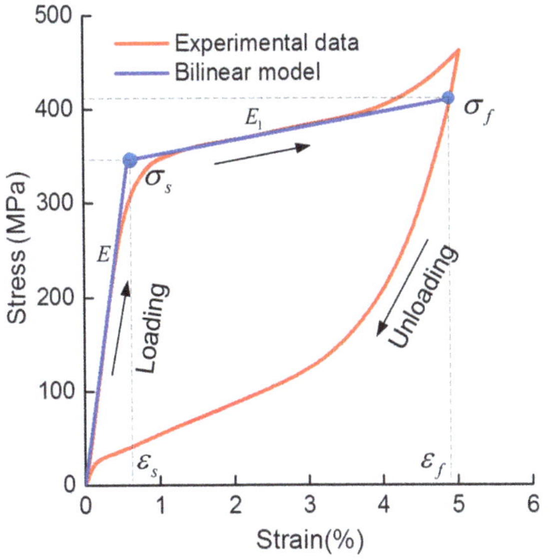

Figure 6. Experimental data of Nitinol uniaxial tensile behavior and the bilinear constitutive model.

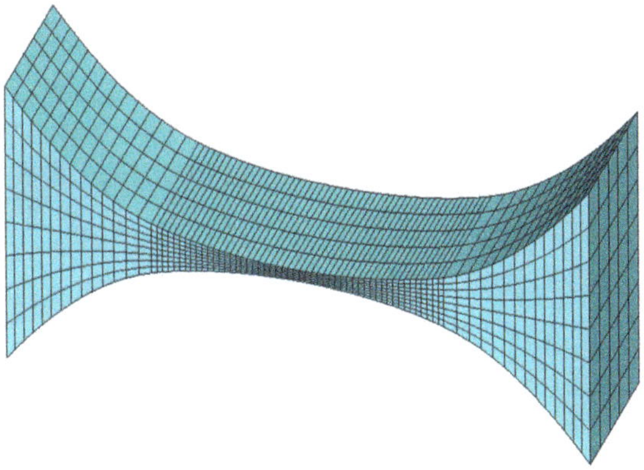

Figure 7. The finite element model in ANSYS.

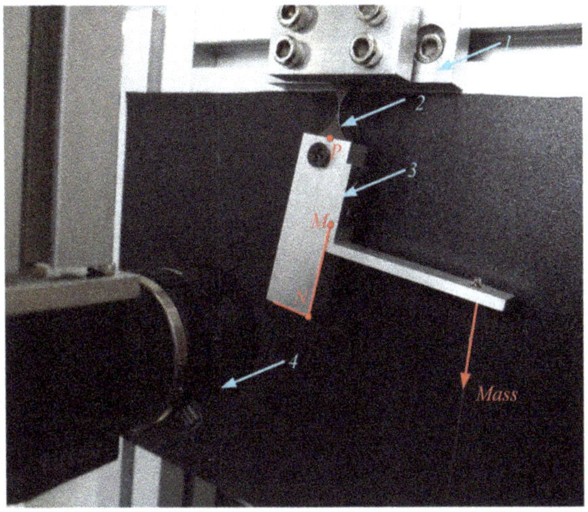

Figure 8. The experiment setup. (1: the base; 2: the flexure hinge sample; 3: the loading device; 4: the CCD camera)

And the local tangent stiffness matrix is than obtained by taking differential of the local internal force with respect to the components of $\boldsymbol{q}_1$ respectively

$$\mathbf{K}_{1_{i,j}} = \frac{\partial \boldsymbol{f}_{1i}}{\partial \boldsymbol{q}_{1j}} (i = 1 \ldots 3, j = 1 \ldots 3). \quad (17)$$

Combining Eqs. (15)–(17) with Eq. (9), the global tangent stiffness matrix of the element $\mathbf{K}_g$ and the global internal force vector $\boldsymbol{F}_g$ of the beam element can be obtained.

The derived global force vector $\boldsymbol{F}_g$ and global tangent stiffness matrix $\mathbf{K}_g$ of the element are then assembled to construct equilibrium equations of the structure. The system of nonlinear equations is solved by a load control algorithm. And the Newton–Raphson method is used to obtain the nodal displacements that minimize the residual stress during every incremental load step. A program implementing this algorithm has been written in Matlab. The end displacements and the stress and strain on the structure can be easily acquired by the program.

4 Verification

To validate the proposed co-rotational beam element based model (CRM), both finite element analysis by ANSYS (AFE) and experimental tests (EXP) are conducted. The height and the width of the flexure hinge samples are designed as $h = 10$ mm and $d = 5$ mm. The corresponding parameters of the material obtained from experimental measurement are as follow: $E = 58.51$ GPa, $E_1 = 1.5$ GPa, $\sigma_s = 346$ MPa, $\sigma_f = 410$ MPa.

The AFE model of the ellipse-parabola shaped superelastic flexure hinge is established via the APDL language and discretized by 20-node solid 186 elements in a mapped meshing way as shown in Fig. 7. The bilinear constitutive model is introduced into the model by the TB command. And the flexure hinge is fixed on one end, and loaded with a moment M_y on the other end.

The photograph of the experiment setup is illustrated in Fig. 8. The deformation of the flexure hinge samples are obtained by a computer vision measurement system. The CCD camera (PointGrey BFLY-PGE-50H5M-C) with 2448 × 2048 pixels is used and the measurement area is about 20 mm × 16 mm. Thus the resolution of the vision measurement is better than 8 μm and 1 mrad for transverse displacement and rotational angle respectively.

Three sample flexure hinges are manufactured by WEDM and deformed by standard weights via the loading device. The moment introduced by the loading device is also taken into consideration to calculate the deformation of the flexure hinge. Transverse displacement of point N and slope of line MN are obtained by the computer vision measurement system. For easy to comparison, the transverse and ro-

tational displacement of point P obtained by CRM and AFE are transformed into the deformation information of point N through a geometric transformation.

The result obtained by different methods are listed in Table 1, where w is the transverse displacements at point N, and θ is the rotational angle. It can be seen that the results calculated by CRM agree well with those by AFE and EXP, and the maximum relative errors are within 1.5 and 7 % respectively. The deviation between the CRM and the EXP may be caused by the approximation of the constitutive model and the manufacturing error of the flexure hinge.

It should be pointed out that the AFE model consists of approximately 3200 elements, and consumes more than 350 s to calculate the results, while the CRM contains only 20 elements and complete the solution within 5 s which is almost 70 times faster than AFE. It will be really helpful for the optimization of superelastic flexure hinges which will be discussed in the following section.

5 Multi-objective optimization of the superelastic flexure hinge

5.1 Optimization objectives

The flexure hinge is designed to serve as a rotational joint with its rotation center is located at the geometric center of the thinnest region. But it does not behave exactly as the ideal rotational joint. Considering the characteristics of the ellipse-parabola shaped superelastic flexure hinge, three performance indexes have been defined.

Motion range θ_{max} The motion range θ_{max} of an ellipse-parabola shaped superelastic flexure hinge is defined as the rotational angle between its two end sections, when the maximum strain on the structure up to the admissible strain. Though the maximum recoverable strain of the Nitinol used in this paper is about 4.9 % as shown in Fig. 6, we set the allowable strain as 3 % in this paper to obtain a long life cycle.

Relative rotational error γ The trace at the free end of the ellipse-parabola shaped superelastic flexure hinge is not a perfectly circular arc, because the rotation center of the flexure hinge significantly shifts during the deformation. This may introduce undesirable errors for transmission. The relative rotational error is defined as,

$$\gamma = \frac{\Delta r}{\theta_{max}}, \tag{18}$$

where Δr is the deviation between the flexure hinge and an ideal rotational joint at the maximum rotation angle θ_{max}, and it is given by

$$\Delta r = \sqrt{\left[u + \frac{l}{2}(1 - \cos\theta_{max})\right]^2 + \left(w - \frac{l}{2}\sin\theta_{max}\right)^2}.$$

Relative compliance ρ The flexure hinges should be very compliant in the motion axis and relative stiff in other directions, therefore the relative compliance is defined as:

$$\rho = \frac{C_m}{C_f}, \tag{19}$$

where C_m and C_f are the angular compliance about y axis and the linear compliance along x axis respectively, and they can be defined as

$$C_m = \frac{\theta_{max}}{M_{ymax}}, C_f = \frac{1}{dE}\int_0^l \frac{1}{2z(x)}dx,$$

in which M_{ymax} is the moment loaded at the free end of the superelastic flexure hinge when θ_{max} is reached.

The optimization problem aims to determine a set of optimal geometric parameters of the superelastic flexure hinge that minimize the difference between a superelastic flexure hinge and an ideal rotational joint, i.e. maximize the motion range and the relative compliance and minimize the relative rotation error during the deformation as well. Thus, the optimization objectives can be described as:

$$f(\boldsymbol{x}) = \left[-\theta_{max}, \gamma, -\rho\right] \rightarrow \text{min}. \tag{20}$$

5.2 Design variables and optimization constraints

The ellipse-parabola shaped superelastic flexure hinge investigated in this paper is made of a rectangular Nitinol strip by removing two symmetric cutouts. Among the parameters that define the geometry of the flexure hinge, the height h and width d are determined by the Nitinol strip and the minimum thickness of the hinge t is determined by the processing method. In order to optimize the ellipse-parabola shaped superelastic flexure hinge, three design variables are introduced: the length of flexure hinge l, the ellipticity of the ellipse arc e and the eccentric angle φ_t at p_2. Hence, the design parameters vector $\boldsymbol{x}$ is given by:

$$\boldsymbol{x} = [l, e, \varphi_t]. \tag{21}$$

According the definition of the flexure hinge profile two geometric constraints have to be added. The first one is related to the x coordinate of point p_2,

$$g_1(\boldsymbol{x}) = ne\cos\varphi_t - \frac{l}{2} \leq 0. \tag{22}$$

And the second constraint is to make sure that the curve is monotonous, thus the slop of the profile at p_3 must be positive

$$g_2(\boldsymbol{x}) = -2a\left(\frac{l}{2} - ne\cos\varphi_t\right) - b \leq 0. \tag{23}$$

The boundary constraints in this optimization problem are defined as below

$$5 \leq l \leq 40; \quad 1 \leq e \leq 5; \quad -\frac{\pi}{2} \leq \varphi_t \leq 0. \tag{24}$$

Table 1. Displacements calculated by CRM, AFE and EXP.

l (mm)	e	φ_t (rad)	M_y (Nmm)	Method	w (mm)	Error (%)	θ (°)	Error (%)
				CRM	17.6575	–	18.3003	–
15	3	$-\frac{7}{16}\pi$	71.024	FEM	17.8665	1.1695	18.5121	1.1440
				EXP	18.8895	6.5217	19.6255	6.7528
				CRM	13.1756	–	12.8801	–
20	2	$-\frac{1}{3}\pi$	71.488	FEM	13.2491	0.5550	12.9546	0.5750
				EXP	13.7327	4.0567	13.4359	4.1365
				CRM	11.0124	–	10.2674	–
25	1.5	$-\frac{1}{4}\pi$	72.231	FEM	11.0842	0.6479	10.3362	0.6652
				EXP	11.4332	3.6808	10.6627	3.7077

5.3 Optimization and results

The problem defined by Eqs. (20)–(24) is an optimization problem with nonlinear constraints. In the traditional methodology, the objectives defined in Eq. (20) are generally converted into a single objective function by using different weighting factors. However, this methodology can only obtain a single optimal solution, and the determination of the weighting factors depends heavily on experiences. While the multi-objective optimization treats all the objectives separately and delivers a set of optimal solutions form the Pareto frontier. A Pareto optimal solution is one that any improvement in an objective will cause degradation in other objectives, and all Pareto optimal solutions are considered equally good.

In this paper, NSGA-II is adopted to find the Pareto frontier of the multi-objective optimization problem. NSGA-II is a multi-objective exploratory technique, which has been successfully used in many engineering optimization jobs (Nikkhah Kashani and Rafiei, 2014; Ma et al., 2013; Kelaiaia et al., 2012) and is becoming one of the most popular methods to multi-objective problems. For the current work, we use a population size of 100, a number of generations of 100, and a mutation probability of 0.3. All possible and optimal solutions which considering the three objectives simultaneously are shown in Fig. 9. The results indicate that the algorithm was able to find the Pareto front with good distribution.

Table 2 lists 6 representative optimal solutions ranked by motion range θ_{max}, where the index k is the rank of θ_{max} in the Pareto solution set. At the extreme point f_1, the ellipse-parabola shaped flexure hinge has the minimum motion range but higher rotation precision and relative compliance. While at the other extreme point f_6, the flexure hinge has the maximum motion range but lower rotation precision and relative compliance. All the other points are intermediate optimal solutions. The designers can select the optimal geometric parameters from the obtained Pareto solution set based on practical requirements.

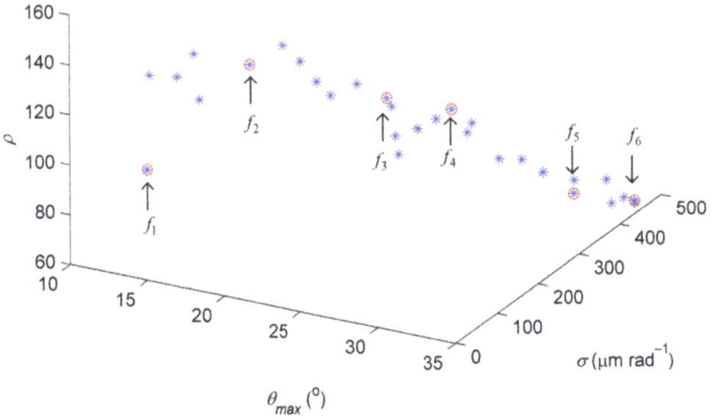

Figure 9. Pareto front obtained by NSGA-II.

6 Conclusion

A novel type of superelastic flexure hinge, called ellipse-parabola shaped flexure hinge, is proposed in this paper to achieve optimal comprehensive performance. The ellipse-parabola shaped flexure hinge is constructed by an ellipse arc and a parabola curve and smoothly connected at the intersection. The static response of the hinge is solved by non-prismatic beam elements which considered the geometric and material nonlinearities. Finite element analysis and experiment tests proved the accuracy and efficiency of the proposed method. The maximum relative error of CRM is within 1.5 % and 7 % compared to the AFE and the EXP respectively. A multi-objective optimization aims to maximize the motion range and the relative compliance of the ellipse-parabola shaped flexure hinge as well as to minimize the relative rotation error during the deformation is conducted. And the Pareto frontier is found by using the NSGA-II algorithm. The optimization methodology presented in this article can be used and extended in the design process of other shaped superelastic flexure hinges.

Table 2. The optimal solutions.

Point	k	Geometric parameter			Performance		
		l (mm)	e (mm)	φ_t (rad)	θ_{max} (°)	γ (μm rad^{-1})	ρ
f_1	1	10.179	2.673	−1.498	13.810	45.738	97.296
f_2	6	14.899	4.912	−1.393	19.645	77.875	142.900
f_3	12	32.484	4.981	−1.146	26.347	161.207	127.878
f_4	18	11.845	4.969	−1.456	29.093	214.126	120.615
f_5	24	8.673	4.950	−1.401	32.747	372.203	72.782
f_6	30	20.893	4.981	−1.236	34.399	457.984	61.899

Acknowledgements. This work was supported by National Natural Science Foundation of China under Grant No. 51475113, Natural Science Foundation of Heilongjiang Province under Grant No. E2015006, the State Key Lab of Self-planned Project under Grant No. SKLRS201501A03, and the Programme of Introducing Talents of Discipline to Universities under Grant No. B07018.

References

Chen, G., Shao, X., and Huang, X.: A new generalized model for elliptical arc flexure hinges, Rev. Sci. Instrum., 79, 095103, doi:10.1063/1.2976756, 2008.

Chen, G., Liu, X., Gao, H., and Jia, J.: A generalized model for conic flexure hinges, Rev. Sci. Instrum., 80, 055106, doi:10.1063/1.3137074, 2009.

Chen, G., Liu, X., and Du, Y.: Elliptical-Arc-Fillet Flexure Hinges: Toward a Generalized Model for Commonly Used Flexure Hinges, J. Mech. Design, 133, 81002, doi:10.1115/1.4004441, 2011.

Criesfield, M. A.: Nonlinear finite element analysis of solids and structures, Volume 1: Essentials, Wiley, Location: New York, 1991.

De Bona, F. and Gh Munteanu, M.: Optimized Flexural Hinges for Compliant Micromechanisms, Analog Integr. Circ. S., 44, 163–174, 2005.

Desroches, R. and Delemont, M.: Seismic retrofit of simply supported bridges using shape memory alloys, Eng. Struct., 24, 325–332, 2002.

Du, Z., Shi, R., and Dong, W.: A Piezo-Actuated High-Precision Flexible Parallel Pointing Mechanism: Conceptual Design, Development, and Experiments, IEEE T. Robot., 30, 131–137, 2014.

Friedrich, R., Lammering, R., and Rösner, M.: On the modeling of flexure hinge mechanisms with finite beam elements of variable cross section, Precis. Eng., 38, 915–920, 2014.

Hesselbach, J. and Raatz, A.: Pseudo-elastic Flexure-Hinges in Robots for Micro Assembly, in: Proc. SPIE Microrobotics and Microassembly II, Boston, USA, 4194, 157–167, doi:10.1117/12.403696, 2000.

Kelaiaia, R., Company, O., and Zaatri, A.: Multiobjective optimization of parallel kinematic mechanisms by the genetic algorithms, Robotica, 30, 783–797, 2012.

Kim, H., Kim, J., Ahn, D., and Gweon, D.: Development of a Nanoprecision 3-DOF Vertical Positioning System With a Flexure Hinge, IEEE T. Nanotechnol., 12, 234–245, 2013.

Liew, K. M., Ren, J., and Kitipornchai, S.: Analysis of the pseudoelastic behavior of a SMA beam by the element-free Galerkin method, Eng. Anal. Bound. Elem., 28, 497–507, 2004.

Lobontiu, N., Paine, J. S. N., Malley, E. O., and Samuelson, M.: Parabolic and hyperbolic flexure hinges: flexibility, motion precision and stress characterization based on compliance closed-form equations, Precis. Eng., 26, 183–192, 2002a.

Lobontiu, N., Paine, J. S. N., Garcia, E., and Goldfarb, M.: Design of symmetric conic-section flexure hinges based on closed-form compliance equations, Mech. Mach. Theory, 37, 477–498, 2002b.

Ma, M., Sun, B., Wang, J., and Shi, J.: Multi-objective optimization design for leg mechanism of hydraulic-actuated quadruped robot, Journal of Beijing Institute of Technology (English Edition), 22, 12–19, 2013.

Nikkhah Kashani, H. and Rafiei, S. M. R.: Optimal Control of Active Power Filters using Fractional Order Controllers Based on NSGA-II Optimization Method, Int. J. Elec. Power, 63, 1008–1014, 2014.

Pacoste, C. and Eriksson, A.: Beam elements in instability problems, Comput. Method. Appl. M., 144, 163–197, 1997.

Paros, J. M. and Weisbord, L.: How to design flexure hinges, Mach. Des., 37, 151–156, 1965.

Pham, H. and Chen, I.: Stiffness modeling of flexure parallel mechanism, Precis. Eng., 29, 467–478, 2005.

Shi, R. C., Dong, W., and Du, Z. J.: Design methodology and performance analysis of application-oriented flexure hinges, Rev. Sci. Instrum., 84, 075005, doi:10.1063/1.4813252, 2013.

Tian, Y., Shirinzadeh, B., Zhang, D., and Zhong, Y.: Three flexure hinges for compliant mechanism designs based on dimensionless graph analysis, Precis. Eng., 34, 92–100, 2010.

Wang, R., Zhou, X., and Zhu, Z.: Development of a novel sort of exponent-sine-shaped flexure hinges, Rev. Sci. Instrum., 84, 095008, doi:10.1063/1.4821940, 2013.

Wu, Y. and Zhou, Z.: Design calculations for flexure hinges, Rev. Sci. Instrum., 73, 3101–3101, 2002.

Yong, Y. K., Lu, T., and Handley, D. C.: Review of circular flexure hinge design equations and derivation of empirical formulations, Precis. Eng., 32, 63–70, 2008.

Permissions

All chapters in this book were first published in MS, by Copernicus Publications; hereby published with permission under the Creative Commons Attribution License or equivalent. Every chapter published in this book has been scrutinized by our experts. Their significance has been extensively debated. The topics covered herein carry significant findings which will fuel the growth of the discipline. They may even be implemented as practical applications or may be referred to as a beginning point for another development.

The contributors of this book come from diverse backgrounds, making this book a truly international effort. This book will bring forth new frontiers with its revolutionizing research information and detailed analysisof the nascent developments around the world.

We would like to thank all the contributing authors for lending their expertise to make the book truly unique. They have played a crucial role in the development of this book. Without their invaluable contributions this book wouldn't have been possible. They have made vital efforts to compile up to date information on the varied aspects of this subject to make this book a valuable addition to the collection of many professionals and students.

This book was conceptualized with the vision of imparting up-to-date information and advanced data inthis field. To ensure the same, a matchless editorial board was set up. Every individual on the board wentthrough rigorous rounds of assessment to prove their worth. After which they invested a large part of theirtime researching and compiling the most relevant data for our readers.

The editorial board has been involved in producing this book since its inception. They have spent rigorous hours researching and exploring the diverse topics which have resulted in the successful publishing of this book. They have passed on their knowledge of decades through this book. To expedite this challenging task, the publisher supported the team at every step. A small team of assistant editors was also appointed to further simplify the editing procedure and attain best results for the readers.

Apart from the editorial board, the designing team has also invested a significant amount of their time in understanding the subject and creating the most relevant covers. They scrutinized every image to scout for the most suitable representation of the subject and create an appropriate cover for the book.

The publishing team has been an ardent support to the editorial, designing and production team. Theirendless efforts to recruit the best for this project, has resulted in the accomplishment of this book. They are a veteran in the field of academics and their pool of knowledge is as vast as their experience in printing. Their expertise and guidance has proved useful at every step. Their uncompromising quality standards have made this book an exceptional effort. Their encouragement from time to time has been an inspiration for everyone.

The publisher and the editorial board hope that this book will prove to be a valuable piece of knowledge for researchers, students, practitioners and scholars across the globe.

List of Contributors

M. Meyer and R. Lupoi
The University of Dublin, Trinity College, Department of Mechanical & Manufacturing Engineering, Parsons Building, Dublin 2, Ireland

M. Oberherber, H. Gattringer and A. Müller
Institute of Robotics, Johannes Kepler University Linz, Altenbergerstr. 69, 4040 Linz, Austria

K. Gok
Dumlupinar University, Kütahya Vocational School of Technical Sciences, Germiyan Campus, 43110 Kütahya, Turkey

S. Inal
Dumlupinar University, School of Medicine, Department of Orthopaedic Surgery, Campus of Evliya Celebi, 43100 Kütahya, Turkey

J. Zhang, Z. Liang, C. J. Han and H. Zhang
School of Mechatronic Engineering, Southwest Petroleum University, Chengdu, 610500, China

M. J. Hosseini and M. Rahimi
Department of Mechanical Engineering, Golestan University, P.O. Box 155, Gorgan, Iran

R. Bahrampoury
Department of Mechanical Engineering, K. N. Toosi University of Technology, Tehran, Iran

A. Gok and M. B. Bilgin
Amasya University, Technology Faculty, Department of Mechanical Engineering, 05000 Amasya, Turkey

K. Gok
Dumlupınar University, Kütahya Vocational School of Technical Sciences, Germiyan Campus, 43100 Kütahya, Turkey

A. R. G. Araújo, N. Peixinho, A. Pinho and J. C. P. Claro
Department of Mechanical Engineering, University of Minho, Guimarães, Portugal

Z. F. Zhu, J. Y. Jia, Y. L. Chen, Z. Zeng and D. L. Yu
School of Mechano-Electronic Engineering, Xidian University, Xi'an, China

H. Z. Fu
ZTE Corporation, Shenzhen, China

Michael R. Morgan, Spencer P. Magleby and Larry L. Howell
Department of Mechanical Engineering, Brigham Young University, Provo, UT 84602, USA

Robert J. Lang
Lang Origami, Alamo, CA 94507, USA

E. I. Jassim
Department of Mechanical Engineering, Prince Mohammad Bin Fahd University, PMU, Al-Khobar, Saudi Arabia

A. Müller
Johannes Kepler University, Altenbergerstr. 69, 4040 Linz, Austria

Elias Rezvani
Department of Mechanical Engineering, Faculty of Engineering, Najafabad Branch, Islamic Azad University, Najafabad, Iran

Hamid Ghayour and Masoud Kasiri
Advanced Materials Research Center, Faculty of Materials Engineering, Najafabad Branch, Islamic Azad University, Najafabad, Iran

M. Neubauer, H. Gattringer, A. Müller and A. Steinhauser
Institute of Robotics, Johannes Kepler University Linz, Altenbergestr. 69, 4040 Linz, Austria

W. Höbarth
Bernecker + Rainer Industrie Elektronik Ges.m.b.H., B&R Str. 1, 5142 Eggelsberg, Austria

U. Hanke, M. Zichner, N. Modler, A. Comsa and K.-H. Modler
Faculty of Mechanical Engineering and Machine Science, TU Dresden, Dresden, Germany

E.-C. Lovasz
Department of Mechatronics, Politehnica University of Timisoara, Timisoara, Romania

Y. F. Liu, W. Li, X. F. Yang, Y. Q. Wang and M. B. Fan
School of Mechatronic Engineering, China University of Mining and Technology, Xuzhou, China

G. Ye
School of Mechanical&Electrical Engineering, Jiangsu Normal University, Xuzhou, China

W. Van Loock, G. Pipeleers and J. Swevers
Department of Mechanical Engineering, Division PMA, KU Leuven, 3001 Leuven, Belgium

Christoph Gruber
eurofunk Kappacher GmbH, St. Johann im Pongau, Austria

Michael Hofbaur
Institute of Robotics and Mechatronics, JOANNEUM RESEARCH, Klagenfurt, Austria

Florian Simroth, and Andrés Kecskeméthy
University of Duisburg-Essen, Duisburg, Germany

Huafeng Ding
China University of Geosciences Wuhan, Wuhan, China

Zhijiang Du, Miao Yang and Wei Dong
State Key Laboratory of Robotics and System, Harbin Institute of Technology, 2 Yikuang Street, Harbin, 150080, China